Manufacturing of Natural Fibre Reinforced Polymer Composites

Mohd Sapuan Salit • Mohammad Jawaid
Nukman Bin Yusoff • M. Enamul Hoque
Editors

Manufacturing of Natural Fibre Reinforced Polymer Composites

Springer

Editors
Mohd Sapuan Salit
Faculty of Engineering
Universiti Putra Malaysia
Selangor, Malaysia

Mohammad Jawaid
Laboratory of Biocomposite Technology
Universiti Putra Malaysia
Selangor, Malaysia

Nukman Bin Yusoff
Department of Mechanical Engineering
University of Malaya
Kuala Lumpur, Malaysia

M. Enamul Hoque
Department of Biomedical Engineering
University of King Faisal
Al-Hofuf, Al-Hassa, Saudi Arabia

ISBN 978-3-319-37201-3 ISBN 978-3-319-07944-8 (eBook)
DOI 10.1007/978-3-319-07944-8

Springer Cham Heidelberg New York Dordrecht London

Softcover reprint of the hardcover 1st edition 2015

Printed on acid-free paper

Springer International Publishing AG Switzerland is part of Springer Science+Business Media (www.springer.com)

Editors are honoured to dedicate
this book to the Late Razia Rahman
(Mother of Dr. Mohammad Jawaid) who left
this worldly life on 25th November, 2014

Preface

Natural fibre composite is an emerging material that has great potential to be used in engineering application. Oil palm, sugar palm, bagasse, date palm, coir, banana stem, hemp, jute, sisal, kenaf, roselle, rice husk, betul nut husk, and cocoa pod, etc. are among the natural fibres reported to be used as reinforcing materials in polymer composites. Natural fibre composites were reported to be used in many industries such as automotive, building, furniture, marine and aerospace industries. Among the advantages of natural fibre composites include low cost, renewable, abundance, light weight, less abrasive, and they are suitable to be used in semi- or non-structural engineering components. Research on various aspects of natural fibre composites such as characterization, determination of properties and design has been extensively carried out. However, publications that reported on research of manufacture of natural fibre composites are very limited especially in the form of book. Therefore, realizing the importance of manufacture of natural fibre composites, an edited book is proposed and it is hoped it can fill the gap of knowledge in the field of natural fibre composites. Generally, manufacturing methods of components from natural fibre composites are similar to the manufacturing methods of components from conventional fibre composites such as glass, carbon and Kevlar fibres. Modification of equipment used for conventional fibre composites may be required when manufacturing components from natural fibre composites and the knowledge on these modifications is needed in natural fibre composite research community. Among the methods reported that are being used to produce components from natural fibre composites include hand lay-up, compression moulding, filament winding, injection moulding, resin transfer moulding, pultrusion and vacuum bag moulding. This book is also intended to address some research on secondary processing such as machining and laser welding of natural fibre composites. It is hoped publication of this book will provide the readers new knowledge and understanding on the manufacture of natural fibre composites.

We wish to express our gratitude to all the contributors who accepted our invitations to contribute and share their knowledge. We admire their proficiency on scattered information from diverse fields in composite manufacturing and accepting editorial suggestions to finally produce this venture that we hope will be a success. We greatly appreciate their commitment.
Lastly, we thank Springer for their generous cooperation at every stage of the book production.

Selangor, Malaysia Mohd Sapuan Salit
Selangor, Malaysia Mohammad Jawaid
Kuala Lumpur, Malaysia Nukman Bin Yusoff
Al-Hassa, Saudi Arabia M. Enamul Hoque

Contents

Chapter 1
The Relationship Between Manufacturing and Design for Manufacturing in Product Development of Natural Fibre Composites

S.M. Sapuan and Nukman Bin Yusoff

Abstract In this chapter, a study of design for manufacturing of natural fibre composites is presented. Initially, the link between design for manufacturing and manufacturing itself is discussed in the context of natural fibre composites. An account of the manufacturing processes to produce products from natural fibre composites is given. Various activities of design for manufacturing of natural fibre composites in various industries are reviewed. Design for sustainability, a recent topic within design for manufacturing or concurrent engineering, in relation to natural fibre composites is discussed and reviewed. Current work on development of packaging materials from sugar palm fibre reinforced biopolymer composites encompassing materials selection of biopolymers and some testing is discussed. Finally, a case study on the application of design for sustainability guides for a natural fibre composite product is presented.

Keywords Design for manufacturing • Manufacturing • Natural fibre composites • Design for sustainability • Concurrent engineering

---

S.M. Sapuan (✉)
Department of Mechanical and Manufacturing Engineering, Universiti Putra Malaysia, 43400 UPM Serdang, Selangor, Malaysia

Faculty of Engineering, Aerospace Manufacturing Research Centre (AMRC), Universiti Putra Malaysia, 43400 UPM Serdang, Selangor, Malaysia

Biocomposite Technology Laboratory, Institute of Tropical Forestry and Forest Products (INTROP), Universiti Putra Malaysia, 43400 UPM Serdang, Selangor, Malaysia
e-mail: sapuan@upm.edu.my

N.B. Yusoff
Department of Mechanical Engineering, University of Malaya, Kuala Lumpur, Malaysia

M.S. Salit et al. (eds.), *Manufacturing of Natural Fibre Reinforced Polymer Composites*, DOI 10.1007/978-3-319-07944-8_1

1.1 Introduction

Natural fibre composites have become very important materials in the recent years. They have been used in many industries such as in automotive, aerospace, building and construction, and furniture industries. However, their applications are limited mainly to semi- and nonstructural components due to their inherently inferior mechanical properties such as strength and stiffness, which are generally very dominant in conventional fibre composites. Many different types of natural fibre composites, especially plant-based fibre composites, have been developed and tested and the sources of natural fibres come from sisal, jute, hemp, flax, banana, pineapple leaf, coir, bagasse, oil palm, roselle, sugar palm, and rice husk, just to name a few. Among the advantages of natural-based composites include light weight, posing minimum or no health hazard, abundance, renewable, low cost, the products are aesthetically pleasing, and comparable specific strength and stiffness. However, natural fibre composites also suffer from several drawbacks such high moisture absorption, poor wettability, weak interfacial bonding, inability to withstand high load, application is limited to semi- and nonstructural components, and unproven manufacturing technology. The major problem in using natural fibre composites is the interfacial bonding between hydrophilic fibres and hydrophobic polymer matrices. The problem can be solved by treating the fibres using various established methods such as alkali treatment, silane coupling agent, and compatibilizing agent.

Natural fibre composites can be considered environmentally benign materials compared to conventional fibre composites, but since in many cases, the matrix system is still derived from synthetic polymers, they cannot be considered as totally environmentally benign. At the end of this chapter, a section on design for sustainability (DFS) is included to address this issue.

1.2 The Link Between Design for Manufacturing and Manufacturing

Design for manufacturing is the integration of product design and manufacturing process planning into one activity, and it meant to bridge the gap between design and manufacturing. It is also known as concurrent engineering (Sapuan and Mansor 2014).

Design for manufacturing activity is in fact a design activity. It requires the consideration of all manufacturing and other related issues such as assembly, cost, sales, maintenance, disposal, and recycling early in the design process. It is not manufacturing activity itself but all considerations regarding manufacturing are made during this stage of product development. It has become an accepted fact that decisions made during the early stage of design have a major effect on downstream activities. According to Poli (2001) without embarking any manufacturing tasks, over 70 % of the product manufacturing cost is determined during the early stage of design. This information is applicable to many engineering products including products from natural fibre composites (Mansor et al. 2014).

In practice, despite design for manufacturing philosophy is regarded as an integrated approach in design and manufacture, when coming to discussion on the issues or activities related to this philosophy, such as conceptual design, materials selection, and manufacturing process selection, each issue has to be treated in detail separately. Some experts focus on materials selection, while others on conceptual design, but at the back of their mind, these activities should really be integrated. Integration can be in the form of integrated computer system like knowledge-based system, computer-aided design (CAD), and finite element analysis (FEA) with a common software that can integrate them. Integration can also be done by human through coordination of experts in various fields.

1.3 Manufacturing Process of Natural Fibre Composites

Having known the role of design for manufacturing in manufacturing process, this section is now devoted to the manufacturing processes reported to be used in the development of products or components from natural fibre composites. Ho et al. (2012) performed extensive literature review on composite manufacturing processes that are related to natural fibre composites. They reported that majority of the work related to natural fibre composites used conventional composite manufacturing processes such injection moulding, compression moulding, resin transfer moulding (RTM), vacuum infusion moulding, and hot press processes. In this chapter only selected manufacturing processes are reviewed.

1.3.1 Pultrusion

Pultrusion is a combination of two words: pull and extrusion. The process is similar to extrusion process, but in pultrusion, the product is developed by pulling the materials rather than pushing the materials through the die as in the case of extrusion. Continuous fibre rovings or tapes are pulled by means of a puller through a resin bath of thermosetting polymer. Examples of thermosetting polymers that can be used in this process include epoxy and unsaturated polyester. The resulting impregnated fibre composites leave the resin bath and are being pulled and passed through a series of forming dies. The shape of the final product follows the shape of the cross of the dies and the shape can be circular, rectangular, square, and I-shaped and H-shaped sections. The composite is also cured in one of the dies. The end products are normally in the form of rods and bars. At the end of the pultrusion process, the products are cut to the required lengths.

Fairuz et al. (2014) reported the work on pultruded kenaf fibre reinforced vinyl ester composites using pultrusion process. Kenaf composite rods were produced. In this study, three different parameters were investigated, i.e. speeds, temperatures, and filler loadings. Taguchi's design of experiment was used to determine the number of experiments to be conducted. Specimens in the form of rods were produced

Fig. 1.1 Pultrusion process of kenaf fibre fibre composites (Courtesy of A.M. Fairuz)

at three different speeds by fixing temperature and filler loading. Then the same is repeated for different parameters, temperatures, and filler loadings. The mechanical properties of the composite rod specimens such as tensile, flexural, and impact properties were tested for each parameter. Figure 1.1 shows the pultrusion process of kenaf fibre reinforced composites.

1.3.2 Filament Winding

Filament winding is a composite manufacturing process to produce components that are normally circular in shape. It can be considered an open mould process and the mould is actually a rotating mandrel. In this process, continuous fibre rovings are being drawn by a puller through a resin bath. The preimpregnated composites are in turn being wound around the rotating mandrel. The composites are also being traversed in x-direction so that even distribution of the composites along the length of the mandrel can be achieved. There are three types of winding patterns in the filament winding process, i.e. hoop, helical and polar windings.

Misri et al. (2015) developed kenaf fibre reinforced unsaturated polyester composite hollow shafts and studied their mechanical behaviour. Continuous kenaf fibre rovings were pulled through a drum type resin bath and the preimpregnated fibre composites were wound around a hollow aluminium rotating mandrel. They studied various properties such as tensile and torsion properties of the composites, and such findings were compared with the results of FEA. The major problem with developing filament wound kenaf fibre composites is how to obtain consistently smooth surface of shafts. Secondary processing like milling is required.

1.3.3 Hand Lay-Up

Hand lay-up is an open moulding process to produce polymer composite product. This is a labour-intensive process where a high skill operator is required to perform the fabrication task. A mould release agent is applied onto the mould surface for ease of removing of finished composite products. The laying-up of a component begins with applying gel coat to the mould surface. It is resin-rich layer and the purpose is to prevent the fibres appearing on the mould surface. The composites are then prepared by placing a fibre reinforcement and by adding polymer resin onto it. A roller is used to consolidate the composites and the composites are made layer by layer. The process is repeated with more layers until the desired thickness is obtained. Then the composites are cured at room temperature or in an oven. Thermosetting resins like unsaturated polyester, vinyl ester, and epoxy are among the most commonly used resins.

Hand lay-up or also known as wet lay-up has been used to produce components from natural fibre composites for many decades now. Ford had used hand lay-up process to fabricate automotive body from natural fibre reinforced soya-based polymer composites around a century ago. Misri et al. (2014a) have used hand lay-up process to develop a small boat from hybrid glass-sugar palm fibre reinforced unsaturated polyester composites. The fabrication work was carried out in association with an established boat builder. This process was labour intensive and required high skill operators to fabricate the boat. The mould was made from glass fibre reinforced polyester composite and it was developed in such a way that a very high quality surface was obtained. Then mould release agent was placed onto the mould surface to prevent the component sticking to the mould. Then glass fibres were laid up followed by resin and they were consolidated using a roller. Methyl ethyl ketone peroxide (MEKP) was used as a hardener. The process is repeated with sugar palm fibres and the process continued layer by layer until the desired thickness of the composites was achieved. Figure 1.2 shows the hand lay-up process of hybrid glass-sugar palm fibre composite boat.

Fig. 1.2 Hand lay-up process for hybrid glass-sugar palm composite boat (Courtesy of S. Misri)

1.3.4 Resin Transfer Moulding

RTM is an established composite manufacturing process that is normally used to fabricate automotive and aircraft components. In this process, fibre reinforcement, either long or woven fibres are initially cut out using a template and a knife or scissors. These reinforcements called performs are bound by means of thermoplastic binder and then placed inside the mould cavity in the closed mould process. Resin is being transferred into the mould cavity by means of a variety of equipments such as by pressure or vacuum. Resins that are normally used in this process include polyester, epoxy, vinyl ester, and phenolic.

Salim et al. (2011) used RTM to develop composite specimens for nonwoven kenaf fibre mat reinforced epoxy composites. Kenaf fibre mats were prepared after undergoing a series of processing such as opening, carding, cross lapping, and needle punching processes. Stitching density was optimized to study its effects on the mechanical properties of the composites. In addition, the effects of fibre loading on the mechanical properties were also investigated. The mechanical properties improved when the kanaf mat in composites was stitched compared to unstitched kenaf mat composites.

1.3.5 Vacuum Bagging

Vacuum bagging is one of the processes in the category of prepreg moulding. Another process is called autoclave moulding. The difference between these two processes is on the way curing process was done: one in vacuum bag (oven) and the other in an autoclave. Vacuum bagging is a composite manufacturing process that utilizes a vacuum bag to provide compaction pressure and consolidation of plies within the laminate. It is an extended version of hand or wet lay-up process. It also sometimes called vacuum bag moulding (Mayer 1993). In this process, a mould base is placed horizontally. Then composite prepreg laminates are stacked onto the mould base. Then a series of equipments and materials are covered onto the composite layers such as release film layer, air-bleeder layer, blocked film layer, and breather layer. Finally, a vacuum bag is placed to cover the composite laminate and all the layers. The vacuum bag is then sealed by means of sealers. Polymer materials that can be used in this process include epoxy, phenolic, and polyimide.

Mariatti and Abdul Khalil (2009) used vacuum bagging process to prepare samples of bagasse fibre reinforced unsaturated polyester composites. Two components of bagasse, i.e. rind (outer part) and pith (inner part), were used as reinforcement fibres in composites. The hardener used in this experiment was MEKP. Pressure was applied to the laid-up laminates in order to consolidate the final composites and the bonding between composite plies. Volatiles and trapped air in the laminates were drawn out using vacuum and void contents were minimized. Curing process took place at room temperature. Composite specimens were then tested mechanically to determine tensile, flexural, and impact properties of the composites. Improvements were made on the mechanical properties of bagasse fibre unsaturated polyester composites by fibre chemical treatment using sodium hydroxide (NaOH) and acrylic acid (AA).

1.3.6 Compression Moulding

Compression moulding is a composite manufacturing process normally used to produce composite components in high production volume such as automotive components. There are two types of compression moulding process, i.e. cold compression and hot compression mouldings. In this process, for thermosetting matrices, normally an intermediate material called moulding compound is used. It is really a semicured composite. Sheet moulding compound (SMC) and bulk moulding compound (BMC) are two widely used thermosetting-based moulding compounds. For thermoplastic polymer, the moulding compound normally used is glass mat thermoplastics (GMT). For preparing composite specimens in the laboratory scale compression moulding, composite pellets are normally prepared using internal mixer and twin screw extruder. In compression moulding process, only pressure is applied in the case of cold press and pressure and temperature are applied in hot press after the moulding performs are placed in a mould cavity. In cold compression moulding, curing process takes place at room temperature, while for hot press it takes place by applying heat to the mould and it in turn being transferred to the composites. Matrices used in this process include unsaturated polyester, vinyl ester, and phenolic for thermosetting polymers, while among the thermoplastic polymers, polypropylene, polyamide, and polyetheretherketone (PEEK) can be used.

Wirawan et al. (2011) and El-Shekeil et al. (2012) used laboratory scale compression moulding (Fig. 1.3) to prepare composite specimens from sugar cane bagasse and poly(vinyl chloride) (PVC) and kenaf and thermoplastic polyurethane, respectively. The intermediate materials were made using internal mixer. The intermediate materials were placed into the mould and the heat and pressure were applied to form the composite plates. Hot pressing was then carried out at a temperature around 170 °C for 12.5 min and the mixture was cooled under pressure to room temperature. The mechanical and thermal properties of the composites were then determined.

1.3.7 Injection Moulding

Injection moulding is one of the most widely used manufacturing processes to produce plastic components. Injection moulding can also be used to produce polymer composite products, but the fibres used in the composites are short fibres in the form of particles or powder (Fig. 1.4). Injection moulding compound (IMC), in the form

Fig. 1.3 Hot Compression moulding machine (Courtesy of Y.A. El-Shekeil)

Fig. 1.4 Injection moulding machine (Courtesy of M.D. Azaman)

of granules, is prepared using twin screw extruder. Then, the granules are fed inside the mould through a hopper and a heated barrel. The shearing action of a reciprocating screw and heating process in heated barrel caused the IMC to melt. The molten composites are injected into the mould to form a composite part. The composite part is then ejected from the mould.

Huda et al. (2006) carried out comparative studies of mechanical properties of chopped glass and recycled newspaper cellulose fibre reinforced poly(lactic acid) (PLA) composites fabricated using twin screw extruder and injection moulding process. Tensile and flexural properties of recycled paper composites were superior to virgin polymers. Morphological evaluation using scanning electron microscopy (SEM) analysis showed that both chopped glass and recycled paper fibres were uniformly dispersed in PLA matrices.

Azaman et al. (2013, 2014) carried out mould flow analysis of wood fibre reinforced polypropylene composites using Autodesk MoldFlow Insight 2011 software to simulate the behaviour of composite flow inside the mould to study the behaviour of residual stress, shrinkage, and warpage of thin walled parts used for electronic packaging purposes like mobile phone casing and computer mouse.

1.4 Design for Manufacturing of Natural Fibre Composites

The late Professor Stuart Pugh (1991) is considered one of the authorities in product development, and his work on total design model is accepted worldwide as a tool in design for manufacturing. Total design model comprises several important stages that a designer has to follow to come up with a good design and eventually a product. These stages include market investigation, product design specification, design concept, detail design, and manufacture and sale. A design concept evaluation method called Pugh selection method is one of the well-accepted methods for design concept evaluation and this method has been recognized as an important tool in design for manufacturing or concurrent engineering field. In fact not only Pugh selection method is recognized, but the total design model has also been recognized as the tool to support concurrent engineering.

The author, being a former student from a university where Pugh was employed (not at the same time though), used this method in specific and the total design model in general, in most of his design activities, including design of natural fibre composite products. It is very difficult to find previous published work on the design for manufacturing related to natural fibre composite products and it is equally difficult to write in details about this topic. However the author and his co-workers have attempted to carry out research work related to design for manufacture of natural fibre composites and they are given below, and the focus of the research is mainly on the conceptual design of natural fibre reinforced polymer composite products.

1.4.1 Automotive Components

Mansor et al. (2014) carried out conceptual design study of hybrid glass and kenaf fibre reinforced polypropylene composite automotive brake lever. The development of this seemingly simple component requires a lot of knowledge and information from various sources in order to come up with best design solution. Automotive component vendors, automotive component manufacturers, and composite material experts were consulted in carrying the design task. In concept generation stage, methods like TRIZ and morphological chart were used. It was assisted by the results of FEA (Fig. 1.5). Analytical hierarchy process was the major tool used in this study to select the best design concept and natural fibre for the components.

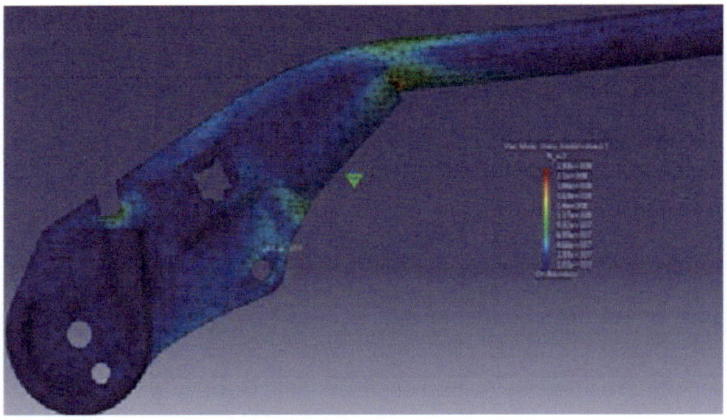

Fig. 1.5 Finite element analysis model for automotive parking brake lever (Courtesy of M.R. Mansor)

Davoodi et al. (2011) used several designs for manufacturing tools in the development hybrid glass-kenaf fibre reinforced polymer composite automotive bumper beam. They carried out concept generation for design concepts and concept selection process using the method that they developed. FEA and experimental testing were conducted to provide supporting data and information that helped the team to come up with the best design concept.

1.4.2 Marine Application

Misri et al. (2014b) used total design to develop a small boat made from hybrid glass-sugar palm fibre reinforced unsaturated polyester composites. Hand lay-up process was employed to fabricate the components. Designers were working in a team comprising a designer, material experts, and boat manufacturers. Discussion session was conducted and various concepts were generated using several tools like morphological chart (Fig. 1.6), questionnaire survey, weighted objective method, and brainstorming. Difficulties arose in deciding the best manufacturing process to develop the boat. Materials selection was another issue. Decision to use hybrid glass-sugar palm polyester composites was based on the fact that sugar palm fibres alone were expected to be unable to support the applied loads.

1.4.3 Household Appliances

Sapuan and Maleque (2005) performed conceptual design studies on the development of a telephone stand. The product was made from woven banana pseudo-stem fibre reinforced epoxy composites using hand lay-up process. Sapuan et al. (2007) used the same method to develop a small multipurpose table also from woven banana pseudo-stem fibre reinforced epoxy composites. In both products, total

Sub Function	Solution				
	1	2	3	4	5
1. Capability on boat structure	no bone	2 bones	3 bones	4 bones	5 bones
2. Dimension of boat	simple	medium	complex		
3. Boat fabrication process	hand lay up	resin transfer moulding (RTM)	resin infusion moulding		
4. Number of passengers	1	2	3	4	5
5. Safety	1 buoy	2 buoys	3 buoys	4 buoys	5 buoys
6. Boat application	lake and small river	river	sea		

Fig. 1.6 Morphology chart for the composite boat (Courtesy of S. Misri)

design model was adopted. Several design concepts were generated using various idea generation techniques. Concept evaluation method (weighted objective method and Pugh selection method) was used to select the most suitable design concept for the products.

1.5 Design for Sustainability of Natural Fibre Composites

Design for manufacturability is just one of the elements of design for X (DFX) where Xs are all abilities or DFX can be used to represent design for excellence. Another name for design for manufacturing is concurrent engineering (CE). In the recent years, whenever the topic of concurrent engineering is discussed, it will invariably involve a new term called design for sustainability (DFS). It is observed that the work on DFS of natural fibre composites is very limited in the literature.

Marques et al. (2012) stated that DFS cannot be achieved through conventional composites utilizing synthetic polymer matrices like thermoplastics and thermosetting polymers and synthetic fibres like glass, carbon, and aramid. These materials are not generally environmentally friendly particularly upon disposal. However, the DFS principle in composite materials is supported by the use of biopolymers as matrices and natural fibres as reinforcement materials. PLA and polyhydroxyalcanoate (PHA) can be used as biopolymer matrices and cork is used as core (reinforcement material) to form a composite for city bus bodies.

Sahari et al. (2013) and Sanyang and Sapuan (2015) were working on research projects that can provide some solution for packaging industry through the development of natural fibre reinforced biopolymer composite products. The final tangible products are still far from reality. The first challenge is on how to develop the materials based on the resources, funding, and constraints that are in hand. The author is the leader for this project. This funded project only focuses on the use of fibre from sugar palm (*Arenga pinnata*) plant. Sahari et al. (2013) developed 100 % biocomposite specimens from sugar palm fibres and sugar palm starch-based biopolymer. Mechanical, physical, and thermal properties of these 100 % biocomposites were tested.

Sanyang and Sapuan (2015) used expert system to select the most suitable biopolymer materials as matrices for the packaging materials from composites. The expert system utilizing rule-based system is used to select the most suitable biopolymers to be used with sugar palm fibres to form composites for packaging industry. Reasoning is carried out using decision rules. Rule-based system is an important tool in materials selection. Several rules were developed in Exsys Corvid software for selecting the most suitable materials. When the conditions of a rule were satisfied, then the conditions were valid. To select the material, the following approach was being followed:

If <X> Then <Y>

where X are the conditions and Y are the conclusions. For example, the selection of material for food packaging could be defined through the following rules:

If		
	(the tensile strength of this material is high)	*and*
	(the stiffness of this material is high)	*and*
	(the elongation of this material is high)	*and*
	(the compostability of is this material is high)	*and*
	(the cost of this material is low)	*and*
Then		
	(this material is candidate for food packaging)	

Any materials that fulfil these constraints are the suitable candidates for food packaging. In the case when there are more than candidates selected, the expert system is able to rank the materials in the sequence of suitability.

The work is now extended to study the mechanical properties of sugar palm fibre reinforced PLA/sugar palm starch biopolymer blend composites. The intention of this research is to design and fabricate a final product from sugar palm biopolymer composites for food packaging.

1.5.1 Design for Sustainability Guides (EcoDesign Foundation 2014) Used for Natural Fibre Composite Product

EcoDesign Foundation (2014) has come up with DFS guides used in the product development. The DFS guides are used to determine a product from natural fibre composites that follows the DFS guides. In the DFS guides, ten guide stages are listed. The most relevant stages are stage 5 (Lean Design) and stage 6 (Manufacturing Issues). In this chapter, only the former is considered. Under Lean Design, designer is required to answer 15 questions. The answers to those questions are really the DFS guides.

Exercises:

1. Choose a product and list the materials that are used in the products.
   A product is chosen (say a chair (e.g. Wool and Sun 2005)) and a designer should imagine it is made from natural fibre reinforced biopolymer composites.
2. Why these materials are used in this product?
   Light weight, low cost, renewable, abundantly available
3. What are the inherent qualities of these materials?
   Fully biodegradable, environmentally friendly
4. What do you think about these materials?
   They are the most suitable materials for the product.
5. Are packaging materials also 'green' materials?
   The packaging materials are also 'green' materials.
6. How the amount of materials can be minimized?
   Parts are normally reduced when designing with composites. This in turn reduces material use.
7. Can the volume be reduced (say by folding) when transporting the product?
   A good designer can design the product with such features.
8. Are these materials toxic and posed health hazard?
   Natural fibre reinforced biopolymer composites are very safe to use.
9. How are natural fibre reinforced biopolymer composites being interfaced with other materials?
   Many ways of doing it which include adhesive joints and mechanical fasteners
10. Can the weight be reduced?
    This is the prime advantage of using natural fibre composites.
11. Can they be recycled?
    They can be recycled but there is no need for doing that as they can be found in abundance at low cost.
12. What are the alternative materials available?
    There are wide varieties of natural fibres and bio-based polymers.
13. Are these materials from renewable sources?
    These materials are renewable.

14. Can this product be made from reclaimed materials?
    It can be made.
15. How this product can be made leaner?
    Use design for manufacturability principle.

By answering 15 questions concerning Lean Design stage in DFS guides, natural fibre reinforced biopolymer composites are the natural choices of materials for 'green' product. These materials are possible solutions for DFS issues.

1.6 Conclusions

From this study, it can be concluded that it is possible to carry out design for manufacturing study for natural fibre composites. Design for manufacturing is closely related to manufacturing. Generally, the successful implementation of design for manufacturing leads to the success of the manufacturing of products from natural fibre composites. These were supported in the development of selected products such as automotive components, household appliances, and a small boat. DFS has emerged as very important philosophy in product development, and attempts have been made to develop products from natural fibre reinforced biopolymer composites, which are truly environmentally benign materials that fit well in the DFS definition.

Acknowledgements The author wishes to thank the Universiti Putra Malaysia for giving him the opportunity through sabbatical leave to carry out research on design for manufacturing of natural fibre composites at the Universiti Malaya from the 1 October 2013 to 30 June 2014. The support of Associate Professor Dr Nukman Yusuff of the Universiti Malaya is highly appreciated. The contributions of M. R. Mansor, M. M. Davoodi, M. L. Sanyang, J. Sahari, M. D. Azaman, A. M. Fairuz, and S. Misri in providing case studies and valuable discussion are highly appreciated.

References

Azaman MD, Sapuan SM, Sulaiman S, Zainudin ES, Khalina A (2013) Shrinkages and warpage in the processability of wood-filled polypropylene composite thin-walled parts formed by injection molding. Mater Des 52:1018–1026

Azaman MD, Sapuan SM, Sulaiman S, Zainudin ES, Khalina A (2014) Numerical simulation analysis of the in-cavity residual stress distribution of lignocellulosic (wood) polymer composites used in shallow thin-walled parts formed by the injection moulding process. Mater Des 55:381–386

Davoodi MM, Sapuan SM, Ahmad D, Aidy A, Khalina A, Jonoobi M (2011) Concept selection of car bumper beam with developed hybrid bio-composite material. Mater Des 32:4857–4865

EcoDesign Foundation (2014) Design for sustainability guides, Sydney. http://www.changedesign.org/DfSGuideWebsite/IntroBackg/IntroFrameset.htm. Accessed in 2014

El-Shekeil YA, Sapuan SM, Abdan K, Zainudin ES (2012) Influence of fiber content on the mechanical and thermal properties of kenaf fiber reinforced thermoplastic polyurethane composites. Mater Des 40:299–303

Fairuz AM, Sapuan SM, Zainudin ES, Aiza CN (2014) Study of pultrusion process parameters. In: Proceedings of the postgraduate symposium on composites science and technology 2014 and the 4th postgraduate seminar on natural fibre composites 2014, 28 January, Putrajaya, pp 116–120

Ho M, Wang H, Lee J, Ho C, Lau K et al (2012) Critical factors on manufacturing processes of natural fibre composites. Compos Part B Eng 43:3549–3562

Huda MS, Drzal LT, Mohanty AK, Misra M (2006) Chopped glass and recycled newspaper as reinforcement fibers in injection molded poly(lactic acid) (PLA) composites: comparative study. Compos Sci Technol 66:1813–1824

Mansor MR, Sapuan SM, Zainudin ES, Nurani AA, Hambali A (2014) Conceptual design of kenaf fibre polymer composite automotive parking brake lever using integrated TRIZ-Morphological Chart-Analytical Hierarchy Process method. Mater Des 54:473–482

Mariatti J, Abdul Khalil S (2009) Properties of bagasse fibre-reinforced unsaturated polymer (USP) composites. In: Salit MS (ed) Research on natural fibre reinforced polymer composites. UPM Press, Serdang, pp 63–83

Marques AT, Esteves JL, Viana J, Loureiro H, Arteiro A (2012) Design for sustainability with composite systems. In: 15th international conference on experimental mechanics, 22–27 July, Porto, pp 1–2

Mayer RM (1993) Design with reinforced plastics: a guide for engineers and designers. The Design Council, London

Misri S, Leman Z, Sapuan SM (2014a) A small boat from woven glass-sugar palm fibre reinforced unsaturated polyester composite. In: S.M. Sapuan (ed) Engineering composites: properties and applications. UPM Press, Serdang, pp 297–312

Misri S, Leman Z, Sapuan SM (2014b) Total design of a small boat using woven glass- sugar palm fibre reinforced unsaturated polyester composite. In: Engineering composites: properties and applications. UPM Press, Serdang, pp 224–247

Misri S, Sapuan SM, Leman Z, Ishak MR (2015) Fabrication and torsional behaviour of filament wound kenaf yarn fibre reinforced unsaturated polyester composite hollow shafts. Mater Des 65:953–960

Poli C (2001) Design for manufacturing: a structured approach. Butterworth-Heinemann, London

Pugh S (1991) Total design: integrated methods for successful product engineering. Addison Wesley, Wokingham

Sahari J, Sapuan SM, Zainudin ES, Maleque MA (2013) Mechanical and thermal properties of environmentally friendly composites derived from sugar palm tree. Mater Des 49:285–289

Salim MS, Ishak ZAM, Hamid SA (2011) Effect of stitching density on non-woven fibre mat towards mechanical properties of kenaf reinforced epoxy composites produced by resin transfer moulding (RTM). Key Eng Mater 471–472:987–992

Sanyang ML, Sapuan SM (2015) Development of expert system for biobased polymer material selection: food packaging application. J Food Sci Technol, in press, doi:10.1007/s13197-015-1759-6

Sapuan SM, Maleque MA (2005) Design and fabrication of natural woven fabric reinforced epoxy composite for household telephone stand. Mater Des 26:65–71

Sapuan SM, Mansor MR (2014) Concurrent engineering approach in the development of composite products: a review. Mater Des 58:161–167

Sapuan SM, Harun N, Abbas KA (2007) Design and fabrication of a multi-purpose table using composite of epoxy and pseudo-stem banana fibres. J Trop Agric 45:66–68

Wirawan R, Sapuan SM, Yunus R, Khalina A (2011) Properties of sugarcane bagasse/poly(vinyl chloride) composites after various treatments. J Compos Mater 45:1667–1674

Wool RP, Sun XS (2005) Bio-based polymers and composites. Elsevier, Burlington

Chapter 2
Introduction to Manufacturing of Natural Fibre-Reinforced Polymer Composites

M. Arifur Rahman, Fahmida Parvin, Mahbub Hasan, and M. Enamul Hoque

Abstract In the recent era, different environmental issues have significantly influenced the innovations in material science and technology. The burgeoning demand for clean environment has led the innovation of green materials and utilization of natural materials. Thus, the urge for the production of high-performance engineering products from natural renewable resource is growing day by day. Composites are among those versatile, high-performance materials which combine the unique mechanical and thermal properties that cannot be achieved in a single material. In the recent decade, scientists continued to explore the potential of natural fibres as the reinforcing phase for polymer composites. The important driving force for such emergence of utilizing natural resources is that they are renewable and biodegradable and impose no adverse effects on environment, whereas petroleum-based products are limited and cause environmental problems. This review gives the state-of-the-art overview on currently developed natural fibre-reinforced polymer composites focusing on structure–property relationship of fibres, different polymer matrices used to develop composites, their mechanical performances, different composite fabrication techniques, and the application of such composites in different areas. Critical issues of biocomposites have also been discussed along with their advantages and disadvantages. This article also summarized the critical issues in the manufacturing of natural fibre composites.

M. Arifur Rahman
Department of Macromolecular Science and Engineering, Case Western Reserve University, Cleveland, OH 44106-7202, USA

F. Parvin
Department of Environmental Sciences, Jahangirnagar University, Savar, Dhaka 1342, Bangladesh

M. Hasan
Department of Materials and Metallurgical Engineering, Bangladesh University of Engineering and Technology, Dhaka 1000, Bangladesh

M. Enamul Hoque (✉)
Department of Biomedical Engineering, University of King Faisal, Al-Hofuf, Al-Hassa, Saudi Arabia
e-mail: enamul1973@gmail.com

M.S. Salit et al. (eds.), *Manufacturing of Natural Fibre Reinforced Polymer Composites*, DOI 10.1007/978-3-319-07944-8_2

Keywords Natural fibre • Composites • Biodegradable • Polymers • Manufacturing • Biocomposites

2.1 Introduction

In the recent decades, scientific and technological interests have been shifting toward the development of stronger materials based on the fibre-reinforced composites. Such idea or innovation of composites is, in fact, an inspiration from biological materials such as human bone where soft tissues are reinforced with fibrous protein and exhibits unique mechanical performance. Many artificial polymer-based composites, reinforced by carbon, glass, or aramid fibres, have been developed in automotive, aerospace, construction, or sporting industries. Among many man-made fibres, glass fibre is the mostly used one for reinforcing plastics because of their comparative low-cost and high mechanical performance. In spite of being highly stable during processing, glass fibres have high density, consume high energy and cost for production, are not biodegradable, and also possess health hazards (Wambua et al. 2003). In addition, environmental concern related to the waste disposal of such nonbiodegradable composite has been growing rapidly. Since the 1990s, natural fibre-reinforced composites are evolving as convincing replacements to glass-reinforced composites in numerous uses. Natural fibres are abundant in nature; they can be grown, have cost-effective processing, and also possess unique physical and mechanical properties. Thus, they can be considered as renewable source of fibres. Plenty of different natural fibres have been used as the reinforcing phase of composites. For example, in automotive industries, hemp fibre epoxy, flax fibre polypropylene (PP), and china reed fibre PP are intensively used due to their low bulk density and lower production cost. Moreover, being thermoplastic in nature, PP- and PE-based composites can be easily recycled, and thus they have drawn a lot of attention in different industries (Mohanty et al. 2000). In addition to the cost-effective processability, NFRCs also offer a number of environmental benefits such as less pollution, less dependence on nonrenewable resources, improved energy recovery, and biodegradability.

Fabrication of natural fibre-reinforced composites has always been a challenge due to their low stability during processing. Generally, traditional manufacturing techniques, which are designed for conventional fibre-reinforced polymer composites and thermoplastics, such as compression molding, injection molding, or vacuum infusion, are still being used for the fabrication of NFRCs. These techniques are still considered to be efficient for the production of high-quality composites. However, the success in the production of NFRCs by utilizing these techniques requires proper understating of the structure—properties of natural fibres which have low thermal stability and low mechanical resistance to shear and compatibility with conventional thermoplastics (Ho et al. 2012a). Thus, pretreatment has always been an important issue for better processability of NFRCs. In many cases, chemical treatments of fibre surface are necessary to improve the interfacial interaction between fibre and matrix polymer. Thus, the correlation among the selection of materials and processing methods to the final properties of the composites is very important.

The versatile applications of NFRCs are widely known. Recent study estimates that in 2010, total global natural fibre composite materials market shipments topped 430.7 million pounds with a value of US$289.3 million, and the market is expected to grow to US$531.3 million in 2016 with an 11 % compound annual growth rate (CAGR) over the next 5 years (Faruk et al. 2014). Moreover, automotive industries are consuming most of the NFRCs and expected to be the largest market for NFRCs till 2016. Automotive industries have already adopted the technology of NFRCs due to their low weight, lower production cost, and recyclability. Thus, various products, not restricted only for interior but also nonstructural applications, have been marketed by the automotive industry. On the other hand, concerning the environmental issues related to the disposal of plastic waste, a remarkable trend of utilizing natural fibre-reinforced bioplastics can be noticed (Shen et al. 2009).

Another interesting fact is that the number of publications as patents, books, or peer-reviewed papers is increasing rapidly which indicates the growing importance of NFRCs. Figure 2.1 shows that the number of patents and publications increased rapidly after year 2000 (Figure 1, Satyanarayana et al. 2009). This chapter introduces different NFRCs and the state-of-the-art review of their manufacturing techniques as well as their potential applications.

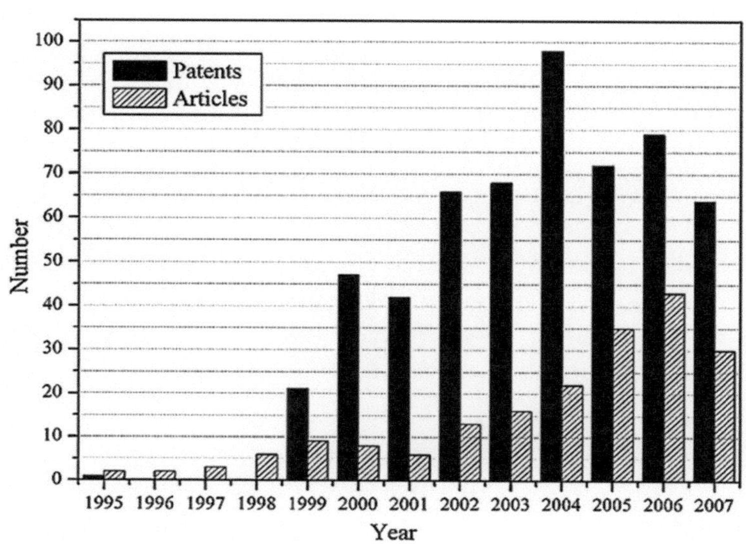

Fig. 2.1 This graph is showing the number of publications produced on natural fibre-reinforced composites each year (1995–2007)

2.2 Structures and Properties of Natural Fibres

2.2.1 Types and Sources of Natural Fibres

Natural environment comprising both flora and fauna is an enormous resource for fibres. However, in this chapter we will only focus on the plant-based fibres. Plant fibres have become a new generation of reinforcing material (Faruk et al. 2014). These fibres can be grouped into different categories based on their source and derivation of plant, animal, and mineral types (Ho et al. 2012b).

Plant-based fibres can be classified as seed hair, bast fibres, and leaf fibres, depending upon the source. Some examples are cotton (seed hairs); hemp, ramie, jute, kenaf, and flax (bast fibres); and sisal, banana, coir, and abaca (leaf fibres). Figure 2.2 shows the photographs of plant-based fibres. It is reported that Brazil and Tanzania produce the largest amount of sisal fibre, whereas Bangladesh, China, and India are the largest producers of jute fibre.

2.2.2 Structure of Natural Fibres

Plant fibres have complicated composite cell structures. Fibres are composed of cellulose microfibrils that reinforce the lignin and hemicellulose matrix. Figure 2.3 shows that the cellulose microfibrils are helically wounded along the hollow fibre axis.

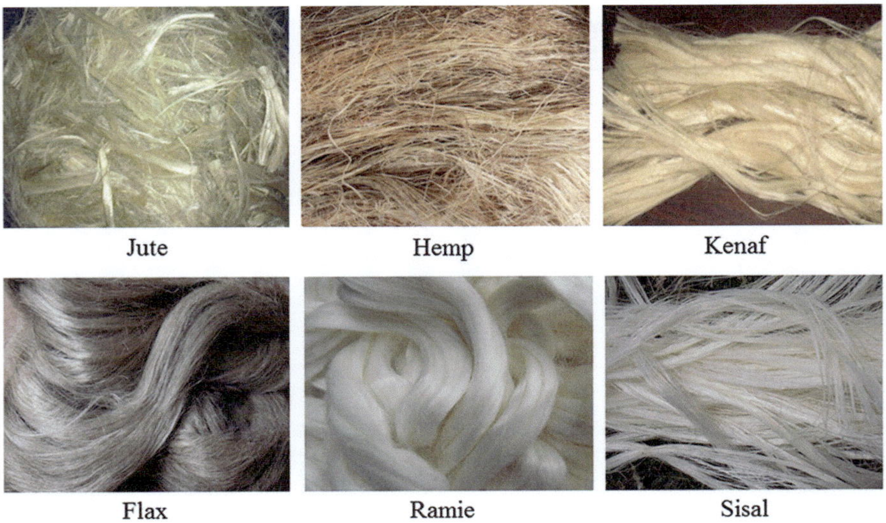

Fig. 2.2 Images of plant-based fibres

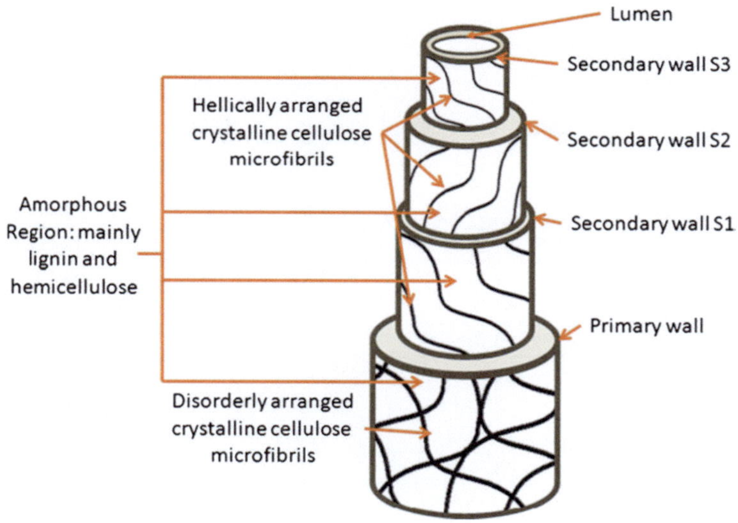

Fig. 2.3 Structural constitution and arrangement of a natural plant fibre cell. (Adopted from John and Thomas 2008)

Such helical conformation gives strength to the fibre because it requires high energy to uncoil the oriented fibrils (Dicker et al. 2014). On the other hand, hemicellulose plays key role for the biodegradation, moisture absorption, and thermal degradation of the fibre. In addition, lignin is responsible for the UV degradation. The percentage composition of each of these components varies for different fibres. Generally, the fibres contain 60–80 % cellulose, 5–20 % lignin, and up to 20 % moisture (Saheb and Jog 1999).

2.2.3 *Properties of Natural Fibres*

Mechanical properties of natural fibres are much lower when compared to those of the most widely used competing reinforcing glass and other man-made fibres (Table 2.1). However, because of their low density, the specific properties (property-to-density ratio), strength, and stiffness of natural fibres are comparable to the values of man-made fibres (Wambua et al. 2003). In case of natural fibre, mechanical properties are determined mainly by the cellulose content and microfibrillar angle. Young's modulus of natural fibre decreases with the increase of diameter. A high cellulose content and low microfibril angle are desirable properties of a fibre to be used as reinforcement in polymer composites (Williams and Wool 2000). The mechanical properties of natural fibre are also significantly related to the degree of polymerization of cellulose in the fibre.

Table 2.1 Mechanical properties of natural and man-made fibres (Taj et al. 2007; Saheb and Jog 1999)

Fibres	Density (g/m^3)	Elongation (%)	Tensile strength (MPa)	Young modulus
Natural fibre				
Jute	1.3	1.5–1.8	393–773	55
Sisal	1.5	2–2.5	510–635	9.4–28
Flax	1.5	2.7–3.2	344	27
Kenaf		1.5	389–930	35–53
Pineapple		2.4	170	62
Cotton	1.5–1.6	7–8	287–597	5.5–12.6
Hemp		1.6	690	
Coir	1.2	30	175	4–6
Ramie		3.6–3.8	400–938	61.4–128
Wool		25–35	120–174	2.3–3.4
Spider silk		17–18	875–972	11–13
Man-made fibre				
E-glass	2.5	2.5	2,000–3,500	70
Aramid	1.4	3.3–3.7	3,000–3,150	63–67
Carbon	1.4	1.4–1.8	4,000	230–240

2.3 Different Polymers as the Reinforcing Phase of NRFCs

2.3.1 Polyolefins

The use of natural fibres in thermoplastics such as polyethylene and polypropylene (PP) has, however, been limited, because of difficulties associated with surface interactions. This is due to the inherent hydrophilic nature of fibres and hydrophobic nature of polyolefin matrices which in consequent gives difficulty in processing. In addition, traditional chemical treatment such as the use of coupling agents is not very effective in improving the interfacial interaction between fibre–polyolefin matrixes. However, the use of functionalized polyolefins, such as MA-*g*-PP or MA-*g*-PE, as compatibilizers has resulted in substantial improvements in the properties of olefin composites (Sanadi et al. 1994). Several research works reported the use of different types of natural fibres (chemically modified or not) as the reinforcing phase for polyolefin-based composites. Sanadi et al. (1994) showed that kenaf fibre-reinforced PP can exhibit significant improvement in tensile and flexible properties over the virgin PP and $CaCO_3$-filled PP. Khan et al. (2002, 2009a, b, 2010a, b, c) explored different ways of modifying jute fibre surfaces in order to enhance PP matrix and fibre interaction that ultimately led to improved mechanical performance. They (2002; 2010a, b, c) also investigated the effect of gamma and UV radiation on the thermal stability and mechanical properties of polyolefin composites. Recently, some works have been reported on the use of oil palm, banana, and coir fibres as the reinforcing agents for PP, LLDPE, LDPE, and HDPE matrices. Ren et al. (2014) used twin screw extruder to produce bamboo pulp fibre-reinforced

HDPE composites and observed improvement in mechanical and thermal properties with 30 % fibre content. Another study (AlMaadeed et al. 2014) showed that the wood content significantly affect the mechanical properties of palm wood/LDPE composites. Increasing wood filler also contributed to the increase in water absorption capacity of the composite. Similar study was done by Darie et al. and they observed pseudoplasticity of LDPE with increasing oak wood content (Darie et al. 2011).

In a recent work, Kakou et al. (2014) developed oil palm fibre-reinforced HDPE composite with 40 % fibre. They used maleated PP as the compatibilizer and thus improved the thermal stability and mechanical properties. Apart from the production of polyolefin composites having natural fibres as the reinforcing phase, the biodegradability is always an important issue with such composites (Joseph et al. 2002; Abu-Sharkh and Hamid 2004; Arutchelvi et al. 2008).

2.3.2 *Glassy Polymers*

Glassy polymers are widely used for their optical clarity, high heat distortion temperature, and high-impact strength. However, such polymers require high temperature for processing which can degrade natural fibre like jute or kenaf. Thus, a limited number of works have been done on natural fibre-reinforced thermoplastic glassy polymers.

2.3.2.1 Polycarbonate

Khan et al. (2005) reported the use of jute fibre treated with HEMA (2-hydroxyethyl methacrylate) to reinforce polycarbonate (PC) matrix. They found that HEMA acted as a coupling agent between fibre and matrix and observed improved adhesion that ultimately improve the storage modulus of the composite significantly. Khan et al. (2006) also showed that similar improvement in interfacial adhesion and mechanical behavior can be achieved by treating the jute fibre with a silane monomer (γ-aminopropyltrimethoxysilane). Recent research also showed that fibre modification is necessary in order to achieve good adhesion between fibre and PC matrix. Threepopnatkul et al. (2009) showed that alkali-treated pineapple leaf fibre modified with silane coupling agents can improve the adhesion between PC matrix and fibre phase and thus improved the thermal stability as well as tensile and impact strength of the composite.

2.3.2.2 Polymethyl Methacrylate

Polymethyl methacrylate (PMMA) is another widely used thermoplastic glassy polymer which is well known for its excellent optical clarity. However, it shows very low fatigue resistance, unsatisfactory transverse strength, and low impact strength. Hence, there is a need for the improvement of its fracture resistance. Very few works have been reported on the use of natural fibres to reinforce the PMMA matrix. John et al. (2014) explored the feasibility of using oil palm fibre as the reinforcement phase for PMMA matrix. They improved the flexural strength and impact strength of PMMA composites. Another work by Sain et al. (2014a) investigated the effect of jute fibre modification on the interfacial compatibility between PMMA matrix and fibre and observed significant improvement in thermal stability and mechanical strength of the composites. Recently, Sain et al. (2014b) reported that by modifying the cellulose nanofibre by maleic anhydride and methyl methacrylate, good biodegradability can be achieved.

2.3.2.3 Polystyrene

Polystyrenes have limited use in engineering applications due to the trade-off between toughness and stiffness. For example, atactic polystyrene is widely regarded as being relatively stiff but suffers from poor toughness. A good number of research works have been done to improve the toughness and impact strength by adding inorganic particles, rubbers, impact modifier particles, and so on. However, also in this case, the use of natural fibre is very limited. Nair et al. (1996) showed that benzoylated short sisal fibre can be used to reinforce the polystyrene (PS) matrix and improve the impact strength. However, the toughness of the composite could not be improved significantly. In a recent work, Oumer and Bachtiar (2014) showed that sugar palm fibre can be used to improve the impact strength and stiffness. However, the toughness was not improved.

2.3.3 Polyamides

Polyamides are important class of engineering thermoplastics and well known for their creep resistance, stiffness, and toughness. Polyamide composites have been used in automotive industries intensively because of their relatively lightweight, durability, corrosion resistant, and attractive appearance (Johnson et al. 2004). A good number of literature can be found on the natural fibre-reinforced nylon composites. This is because nylons possess higher mechanical behavior as the matrix, and they are very compatible with lignocellulosic fibres (Graff 2005). Different natural fibres such as sisal, palm, kenaf, and flax have been used to produce nylon composites. Thitithanasarn et al. (2012) reported the development of nylon 6 composites with epoxy treated jute and observed improved interfacial compatibility that led to improved mechanical performance. Attempts have also been made to optimize the processing condition for nylon 6 composites reinforced with blend of different natural fibres (Ozen et al. 2013). Some research works have also reported the development of bio-based polyamides reinforced with fibres (Feldmann and Bledzki 2014). The advantages of such composite are their biodegradability and recyclability.

2.3.4 *Bioplastics*

Reinforced bioplastics are relatively new class of composites that have attracted a lot of attention in the recent era due to the increasing environmental awareness concerning the waste management of plastics. Although bioplastics are environment-friendly, they still have poor thermal and mechanical stability. However, recent studies showed that biocomposites made of bioplastics and biofibres can compete with the conventional engineering plastics. Among different bioplastics, PLLA, PCL, PVA, PBS, and PHAs and some waste proteins like gelatin are widely used. Table 2.2 shows a list of different biodegradable polymers.

Bledzki and Jaszkiewicz (2010) performed a comparative study on the mechanical properties of PLA, PHB, and PP reinforced with cellulose, abaca, and jute fibres. Figure 2.4 shows that bio-based polymer composites revealed higher tensile strength and modulus compared to petrochemical-based composites. They (Bledzki and Jaszkiewicz 2010) also observed that biofibre-reinforced PLA composites showed improved properties than reinforced PP composites.

Kim et al. (2011) compared the odor emission from PLA and PBS composites with PP- and PE-based composites. They found that PLA and PBS composites emit less odor than PE- and PP-based composites. This indicates that PLA and PBS composites are suitable for interior application and can replace PP- or PE-based composites. It is important to note that many lignocellulosic fibres and man-made fibres have been used to produce PLA composites by using traditional manufacturing technologies, and excellent thermal and mechanical biodegradability was observed as reviewed by Ganster and Fink (2010).

2.4 Critical Issues in the Processing of Natural Fibre-Reinforced Composites

There are several important issues in processing natural fibre-reinforced composites. The issues are mostly related to the structures and properties of the fibres. Thus, the properties of the composites depend on the proper selection of matrix polymer, pretreatment of fibres, and thermal stability of fibre during the processing. Such issues are discussed below.

2.4.1 *Thermal Stability of Natural Fibres*

The thermal degradation of natural fibres is an important concern for the processing of NFRCs. It is well known that all lignocellulosic fibres undergo chemical and physical changes when they are heated at a temperature range of 100–250 °C (Gassan and Bledzki 2001). Some of the physical changes involved the alterations

Table 2.2 Different types of biopolymers, their manufacturers, and trade names (Adopted from Mohanty et al. 2000)

Material class	Manufacturer	Product name
Cellulose acetate	Mazzucchelli	BIOCETA®
	Planet polymer	EnviroPlastic®-Z
Copolyester	BASF	Ecoflex
	Eastman	Easter Bio™
Polycaprolactone (PCL)	Birmingham polymers	Poly(caprolactone)
	Planet polymer	EnviroPlastic®-C
	Solvay	CAPA®
	Union carbide	TONE®
Poly(ester amide)	Bayer	BAK 1095
		BAK 2195
Polyethylene terephthalate (PET) modified	DuPont	Biomax®
Polyglycolide (PGA)	Alkermes	Medisorb®
	Birmingham polymers	Poly(glycolide)
	Boehringer ingelheim	Resomer®
	PURAC	PURASORB® PG
Polyhydroxyalkanoates (PHA)	Metabolix	PHA
	Biomer	Biomer™
	Monsanto	Biopol®
Poly(lactic acid) (PLA)	Alkermers	Medisorb®
	Birmingham polymers	Poly(L-lactide) and poly(D,L-lactide)
	Boehringer ingelheim	Resomer®
	PURAC	PURASORB® PL/PD/PDL
Poly(vinyl alcohol) (PVOH)	Idroplast	Hydrolene®
	Novon Planet polymer	Aqua-NOVON® Aquadro™
	Texas polymer	Vinex™
Starch and starch blends	AVEBE	Paragon™
	BioPlastic (Michigan)	Envar™
	BIOTEC	Bioplast®, Bioflex®, Biopur®
	Earth Shell	Starch-based composite
	Novamont	Mater-Bi™
	Novon	Poly-NOVON®
	Starch Tech	ST1, ST2, ST3
Other blends	Alkermers	Medisorb®
	BioPlastic (Colorado)	Biocomposite material
	Birmingham polymers	Poly(D,L-lactide-*co-caprolactone*)
		Poly(D,L-lactide-*co*-glycolide)

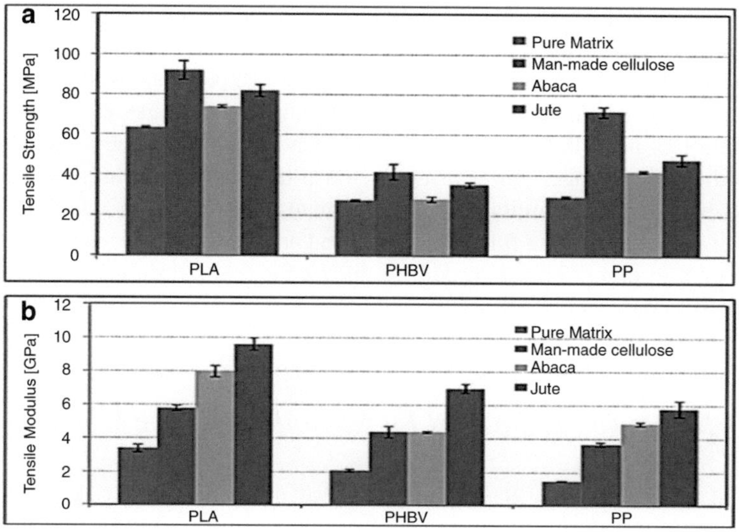

Fig. 2.4 Tensile strength (**a**) and tensile modulus (**b**) of PLA, PHBV, and PP composites with man-made cellulose, abaca, and jute fibres

in physical or chemical structures such as depolymerization, hydrolysis, oxidation, dehydration, decarboxylation, and recrystallization. For an example, Zeronian (1977) reported that the depolymerization takes place in cotton fibre when they are heated from 165 to 240 °C. The degree in the depolymerization decreased from 5,360 to 320 and also recrystallization took place. Similar observation was reported by Rusznák and Zimmer (1971). However, heating (at 160 °C) a composite reinforced with cotton fibre can cause rapid deterioration of strength (Brushwood 1988). Burger et al. (1995) observed similar loss in strength and tenacity of flax fibre. There are other reports that also showed the detrimental effect of processing temperature on the mechanical properties of natural fibre-reinforced composites. Thus, it is important to improve the thermal stability of natural fibre that will also improve the mechanical performances of the composites.

A good number of chemical and physical treatments of natural fibres, as a crucial preprocessing step of composite fabrication, have been reported in the literature. Grafting of functional monomer on the fibre surface is a popular way to improve the thermal stability of natural fibres. For example, Mohanty et al. (1989) reported the improvement in thermal stability of jute fibre by grafting acrylonitrile on the fibre surface. Khan et al. (2002, 2009a, b, 2010a, b, c) reported quite a good number of research works on the fibre surface modification by photoinduced grafting of functional monomers on the fibre surface. They also reported several work on the gamma radiation-treated jute fibres. They stated that the thermal stability of jute fibre can be improved by gamma irradiation (at a certain dose). Recent study by George et al. (2014) showed an inexpensive and environment-friendly method to improve the thermal stability of natural fibres. They improved the thermal stability of flex and hemp fibres by treating them with different enzymes such as hemicellulases, pectinases, and oxidoreductase. Treatment with hemicellulose and pectinase improved the thermal properties for both fibres. Kalia et al. (2013) wrote a nice review on the surface modification of plant fibres by environment-friendly methods. They stated that surface modification of plant fibres using chemical treatments becomes less attractive because of a number of limitations. Environment-friendly methods such as plasma treatment, treatments using fungi, enzymes, and bacteria, can be used for the surface modification of plant fibres on order to improve the thermal stability as well as the other physico-mechanical properties.

2.4.2 Dispersion of Natural Fibres in the Polymer Matrix

It is well known that the dispersion of natural fibres determines the mechanical performances of the composites. Most often bad dispersion of fibres can lead to the deterioration of the mechanical properties of the composites. Although the dispersion is a fibre size-dependent phenomenon, it is not easy to achieve good dispersion of natural fibres in the matrix. Different physical and chemical treatments of fibres can be carried out to improve the fibre dispersion in the matrix. Most often pretreatment of fibres is done in order to improve the dispersion in the matrix. Different coupling agents, such as maleic anhydride-grafted PE, mineral oil, or stearic acid, have been used to improve the dispersion. Stearic acid improves the dispersion by reducing the fibre to fibre interaction, whereas mineral oil acts as a lubricant that induce disentanglement of fibres and thus improves the dispersion.

2.4.3 Hydrophilicity of Natural Fibres

Most of the lignocellulosic fibres are hydrophilic due to the presence of hydroxyl groups, and thus they are vulnerable to microbial degradation. In addition, these fibres are incompatible to most of the hydrophobic polymer matrices. The incompatibility of fibres to the matrix contributed to poor mechanical performances of the composites. Moreover, poor wetting fibres cause difficulty in mixing. Thus, chemical and physical treatments are necessary to reduce the moisture absorption by fibres (John and Anandjiwala 2008). One of the most conventional ways of reducing moisture absorption by fibres is the hydrothermal treatment of natural fibres. Due to the increase in crystallinity of cellulose and separating portion of hemicellulose by hydrothermal treatment, water absorption is reduced (Pott et al. 1999, 2000). Photocuring of monomers on fibre surface is a popular chemical treatment to reduce the water absorption (Khan et al. 2001, 2005, 2008, 2010a, b, c).

2.4.4 *Interfacial Adhesion Between Natural Fibre and Polymer Matrix*

Most of the polymers are hydrophobic in nature, and thus good adhesion between fibre and polymers is the most important issue in achieving desired properties of the matrix (Tang and Kardos 1997). In addition, interface in composites plays key role in transferring the stress through the fibres to matrix. There are several ways to improve the interfacial adhesion between fibres and matrix (Herrera-Franco and Drzal 1992; Qaiss et al. 2014). For example, fibre surface can be treated with different functionalized monomers that act as coupling agents to matrix, or functionalized polymer matrix can also be used (Petinakis et al. 2014). In addition, radiation treatment of fibres such as gamma irradiation or plasma treatment of fibre surface can be useful to functionalize the surface of polymer as well as the fibres. In addition to the surface treatment of fibres, the use of a compatibilizer or coupling agents can also improve the effective stress transfer across the interface (Tang and Kardos 1997; Wang 2006). The compatibilizer can be polymers with functional groups grafted onto the chain of the polymer such as tetrafunctional organometallic compounds based on silicon, titanium, and zirconium and are commonly known as silane, zirconate, or titanate coupling agents (Tang and Kardos 1997).

2.5 Different Processing Techniques to Fabricate NFRCs

Appropriate manufacturing process is necessary in order to alter materials into the right shape without producing any imperfection into the product. There are several important criteria to find out the right processing technique. The size, shape, and desired properties of the composites are the three most important criteria for the production of composites. In addition, speed of the production and manufacturing cost are fairly more important criteria in the industries. The preliminary assessment to choose appropriate processing techniques depends on the size and shape of the composites. In addition, the properties of raw materials also determine further modification to the processing technique. Several manufacturing techniques have been discussed below.

2.5.1 *Fabrication of NFRCs by Injection Molding*

A great deal of research works have been reported on the fabrication of NFRCs by injection molding (Huda et al. 2005a, b, 2006a, b; Birgitha 2007). Usually polymer pellets containing chopped fibres are fed through the hopper into feed block with rotating screws. Then the viscous flow of the polymer melt carrying the fibres is forced into the mold cavities. The viscous melt then solidifies at certain pressure and temperature.

The orientation and distribution of the fibres are determined in this stage. After sufficient cooling, the composite is removed with a desired shape.

The traditional injection molding process has several limitations. For example, the fibre length is usually shorter in the final product than it is expected due to the solidification of the melt in the mold cavities. There are several other issues that need to be addressed properly in order to obtain good composites, and the issues are residual stress development during the molding that may cause cracking and long-term deformation of the composites.

Process, material, and geometric parameters should be optimized to minimize these problems to happen. Process parameters include the melt temperature, injection and screw speeds, injection pressure, and the mold temperature that can be controlled on the injection units. Increasing mold temperature results in a decreasing overall stress level, while the compressive stress region is shifted onto the surface (White 1985). According to recent research and investigations, for biocomposites, the machine temperature of biodegradable polymer, such as poly-lactic-acid (PLA) composites, was made by using injection molding process that should be restricted in the range of 150–210 °C depending on the type of PLAs and their crystallinity from diverse manufacturers.

2.5.2 *Compression Molding*

Fabrication of NFRCs by compression molding can be done in two different ways: hot press or autoclave process. In an autoclave process, the compounded materials are kept inside the pressure chamber under vacuum at a certain temperature, certainly above the softening point of the polymer matrix to allow sufficient flow of the fibres and matrix. After several heat–pressure cycles, a composite is formed (Mallick 1993). On the other hand, for the hot-press processing, a close mold may or may not be necessary. It is important to make sure that fibres do not break under a certain load in the hot press. Sheet molding compounds (SMCs) and bulk molding compounds (BMCs) are traditional initial charges for compression molding process.

2.5.3 *Resin Transfer Molding*

Liquid composite molding processes encompass resin transfer molding (RTM), vacuum-assisted resin transfer molding (VARTM), structural reaction injection molding (S-RIM), co-injection resin transfer molding (CIRTM), and other subsets where the basic approach is to separately inject the liquid resin into a bed of stationary preforms. The RTM process has become a popular composite manufacturing process due to its capability for high-volume production and cost-effectiveness. Many studies have been made on the potential of natural fibres as reinforcement with renewable polymers as matrix through RTM (Sreekumar et al. 2007; Williams and Wool 2000; Ferland et al. 1996; Kim and Daniel 2003; Ikegawa et al. 1996;

Warrior et al. 2003). In the RTM process, dry fibre preform (impregnating) or porous fibrous preform is placed into the mold cavity. Two matching mold halves are clamped tightly to avoid leakage of resin during injection process. Then, using dispensing equipment, a pressurized molten plastic is injected into the heated mold using single or multiple inlet ports in the mold depending on the complexity of the shape of a final product until the mold is filled with resin. After cooling, the part is then removed from the mold. Post-curing normally is needed to ensure the resin is fully cured (chemically reacted between the resin and its catalyst).

2.6 Structure and Properties of Various NFRCs

Natural fibre-reinforced polymer composites are gaining paramount importance day by day in the composite research field and industry due to their versatile diversified nature and renewability. Structure and properties of lignocellulosic natural fibre-reinforced polymer composites are mainly affected by the type of matrix, the content and properties of the reinforcing fibre, and the fibre–matrix interaction. Some of those properties and their related structures are described below.

2.6.1 Structural Properties

Scanning electron microscope (SEM) is one of the most important and frequently used tools for structural analysis of natural fibre-reinforced polymer composites. Figure 2.5 shows SEM micrographs of jute, coir, abaca, bagasse, bamboo, and palm fibre-reinforced polypropylene (PP) composites. The SEM image of jute fibre-reinforced PP composites (Fig. 2.5a) shows agglomeration of untreated jute in the PP matrix. This feature suggests weak interfacial bonding between the fibre and matrix. Fibre agglomeration is also seen in the SEM images of coir, abaca, and palm-reinforced PP composites (Fig. 2.5b, c, f) indicating weak interfacial adhesion between the fibre and matrix. SEM micrograph of bagasse-reinforced PP composites (Fig. 2.5d) shows fibre pullout due to the lack of interfacial adhesion between raw bagasse fibre and PP matrix. On the other hand, bamboo fibre-reinforced PP composite had broken fibre in their microstructure (Fig. 2.5e). One of the ways of improving the interfacial adhesion between the natural fibre and polymer matrix in composites is chemical treatment of the fibre prior or after composite manufacturing (Rahman et al. 2008; Haque et al. 2010a; Karim et al. 2013; Nahar 2014).

2.6.2 Mechanical Properties

Mechanical properties of natural fibre-reinforced polymer composites are commonly characterized using tensile, flexural, impact, and hardness tests. Those properties of various natural fibre polymer composites are described below.

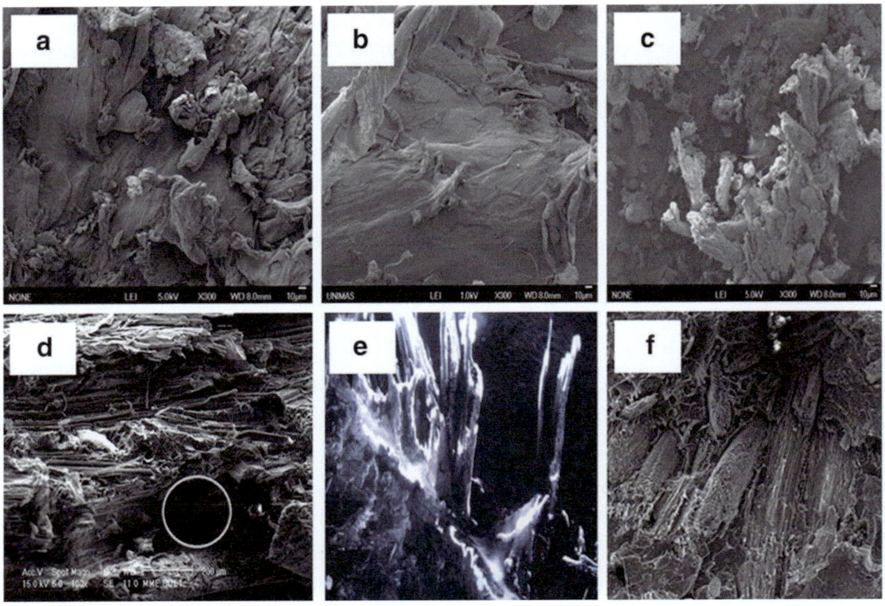

Fig. 2.5 Scanning electron micrographs of (**a**) jute, (**b**) coir, (**c**) abaca, (**d**) bagasse, (**e**) bamboo, and (**f**) palm fibre-reinforced polypropylene composites

2.6.2.1 Tensile Properties

The ultimate tensile strength and Young's modulus are the main two properties evaluated during the tensile test. Those two properties of various natural fibre-reinforced polymer composites are mentioned in Table 2.3. The tensile strength generally decreases with increase in fibre content. As the fibre load increases, the weak interfacial area between the fibre and matrix increases. This in turn decreases the tensile strength (Rahman et al. 2008; Haque et al. 2010a, b; Karim et al. 2013). Banana fibre-reinforced composites had the highest tensile strength, while palm composites had the lowest ones. On the other hand, the Young's modulus generally increases with fibre loading. During tensile loading, partially separated micro-spaces are created, which obstructs stress propagation between the fibre and matrix. As the fibre load increases, the degree of obstruction increases, which consequently increases the stiffness (Rahman et al. 2008; Haque et al. 2010a; Karim et al. 2013; Haque et al. 2010b). The coir fibre- and banana fibre-reinforced composites had the highest and lowest Young's modulus values, respectively.

2.6.2.2 Flexural Properties

Flexural properties are some of the important parameters influencing the performance of hybrid green composites in actual applications. The variation of flexural properties of various natural fibre polymer composites is mentioned in Table 2.3.

Table 2.3 Variation of mechanical properties of various natural fibre-reinforced polymer composites (Rahman et al. 2008; Haque et al. 2010a; Karim et al. 2013; Nahar 2014)

Type of composite	Tensile strength (MPa)	Young's modulus (GPa)	Flexural strength (MPa)	Flexural modulus (MPa)	Impact strength	Hardness (R_L)
Jute-PP	23–29	1.6–2.4	45–54	1.7–2.8	30–51 J/m	76–91
Coir-PP	25–28	1.7–2.7	47–49	1.6–2.8	41–54 J/m	85–87
Abaca-PP	23–27	1.6–2.6	46–48	1.4–2.6	39–46 J/m	79–86
Bagasse-PP	17–22	1.2–1.4	21–34	0.8–1.6	3.3–6.2 kJ/mm^2	–
Banana-PP	36–41	0.82–0.98	–	–	10.2–12.8 kJ/m^2	–
Hemp-PP	27–29	1.6–1.8	–	–	–	–
Palm-PP	21–30	1.1–1.6	44–55	1.6–2.6	39–53 J/m	92–96

The flexural strength and flexural modulus generally increase with increase in fibre loading. Since natural fibres are high modulus materials, higher fibre concentration demands higher stress for the same deformation. Increased fibre–matrix adhesion provides increased stress transfer between them (Rahman et al. 2008; Haque et al. 2010a; Karim et al. 2013; Haque et al. 2010b). Jute fibre-reinforced composites had the highest flexural strength, while bagasse composites had the lowest ones. On the other hand, coir fibre- and bagasse fibre-reinforced composites had the highest and lowest flexural modulus values, respectively.

2.6.2.3 Impact Strength

The impact strength is the ability of a material to withstand fracture or the amount of energy required to propagate a crack. The impact strength of different natural fibre-reinforced polymer composites is reported in Table 2.3. Generally impact strength of natural fibre-reinforced polymer composites increases with fibre addition. The impact strength mainly depends on the nature of the fibre, polymer, and fibre–matrix interfacial bonding. High fibre content increases the probability of fibre agglomeration which results in regions of stress concentration requiring less energy for crack propagation. Another factor of impact failure of composite is fibre pullout. With increase in fibre loading, bigger force is required to pull out the fibres. This consequently increases the impact strength (Rahman et al. 2008; Haque et al. 2010a; Karim et al. 2013; Haque et al. 2010b).

2.6.2.4 Hardness

Table 2.3 shows the hardness values of some natural fibre-reinforced polymer composites. Hardness of natural fibre polymer composites generally increases with increase in fibre loading. The increase in hardness is due to the increase of stiffness of the composites with fibre loading (Rahman et al. 2008; Haque et al. 2010a; Karim et al. 2013; Haque et al. 2010b). According to Table 2.3, palm and jute fibre-reinforced composites had the highest and lowest hardness values, respectively.

2.6.3 Dynamic Mechanical Properties

Dynamic mechanical analyzer (DMA) is a frequently used characterization machine for evaluating dynamic mechanical properties of natural fibre-reinforced polymer composites. Variation of the dynamic mechanical properties of various natural fibre-reinforced polymer composites is shown in Table 2.4. The storage modulus increased with ammonium polyphosphate (APP) addition and NaOH treatment in kenaf fibre-reinforced PLA composites. This was due to better interfacial adhesion and bond strength among the matrix and fibre. NaOH treatment also reduced the loss factor values due to the same reason mentioned above (Shukor et al. 2014a). The storage modulus increased, while the loss factor decreased with increase in frequency in doum fibre-reinforced PP composites. This was due to relatively high time response of the matrix chains. Hence, at high frequencies, the composite behaved like a solid. However, at low frequencies, the polymer chains had time to relax that decreased the loss factor (Essabir et al. 2013). In kenaf fibre-reinforced polyethylene (PE) composites, the storage modulus decreased with temperature due to the softening of the matrix. On the other hand, the same property increased with fibre content due to the stiffening of the overall composite. The loss modulus and loss factor also increased with increase in kenaf fibre content in the composite.

2.6.4 Thermal Properties

Thermogravimetric analysis (TGA) and differential scanning calorimetry (DSC) are the two main tools that determine the thermal properties of natural fibre-reinforced polymer composites. Variation of thermal properties of various natural fibre-reinforced polymer composites is shown in Table 2.4. The glass transition temperature increased with APP addition and NaOH treatment in kenaf fibre-reinforced

Table 2.4 Variation of dynamic mechanical and thermal properties of various natural fibre-reinforced polymer composites (Elkhaoulani et al. 2013; Shukor et al. 2014a; Essabir et al. 2013; Salleh et al. 2014; Shukor et al. 2014b; Shekeil et al. 2014)

	Dynamic mechanical properties			Thermal properties	
Type of composites	Maximum storage modulus (MPa)	Maximum loss modulus (GPa)	Maximum tan delta	Maximum glass transition temperature (°C)	Maximum thermal degradation temperature (°C)
Kenaf-PLA	5	1	1.7	67	305
Doum-PP	–	–	0.16	–	–
Kenaf-PE	8	0.7	0.24	–	–
Hemp-PP	–	–	–	–	376
Kenaf-PVC/TPU	–	–	–	–	281

PLA composites. This was due to the decrease in mobility of the matrix chains, which indicated improved interfacial adhesion between the fibre and matrix (Shukor et al. 2014a). The thermal stability of the same composites also increased with APP addition and NaOH treatment (Shukor et al. 2014b). In hemp fibre-reinforced PP composites, the thermal stability increased with increase in hemp fibre loading due to the creation of higher barrier to polymer exhaust gas formation (Elkhaoulani et al. 2013). The maximum degradation peak increased with increase in kenaf fibre content in kenaf fibre-reinforced polyvinyl chloride/thermoplastic polyurethane polymer blend composites (Shekeil et al. 2014).

2.7 Application of NFRCs

Being lightweight, economically cost-effective, highly abundant, and environment-friendly, NFRCs are useful for various applications such as products of commodity to aerospace, examples including electroactive papers, fuel cell membranes, controlled drug release mechanisms, and biosensors. Several useful applications of NFRCs in different areas are described below.

2.7.1 Automotive Applications

There is an increasing trend of utilizing NFRCs in automotive industries. Most of the major vehicle manufacturers around the world are utilizing NFRCs in various applications such as those listed in Table 2.5 (Faruk et al. 2014).

In the manufacturing of interior and exterior of vehicles, weight reduction is a key issue. Materials have to be lightweight but strong, i.e., possess good impact performance in order to avoid any occupational damage. It was reported that the reduction of vehicle weight not only reduces the fuel requirements but also reduces the greenhouse gas emission. Pandey et al. reported that a 25 % reduction in vehicle weight is equivalent to a saving of 250 million barrels of crude oil and a reduction in CO_2 emissions of 220 billion pounds per year (Pandey et al. 2010; Gejo et al. 2010; Błędzki et al. 2014).

Among various plastics, PP is the mostly used thermoplastic resin in automotive industries due to their low density, excellent mechanical properties, and processability. Several other thermoplastics such as polyethylene, polystyrene, and polyamides (nylon 6 and 6, 6) have also been used intensively in automotive industries. However, due to the raised awareness on environmental issues with the nonbiodegradable and non-recyclable polymers, most of the plastic manufacturers as well as automobile manufacturers are trying to use bioplastics as the alternative sustainable solutions. Several research works have also reported the manufacturing of bioplastic-based composites for the automotive applications. Graupner et al. (2009) reported a good number of PLA-based composites reinforced with hemp, cotton, kenaf, and

Table 2.5 Automotive models, manufacturers, and components using natural fibre-reinforced composites (Adopted from Faruk et al. 2014)

Models	Manufacturers	Components
A2, A3, A4, A4 Avant, A6, A8	Audi	Seat back, side and back door panel, boot lining, hat rack, spare tire lining
Roadstar, Coupe		
C5	Citroen	Interior door paneling
3, 5, 7 series	BMW	Door panels, headliner panel, boot lining, seat back, noise insulation panels, molded foot well linings
Eco Elise	Lotus	Body panels, spoiler, seats, interior carpets
Punto, Brava, Marea, Alfa	Fiat	Door panel
Romeo 146, 156		
Astra, Vectra, Zafira	Opel	Instrumental panel, headliner panel, door panels, pillar cover panel
406	Peugeot	Front and rear door panels
2000 and others	Rover	Insulation, rear storage shelf/panel
Raum, Brevis, Harrier, Celsior	Toyota	Door panels, seat backs, floor mats, spare tire cover
Golf A4, Passat Variant, Bora	Volkswagen	Door panel, seat back, boot-lid finish panel, boot liner
Space star, Colt	Mitsubishi	Cargo area floor, door panels, instrumental panels
Clio, Twingo	Renault	Rear parcel shelf
Mercedes A, C, E, S class, Trucks, EvoBus (exterior)	Daimler-Benz	Door panels, windshield/dashboard, business table, piller cover panel, glove box, instrumental panel support, insulation, molding rod/apertures, seat backrest panel, trunk panel, seat surface/backrest, internal engine cover, engine insulation, sun visor, bumper, wheel box, roof cover
Pilot	Honda	Cargo area
C70, V70	Volvo	Seat padding, natural foams, cargo floor tray
Cadillac Deville, Chevrolet TrailBlazer	General Motors	Seat backs, cargo area floor
L3000	Saturn	Package trays and door panel
Mondeo CD 162, Focus, Freestar	Ford	Floor trays, door panels, B-pillar, boot liner

cellulose fibres. They investigated the effect of fibre content and loading direction on their mechanical performance and found that PLA reinforced with man-made cellulose fibres showed highest tensile strength followed by hemp- and kenaf-reinforced systems.

Several other bioplastics have been commercialized by different chemical companies around the globe to replace the nonbiodegradable commodity plastics. For example, PTAT or Eastar Bio from Eastman Chemical Company, BioPET or Biomax from Dupont, Ecoflex from BASF, PLA from Dow-Cargill, and PHBV from Metabolix are other examples of environment-friendly material for automobiles.

Most of this material contains biodegradable materials, tested under composting conditions, which can degrade in the natural condition leaving behind biomass and carbon dioxide.

2.7.2 Packaging Applications

Polymers alone do not possess good barrier to most common gases like N_2, O_2, CO_2, or water vapor. This is why polymers are hardly used in perishable food packaging. In order to improve their barrier properties, the phase morphology can be engineered or some fillers or additives can be used. In most NFRCs, fibre phase may play significant role in improving the barrier properties of the polymers. Several research works reported that the presence of crystalline segments in lignocellulosic fibres leads to slower diffusion of gases or water molecule through the composites (Pandey et al. 2010). Lee and Wang (2006) recently reviewed the use of biofibres for packaging applications. They reported several limitations of fibre-filled plastics for packaging applications such as the hydrophilic nature of the fibres, degradability, lower mechanical properties in humid environment, and so on. However, they mentioned that such composites can be utilized in areas where plastics cannot be recycled or reused or do not have sustainable waste disposal facilities.

Natural fibre-based packaging materials have many advantages over synthetic packaging materials such as stiffness vs. weight ratio and recyclability. A good number of research works reported the production of high-performance bioplastic composites filled with fibres, and they can be applied in food and pharmaceutical packaging. In a recent study, it was found that the environmental degradation of packaging materials, which depends on the amount of organic but degradable content, is very important concerning the development of short-life disposable packaging plastics. In recent years several companies took initiatives to produce recycled packaging materials, for example, Placon Corp. has introduced EcoStar™ HS 1000, a PET-based eco-friendly recyclable material, and Toray Plastics America has developed solvent-free polyester LumiLid lidding films.

2.7.3 Bio-Based Electronics

Bio-based electronics is an emerging area that emphasizes the use of bio-based polymers and composites for electronic applications. The aim of such emerging research area is to create and innovate human- and environment-friendly electronics (Irimia-Valdu et al. 2014). This research area may also be useful in fulfilling the creation of low-cost, energy-efficient organic electronics. Cellulose nanocomposites with semiconductive polymer matrixes have been studied intensively for actuator and sensor applications. For example, polyaniline (PAN) reinforced with cellulose nanofibres can act as acid or base sensors or as an actuator and can be applied in heating devices (Bhat et al 2006; Nalwa 1997). Several research works reported the tuning of conductivity by modifying the surface of fibres. Cellulose fibres can be coated with conductive polymers such as PAN or polypyrrole (PPy) to improve the conductivity (Sapurina et al. 2005; Konyushenko et al. 2008). In addition, cellulosic composites have also been used as a green alternative to polyolefin-based separators in rechargeable lithium ion batteries (Kuribayashi et al. 1995; Cámer et al. 2008). Most of the cellulose-based separators are thin composite porous membranes composed of cellulose microfibres and a polymeric matrix component such as ethyl acrylate or silicon resin.

2.7.4 *Other Applications of NFRCs*

NFRCs have long been used in the civil infrastructures. Their applications in civil infrastructures include heat insulating boards, windows, pans, and so on. Natural fibres have also been used as the reinforcement for cements to produce low-cost building materials. Fibre cement is one such product which possesses higher toughness and crack resistance in comparison with conventional cement. In addition, fibre-reinforced composites can be utilized as the low-cost building materials in developing countries. Khan et al. developed corrugated sheets made of jute fibre-reinforced unsaturated polyester matrix (Bangladesh Economic News, April 9, 2010). They proposed that such green composite (JUTIN) also possesses good heat insulation and saltwater-resistant properties and could be best useful in coastal areas of Bangladesh. Similar initiatives have been taken in India in order to produce NFRC-based composites for building materials.

NFRCs have also been applied as fire-resistant materials. Due to the porous microstructure as well as the crystalline fractions, most natural fibres have fire-resistant properties (Chapple et al. 2010). In addition the presence of lignin in fibres also induces fire resistance to most NFRCs.

2.8 Summary and Future Prospects

Natural fibres have the potential to replace the traditional man-made fibres. They are sustainable resources for renewable natural materials. Such fibres are useful to produce lightweight and strong composites. Polymer composites with natural fibres offer a unique opportunity to maintain the sustainable development of economically and ecologically attractive technology. It is still necessary to overcome problems associated with the traditional manufacturing techniques of natural fibre-based composites. The burgeoning demand for more environment-friendly materials has already raised the interests among the scientists of different disciplines to develop a flexible and sustainable approach of composite manufacturing technology. Natural fibre-reinforced composites have already achieved wide acceptance in automotive, aerospace, packaging, construction, and other industries.

Numerous research works on bio-based polymers give the indication to the fact that the future materials will be more environment-friendly. Green chemistry has already revealed ways of producing novel biopolymers that can open new windows of thinking about polymers. Natural fibre-reinforced biocomposites can substantially replace the traditional commodity polymers, although current bioplastics still need to overcome some engineering challenges. On the other hand, recent promising works in understanding the genetic sequence of fibrous plants may offer a unique way to produce fibres with the desired properties for flexible manufacturing of composites. In fine, more bottom-up approaches are necessary to fruitfully utilize such green materials for our socio-environmental benefits.

References

Abu-Sharkh BF, Hamid H (2004) Degradation study of date palm fiber/polypropylene composites in natural and artificial weathering: mechanical and thermal analysis. Polym Degrad Stab 85(3):967–973

AlMaadeed MA, Nogellova Z, Mičušík M, Novak I, Krupa I (2014) Mechanical, sorption and adhesive properties of composites based on low density polyethylene filled with date palm wood powder. Mater Des 53:29–37

Arutchelvi J, Sudhakar M, Arkatkar A, Doble M, Bhaduri S, Uppara PV (2008) Biodegradation of polyethylene and polypropylene. Indian J Biotechnol 7(1):9

Bhat NV, Seshadri DT, Nate MM, Gore AV (2006) Development of conductive cotton fabrics for heating devices. J Appl Polym Sci 102(5):4690–4695

Bledzki AK, Jaszkiewicz A (2010) Mechanical performance of biocomposites based on PLA and PHBV reinforced with natural fibers—a comparative study to PP. Compos Sci Technol 70(12):1687–1696

Błędzki A, Urbaniak M, Jaszkiewicz A, Feldmann M (2014) Włóknacelulozowejakoalternatywad lawłókienszklanych w kompozytachpolimerowych. Polimery 59

Birgitha N (2007) Natural fiber composites optimization of microstructure and processing parameters. Thesis

Brushwood DE (1988) Effects of heating on chemical and physical properties and processing quality of cotton. Text Res J 58(6):309–317

Burger H, Koine A, Maron R, Mieck KP (1995) Gummi FaserKunststoffe 48:475

Cámer JG, Morales J, Sánchez L (2008) Nano-Si/cellulose composites as anode materials for lithium-ion batteries. Electrochem Solid State Lett 11(6):A101–A104

Chapple S, Anandjiwala R (2010) Flammability of natural fiber-reinforced composites and strategies for fire retardancy: a review. J Thermoplast Compos Mater 23(6):871–893

Darie RN, Bercea M, Kozlowski M, Spiridon I (2011) Evaluation of properties of LDPE/oak wood composites exposed to artificial ageing. Cellulose Chem Technol 45(1–2):127–135

Dicker MPM, Duckworth PF, Baker AB, Francois G, Hazzard MK, Weaver PM (2014) Green composites: a review of material attributes and complementary applications. Compos Part A 56:280–289

Elkhaoulani A, Arrakhiz FZ, Benmoussa K, Bouhfid R, Qaiss A (2013) Mechanical and thermal properties of polymer composite based on natural fibers: Moroccan hemp fibers/polypropylene. Mater Des 49:203–208; Engineering. 2014(53):760–766

Essabir E, Elkhaoulani A, Benmoussa K, Bouhfid R, Arrakhiz FZ, Qaiss A (2013) Dynamic mechanical behavior analysis of doum fibers reinforced polypropylene composites. Mater Des 51:780–788

Faruk O, Bledzki AK, Fink HP, Sain M (2014) Progress report on natural fiber reinforced composites. Macromol Mater Eng 299(1):9–26

Feldmann M, Bledzki AK (2014) Bio-based polyamides reinforced with cellulosic fibers—processing and properties. Compos Sci Technol 100:113–120
Ferland P, Guittard D, Trochu F (1996) Concurrent methods for permeability measurement in resin transfer molding. Polym Compos 17(1):149–158
Ganster J, Fink HP (2010) PLA-based bio-and nanocomposites. Nano Biocompos 275–290
Gassan J, Bledzki AK (2001) Thermal degradation of flax and jute fibers. J Appl Polym Sci 82(6):1417–1422
Gejo G, Kuruvilla J, Boudenne A, Sabu T (2010) Recent advances in green composites. Key Eng Mater 425:107–166
George M, Mussone PG, Bressler DC (2014) Surface and thermal characterization of natural fibers treated with enzymes. Ind Crops Prod 53:365–373
Graff G (2005) Under-hood applications of nylon accelerate. Retrieved from Omnexus Web: http://www.omnexus.com/resources/articles/article.aspx?id=9660
Graupner N, Herrmann AS, Müssig J (2009) Natural and man-made cellulose fiber-reinforced poly (lactic acid) (PLA) composites: an overview about mechanical characteristics and application areas. Compos Part A Appl Sci Manuf 40(6):810–821
Haque MM, Islam MS, Islam MN, Huque MM, Hasan M (2010a) Physico-mechanical properties of chemically treated palm fiber reinforced polypropylene composites. J Reinforc Plast Compos 29(11):1734–1742
Haque MM, Rahman MR, Islam MN, Huque MM, Hasan M (2010b) Physico-mechanical properties of polypropylene composites reinforced with chemically treated coir and abaca fiber. J Reinforc Plast Compos 29(15):2253–2261
Herrera-Franco PJ, Drzal LT (1992) Comparison of methods for the measurement of fibre/matrix adhesion in composites. Composites 23(1):2–27
Ho MP, Wang H, Lee JH, Ho CK, Lau KT, Leng J, Hui D (2012a) Critical factors on manufacturing processes of natural fiber composites. Compos Part B Eng 43(8):3549–3562
Ho M, Wang H, Lee JH, Ho CK, Lau K, Leng J, Hui D (2012b) Critical factors on manufacturing processes of natural fiber composites. Compos Part B 43:3549–3562. https://bangladeshecon-omy.wordpress.com/2010/04/09/local-researchers-develop-jute-made-substitute-for-ci-sheet/
Huda MS, Drzal LT, Misra M, Mohanty AK, Williams K, Mielewski DF (2005a) A study on bio-composites from recycled newspaper fiber and poly (lactic acid). Ind Eng Chem Res 44(15):5593–5601
Huda MS, Mohanty AK, Drzal LT, Schut E, Misra M (2005b) "Green" composites from recycled cellulose and poly (lactic acid): physico-mechanical and morphological properties evaluation. J Mater Sci 40(16):4221–4229
Huda MS, Drzal LT, Misra M, Mohanty AK (2006a) Wood-fiber-reinforced poly (lactic acid) composites: evaluation of the physicomechanical and morphological properties. J Appl Polym Sci 102(5):4856–4869
Huda MS, Drzal LT, Mohanty AK, Misra M (2006b) Chopped glass and recycled newspaper as reinforcement fibers in injection molded poly (lactic acid) (PLA) composites: a comparative study. Compos Sci Technol 66(11):1813–1824
Ikegawa N, Hamada H, Maekawa Z (1996) Effect of compression process on void behavior in structural resin transfer molding. Polym Eng Sci 36(7):953–962
Irimia-Vladu M (2014) "Green" electronics: biodegradable and biocompatible materials and devices for sustainable future. Chem Soc Rev 43(2):588–610
John MJ, Anandjiwala RD (2008) Recent developments in chemical modification and characterization of natural fiber-reinforced composites. Polym Compos 29(2):187–207
John M, Thomas S (2008) Biofibres and biocomposites. Carbohydr Polym 71:343–364
John J, Ann Mani S, Palaniswamy K, Ramanathan A, Razak AAA (2014) Flexural properties of poly (methyl methacrylate) resin reinforced with oil palm empty fruit bunch fibers: a preliminary finding. J Prosthodont
Johnson RW, Evans JL, Jacobsen P, Thompson JR, Christopher M (2004) The changing automotive environment: high-temperature electronics. IEEE Trans Electr Packaging Manuf 27(3): 164–176

Joseph PV, Rabello MS, Mattoso LHC, Joseph K, Thomas S (2002) Environmental effects on the degradation behaviour of sisal fiber reinforced polypropylene composites. Compos Sci Technol 62(10):1357–1372

Kakou CA, Arrakhiz FZ, Trokourey A, Bouhfid R, Qaiss A, Rodrigue D (2014) Influence of coupling agent content on the properties of high density polyethylene composites reinforced with oil palm fibers. Mater Des 63:641–649

Kalia S, Thakur K, Celli A, Kiechel MA, Schauer CL (2013) Surface modification of plant fibers using environment friendly methods for their application in polymer composites, textile industry and antimicrobial activities: a review. J Environ Chem Eng 1(3):97–112

Karim R, Rahman Md, Hasan FM, Islam MdS, Hassan A (2013) Effect of fiber loading and alkali treatment on physical and mechanical properties of chemically treated bagasse fiber reinforced polypropylene composites. J Polym Mater 30(4):423–433

Khan MA, Hassan MM (2006) Effect of γ-aminopropyltrimethoxysilane on the performance of jute–polycarbonate composites. J Appl Polym Sci 100(5):4142–4154

Khan MA, Hinrichsen G, Drzal LT (2001) Influence of novel coupling agents on mechanical properties of jute reinforced polypropylene composite. J Mater Sci Lett 20(18):1711–1713

Khan MA, Rahman MM, Akhunzada KS (2002) Grafting of different monomers onto jute yarn by in situ UV-radiation method: effect of additives. Polym Plast Technol Eng 41(4):677–689

Khan MA, Masudul Hassan M, Drzal LT (2005) Effect of 2-hydroxyethyl methacrylate (HEMA) on the mechanical and thermal properties of jute-polycarbonate composite. Compos Part A Appl Sci Manuf 36(1):71–81

Khan MA, Khan RA, Hossain A, Khan AH (2008) Effect of gamma radiation on the physico-mechanical and electrical properties of jute fiber-reinforced polypropylene composites. J Reinforc Plast Compos

Khan RA, Khan MA, Khan AH, Hossain MA (2009a) Effect of gamma radiation on the performance of jute fabrics-reinforced polypropylene composites. Radiat Phys Chem 78(11):986–993

Khan RA, Khan MA, Sultana S, Khan MN, Shubhra QT, Noor FG (2009b) Mechanical, degradation, and interfacial properties of synthetic degradable fiber reinforced polypropylene composites. J Reinforc Plast Compos

Khan MA, Khan RA, Haydar UZ, Noor-A-Alam M, Hoque MA (2010a) Effect of surface modification of jute with acrylic monomers on the performance of polypropylene composites. J Reinforc Plast Compos 29(8):1195–1205

Khan AH, Hossain MA, Khan MA, Khan RA, Hakim MA (2010b) Fabrication and characterization of jute reinforced polypropylene composite: effectiveness of coupling agents. J Compos Mater

Khan RA, Khan MA, Zaman HU, Pervin S, Khan N, Sultana S, Mustafa AI (2010c) Comparative studies of mechanical and interfacial properties between jute and e-glass fiber-reinforced polypropylene composites. J Reinforc Plast Compos 29(7):1078–1088

Kim SK, Daniel IM (2003) Determination of three-dimensional permeability of fiber preforms by the inverse parameter estimation technique. Compos Part A Appl Sci Manuf 34(5):421–429

Kim KW, Lee BH, Kim S, Kim HJ, Yun JH, Yoo SE, Sohn JR (2011) Reduction of VOC emission from natural flours filled biodegradable bio-composites for automobile interior. J Hazard Mater 187(1):37–43

Konyushenko EN, Kazantseva NE, Stejskal J, Trchová M, Kovářová J, Sapurina I, Prokeš J (2008) Ferromagnetic behaviour of polyaniline-coated multi-wall carbon nanotubes containing nickel nanoparticles. J Magnetism Magn Mater 320(3):231–240

Kuribayashi I, Yokoyama M, Yamashita M (1995) Battery characteristics with various carbonaceous materials. J Power Sources 54(1):1–5

Lee SH, Wang S (2006) Biodegradable polymers/bamboo fiber biocomposite with bio-based coupling agent. Compos Part A Appl Sci Manuf 37(1):80–91

Mallick PK (1993) Fiber-reinforced composites: materials, manufacturing, and design. CRC Press

Mohanty AK, Patnaik S, Singh BC, Misra M (1989) Graft copolymerization of acrylonitrile onto acetylated jute fibers. J Appl Polym Sci 37(5):1171–1181

Mohanty AK, Misra M, Hinrichsen G (2000) Biofibres, biodegradable polymers and biocomposites: an overview. Macromol Mater Eng 276(1):1–24

Nahar S (2014) Modification and characterization of technical bamboo fiber and their polypropylene based composites. Ph.D. thesis, Bangladesh University of Engineering and Technology, Bangladesh

Nair KC, Diwan SM, Thomas S (1996) Tensile properties of short sisal fiber reinforced polystyrene composites. J Appl Polym Sci 60(9):1483–1497

Nalwa HS (1997) Handbook of organic conductive molecules and polymers, vol 4: Conductive polymers: transport, photophysics and applications

Oumer AN, Bachtiar D (2014) Modeling and experimental validation of tensile properties of sugar palm fiber reinforced high impact polystyrene composites. Fibers Polym 15(2):334–339

Ozen E, Kiziltas A, Kiziltas EE, Gardner DJ (2013) Natural fiber blend—nylon 6 composites. Polym Compos 34(4):544–553

Pandey JK, Ahn SH, Lee CS, Mohanty AK, Misra M (2010) Recent advances in the application of natural fiber based composites. Macromol Mater Eng 295(11):975–989

Petinakis E, Yu L, Simon GP, Dai XJ, Chen Z, Dean K (2014) Interfacial adhesion in natural fiber-reinforced polymer composites. Lignocellulosic polymer composites: processing, characterization, and properties, pp 17–39

Pott GT, Van Deursen JH, Hueting DJ, Van der Wooning A (1999) A novel flax upgrading process for industrial applications. In: Proceedings of international symposium, Werksto ffeausnachwachsendenRohstoffen, Erfurt

Pott GT, Hueting DJ, Van Deursen JH (2000) Reduction of moisture sensitivity in wood and natural fibers for polymer composites. In 3rd international wood and natural fiber composites symposium, Kassel

Qaiss AEK, Bouhfid R, Essabir H (2014) Natural fibers reinforced polymeric matrix: thermal, mechanical and interfacial properties. In: Hakeem KR, Jawaid M, Rashid U (eds) Biomass and Bioenergy. Springer, 2014, pp 225–245

Rahman MR, Huque MM, Islam MN, Hasan M (2008) Improvement of physico-mechanical properties of jute fiber reinforced polypropylene composites by post treatment. Compos A Appl Sci Manuf 39:1739–1747

Ren W, Zhang D, Wang G, Cheng H (2014) Mechanical and thermal properties of bamboo pulp fiber reinforced polyethylene composites. BioResources 9(3):4117–4127

Rusznák I, Zimmer K (1971) Proceedings of the 18th Hungarian textile conference, Budapest 2, p 119

Saheb DN, Jog JP (1999) Natural fibre polymer composites: a review. Adv Polym Technol 18(4):351–363

Sain S, Ray D, Mukhopadhyay A (2014a) Improved mechanical and moisture resistance property of in situ polymerized transparent PMMA/cellulose composites. Polym Compos

Sain S, Sengupta S, Kar A, Mukhopadhyay A, Sengupta S, Kar T, Ray D (2014b) Effect of modified cellulose fibers on the biodegradation behaviour of in-situ formed PMMA/cellulose composites in soil environment: isolation and identification of the composite degrading fungus. Polym Degrad Stab 99:156–165

Salleh FM, Hassan A, Yahya R, Azzahari AD (2014) Effects of extrusion temperature on the rheological, dynamic mechanical and tensile properties of Kenaf fiber/HDPE composites. Compos Part B 58:259–266

Sanadi AR, Calufield DF, Rowell RM (1994) Reinforcing polypropylene with natural fibers. Plast Eng (USA) 50(4):27–28

Sapurina I, Kazantseva NE, Ryvkina NG, Prokeš J, Sáha P, Stejskal J (2005) Electromagnetic radiation shielding by composites of conducting polymers and wood. J Appl Polym Sci 95(4):807–814

Satyanarayana KG, Arizaga GG, Wypych F (2009) Biodegradable composites based on lignocellulosic fibers—an overview. Prog Polym Sci 34(9):982–1021

Shekeil YAE, Sapuan SM, Jawaid M, Shuja'a OMA (2014) Influence of fiber content on mechanical, morphological and thermal properties of Kenaf fibers reinforced poly(vinyl chloride)/thermoplastic polyurethane poly-blend composites. Mater Des 58:130–135

Shen L, Haufe J, Patel MK (2009) Product overview and market projection of emerging bio-based plastics, PRO-BIP, Final Report, Utrecht, The Netherlands
Shukor F, Hassan A, Hasan M, Islam MdS, Mokhtar M (2014a) PLA/Kenaf/APP biocomposites: effect of alkali treatment and ammonium polyphosphate (APP) on dynamic mechanical and morphological properties. Polym Plast Technol Eng 53:760–766
Shukor F, Hassan A, Saiful Islam Md, Mokhtar M, Hasan M (2014b) Effect of ammonium polyphosphate on flame retardancy, thermal stability and mechanical properties of alkali treated kenaf fiber filled PLA biocomposites. Mater Des 54:425–429
Sreekumar PA, Joseph K, Unnikrishnan G, Thomas S (2007) A comparative study on mechanical properties of sisal-leaf fiber-reinforced polyester composites prepared by resin transfer and compression moulding techniques. Compos Sci Technol 67(3):453–461
Taj S, Munawar MA, Khan SU (2007) Natural fibre-reinforced polymer composites. Proc Pakistan Acad Sci 44(2):129–144
Tang L-G, Kardos JL (1997) A review of methods for improving the interfacial adhesion between carbon fiber and polymer matrix. Polym Compos 18(1):100–113
Thitithanasarn S, Yamada K, Ishiaku US, Hamada H (2012) The effect of curative concentration on thermal and mechanical properties of flexible epoxy coated jute fabric reinforced polyamide 6 composites. Open J Compos Mater 2(04):133
Threepopnatkul P, Kaerkitcha N, Athipongarporn N (2009) Effect of surface treatment on performance of pineapple leaf fiber–polycarbonate composites. Compos Part B Eng 40(7): 628–632
Wambua P, Ivens J, Verpoest I (2003) Natural fibers: can they replace glass in fiber reinforced plastics? Compos Sci Technol 63(9):1259–1264
Warrior NA, Turner TA, Robitaille F, Rudd CD (2003) Effect of resin properties and processing parameters on crash energy absorbing composite structures made by RTM. Compos Part A 34:543–550
White JR (1985) On the layer removal analysis of residual stress. J Mater Sci 20(7):2377–2387
Williams GI, Wool RP (2000) Composites from natural fibers and soy oil resins. Appl Compos Mater 7(5–6):421–432
Zeronian SH (1977) In: Arthur JC (ed) Cellulose chemistry and technology. American Chemical Society, Washington, DC

Chapter 3
Green Composite Manufacturing via Compression Molding and Thermoforming

Birat KC, Muhammad Pervaiz, Omar Faruk, Jimi Tjong, and Mohini Sain

Abstract Increasing concern over material usage and its impact on the environmental have escalated the growth of green composite materials. There are tremendous opportunities where conventional mineral and synthetic-based materials can be replaced with green composite materials. Before green composites can be used to manufacture various products, it is important to understand their processing steps and optimize process parameters. Past researches on green composites were focused mostly on characterization, and less research can be found in manufacturing of green composites. Common technologies include but are not limited to injection molding, extrusion, thermoforming, and compression molding. In this chapter, manufacturing process of green composites via compression molding and thermoforming is developed based on patents and literature review. Main emphasis is given toward key processing steps, such as material selection, preprocessing, semifinished product manufacturing, and green composite fabrication. Moreover, processing data of some commercially available green composites and biopolymers is summarized.

Keywords Green composites • Compression molding • Thermoforming • Material selection • Semifinished product • Prepregs • Composite fabrication

3.1 Green Composites

There is a growing trend in plastic part manufacturers for processing new class of composite materials, called green composites for reducing the environmental impact and reducing the cost of raw materials (Reddy and Yang 2011). These composites offer practical solution to conventional glass-fibre-reinforced or mineral-filled plastic composites where renewability and sustainability of materials were compromised

Birat KC (✉) • M. Pervaiz • O. Faruk • M. Sain
Faculty of Forestry, Centre for Biocomposites and Biomaterials Processing (CBBP), University of Toronto, 33 Willcocks Street, Toronto, ON, Canada, M5S 3B3
e-mail: b.kc@mail.utoronto.ca

J. Tjong
Faculty of Forestry, Centre for Biocomposites and Biomaterials Processing (CBBP), 33 Willcocks St, Toronto, ON, Canada, M5S 5B3

M.S. Salit et al. (eds.), *Manufacturing of Natural Fibre Reinforced Polymer Composites*, DOI 10.1007/978-3-319-07944-8_3

(Thakur et al. 2013). Green composites can be defined as the combination of two or more materials completely from renewable sources. For example, composites containing natural fibre and biopolymers or the combination of two biopolymers (Ben and Kihara 2007; Gejo et al. 2010) Natural fibre are carbohydrates derived from various plant sources, while biopolymers are usually derived from plant-based sugars, starch, proteins, or vegetable oils (Ben and Kihara 2007). Most of the research and development work in the past focused on composites from natural fibre reinforced with polyolefins and thermosets (Ben and Kihara 2007; Reddy and Yang 2011). With these composites, there are concerns regarding their end-of-life recycling and disposal (Gejo et al. 2010). Hence, the research focus has shifted toward green composites that come from 100 % renewable source and are completely biodegradable after the product life (Takemura 2010). In recent years, few applications of green composites are found in food and packaging, agricultural, automotive, and medical industry. Most commonly used natural fibres are wood, hemp, flax, wheat, jute, sisal, coconut, banana, and bagasse. Recently, micro- and nanocellulose from plants, algae, and bacteria are also being studied as a new source of natural fibre reinforcements (Reddy and Yang 2011). Commonly studied biopolymers are soy-based epoxy, starch, proteins polycaprolactone (PCL), polyhydroxybutyrate (PHB), polylactic acid (PLA), polyester amide 11 (PA11), and polyvinyl alcohol (PVA) (Reddy and Yang 2011; Takemura 2010). Among these, PLA has seen much commercial success while PHB and PA11 are also slowly emerging (U.S. Patent No. US20100170649 A1 2010a).

Most common processing technologies for green composites are injection molding, compression molding, extrusion, thermoforming, pultrusion, and resin transfer molding (RTM) (Faruk et al. 2012). Study by Altun et al. (2013) used injection molding to manufacture wood flour and PLA composites, while Hu et al. (2012) used compression molding to manufacture jute and PLA composites. Other study reported use of thermoforming (Chang et al. 2009) and RTM (Faruk et al. 2012). Higher demand for plastic composites as a substitute for metal parts has led to series of novel methods that combine these technologies such as direct long-fibre thermoplastics (DLFTs) (Faruk et al. 2012). Broadly, selection of these manufacturing technologies depends on the type and the form of material to be processed, volume of production, capital cost requirement, complexity of the part design, and quality of the part (Sain and Pervaiz 2008). In this chapter, we explore the fabrication process for green composites via compression molding (CM) and thermoforming (TF). Basic processing steps and their features are briefly summarized with the emphasis on green composite materials.

3.1.1 Compression Molding and Thermoforming

In compression molding, thermoplastic or thermoset material (in the form of granules, sheet, or prepregs) is melted in the mold cavity under heat and pressure, followed by cooling and part removal after it is cured (solidified) (Faruk et al. 2012; Onal and Adanur 2005). In thermoforming, a thermoplastic sheet is heated to its softening point and stretched over into a single-sided mold and holding in place while it cools and solidifies to form a desired part (Erchiqui et al. 2009; Klein 2009). Schematic of CM and TM process is shown in Fig. 3.1. From the production point of view, CM is usually used for complex part designs and TM for simpler part designs. However more and more complex parts are being thermoformed due its increasing use (Erchiqui et al. 2009). Part quality requirement is also another consideration for the selection of CM and TF. CM is also used in relatively higher-quality parts as compared to TF because of fewer knit lines and less fibre-length degradation (Faruk et al. 2012). However inconsistencies as a result of incomplete mixing of fibre with resin may persist in CM if fibre yarns or fibre mats are used (Cheng 2009). Likewise, stretching process in TF creates nonuniform wall thickness which is undesirable for higher-quality parts with tight geometric tolerances (Klein 2009). Some other features of CM and TF are summarized in Table 3.1.

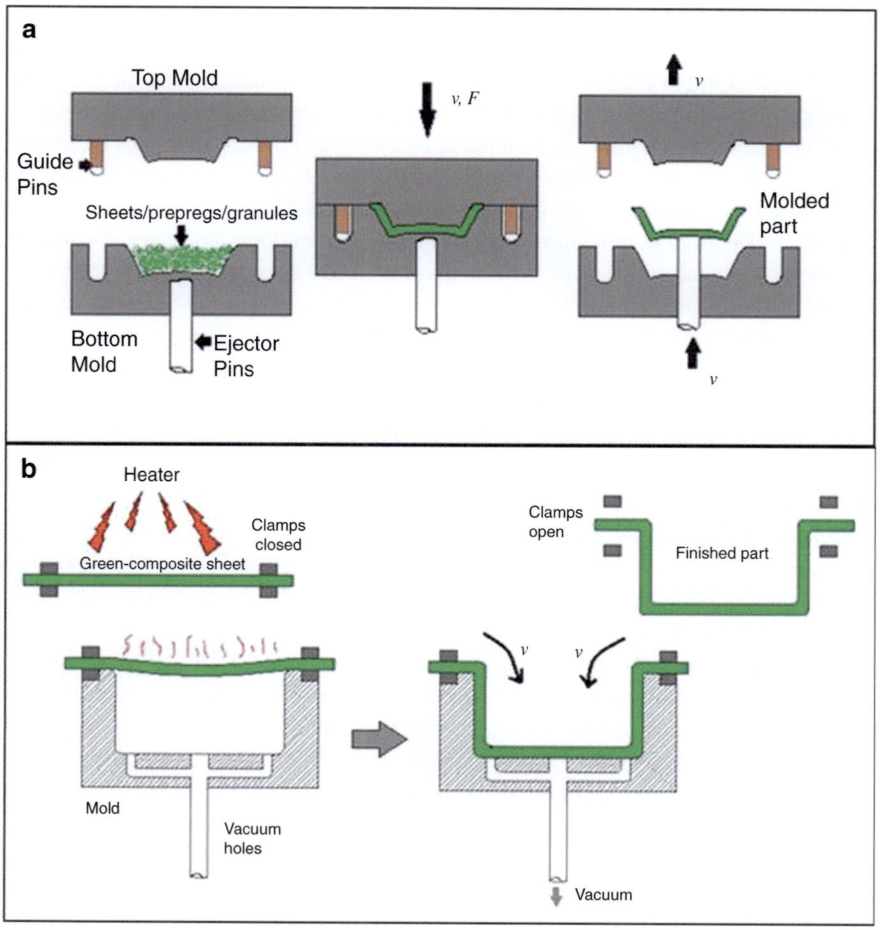

Fig. 3.1 Schematic diagram of (**a**) compression molding and (**b**) thermoforming (via vacuum) process

Table 3.1 Features of compression molding and thermoforming technologies

SN	Features	Compression molding (CM)	Thermoforming (TF)
1	Type of process	Usually discontinuous	Continuous
2	Application type	Low-volume production and high-tolerance parts	High-volume production and low-tolerance parts
3	Acceptable raw material form	Sheets, composite granules, prepregs	Sheets only
4	Material waste	Low	High
5	Application	Automotive hood, fender, door panels, gears, etc.	Food and packaging containers, aircraft windscreens, interior panels, boat hulls, etc.

3.2 Fabrication Process

During fabrication of green composites, processing parameters determine the extent of morphological changes, surface interaction, and chemical cross-linking between constituents of green composite (Gällstedt et al. 2004). Understanding of these phenomena is significant for gaining fundamental knowledge of processing window. Processing of green composite is different to traditional polymer composite due to changes in rheological behavior, thermal instability, tendency to water absorption, and morphological differences within natural fibre types and species (Sain and Pervaiz 2008). Natural fibres and biopolymers exhibit viscoelastic behavior causing a change in its flow characteristics (Sain and Pervaiz 2008). Fewer studies on the flow behavior of natural fibre-filled systems showed increase in viscosity of the composites with fibre content and reduced processability (La Mantia and Morreale 2011). On the other hand, biopolymers like PLA are crystalline and exhibit poor flow properties (Lim et al. 2008). Their degradation starts at temperature above 190 °C and narrows at the processing window (Zampaloni et al. 2007). Moreover, hydrophilic nature of natural fibre reduces fibre-matrix adhesion and fibre dispersion which are keys for processing within preferred output rates (Thakur et al. 2011b). Other challenge is that natural fibres from different plant sources are different in chemical composition and morphology making them even more difficult to process (Faruk et al. 2012).

Compression molding and thermoforming are secondary processes because they both involve preprocessing of selected materials to manufacture semifinished products such as sheets, granules, or prepregs (Du et al. 2010; Klein 2009). Fabrication of green composites via compression molding and thermoforming can be divided into following four basic steps (Fig. 3.2):

1. Material selection
2. Material preprocessing
3. Semifinished product manufacturing
4. Green composite fabrication

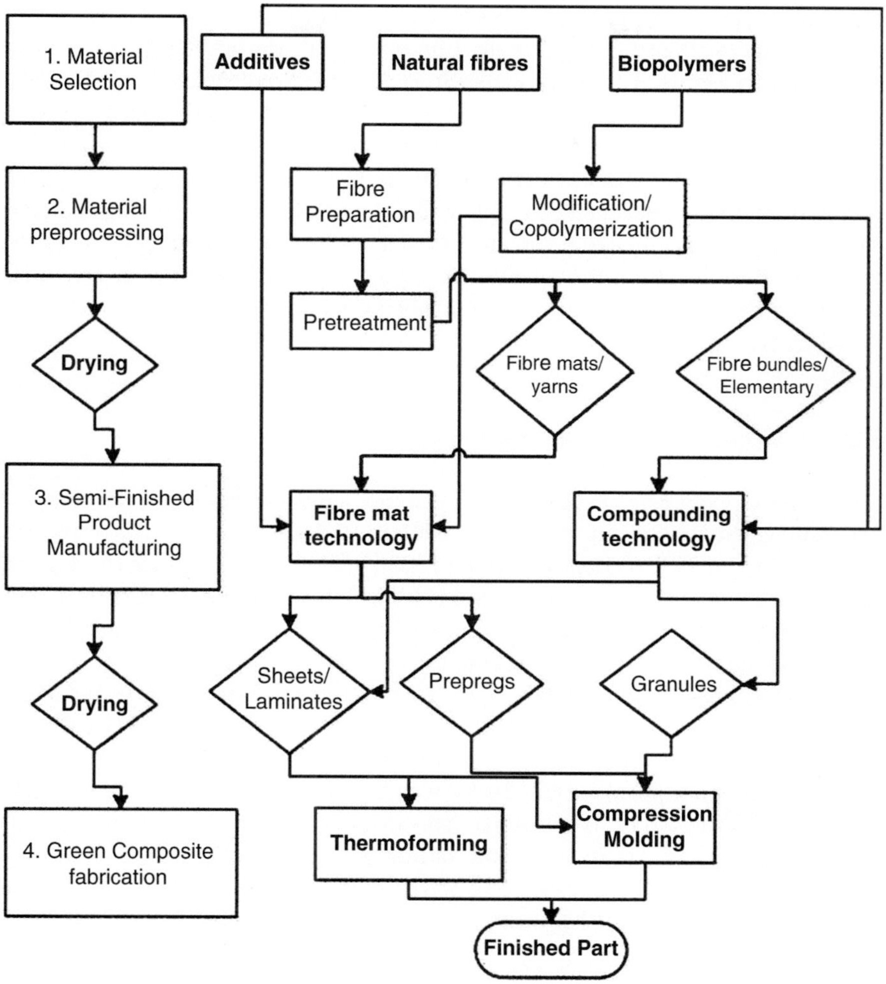

Fig. 3.2 Flowchart of green composite fabrication process using compression molding and thermoforming

3.2.1 Material Selection

3.2.1.1 Material Selection Criteria

Raw material selection starts early in the product development process. Selection is based on considerations such as cost, functional properties, and processability of green composites (Erchiqui et al. 2009). Lower raw-material costs and lower processing temperature for green composites have resulted in overall cost benefit to the manufactures (Nilsson et al. 2012). Next criterion for material selection is to achieve mechanical, physical, and thermal properties as per part specification (Gejo et al. 2010).

Moreover, properties such as melt viscosity, specific heat capacity, thermal conductivity, and crystallinity affect their processing (Lim et al. 2008; Menzel et al. 2013). Higher melt viscosity can lead to problems such as unwanted molecular orientation and internal stress in the downstream processing (Menzel et al. 2013). This property is also critical for maintaining good flow of the melt and reducing the processing time. Besides, specific heat and thermal conductivity of the polymer dictate how much energy is required to heat and cool the mold during compression molding and thermoforming. Another material property specific to thermoforming process is the areal draw ratio (ADR). It is defined as the ratio of area on formed part to area prior to forming. Typical ADR value for commonly thermoformed materials ranges from 3.4 to 8.0 (Klein 2009).

3.2.1.2 Overview of Raw Materials for Green Composites

Natural fibres, polymers, and additives are common raw materials in green composites. Among natural fibre and polymers, cellulose-PLA (Frone et al. 2011; Wang and Drzal 2012), kenaf-PLA (Ben and Kihara 2007), jute fibre-wheat gluten (Reddy and Yang 2011), wheat gluten-cellulose, and PHA with hemp, jute, or flax (U.S. Patent No. US20080160567 2011) were previously used. The composition of natural fibres usually ranged from 40 to 75 % in compression molding and 15 to 50 % in thermoforming (Ayrilmis and Jarusombuti 2010; Takagi 2011). In case of micro- and nanocellulose fibres and PLA, fibre content greatly varied between 2 and 32 % (Cherian et al. 2011; Wang and Drzal 2012). In addition to fibre and matrix, various additives are used to enhance processing by better wetting and dispersing fibres in the polymer matrix (Wang 2011). Types of additives include lubricants, dispersants, plasticizers, release agents, stabilizers, and various processing aids. Based on several studies, recommended compositions of additives were 1–5 % (by weight) for green composites (La Mantia and Morreale 2011; Zampaloni et al. 2007).

3.2.2 Preprocessing

3.2.2.1 Fibre Preparation

Commonly, natural fibres in bales, silos, or wood chips are predigested using mechanical, chemical, thermal, enzymatic, or combination of these methods to separate fibre bundles (Du et al. 2010; Faruk et al. 2012; Sain and Pervaiz 2008). Depending on the source of natural fibre, they can be categorized as core fibre, wood fibre, seed fibre, leaf fibre, bast fibre, etc. (Faruk et al. 2012). Despite their source, basic chemical components are cellulose, hemicellulose, and lignin (Sain and Pervaiz 2008). Among these, cellulose fibres are more commonly used as a reinforcement in composites (Faruk et al. 2012). Most studies use cellulose fibres in fibre bundle form, while some refine further into elementary fibres such as

Fig. 3.3 Wood-based micro-cellulose fibres after acid hydrolysis (courtesy of CBBP, University of Toronto)

microcrystalline cellulose (MCC) and nanocellulose (NC) for higher mechanical properties. An example of acid-hydrolyzed micro-cellulose fibre is shown in Fig. 3.3. In a study by Frone et al. (2011), microfibrils of cellulose were prepared by mechanical grinding and solution-casting technique, while MCC were prepared by acid hydrolysis. In another study, combination of acid hydrolysis and steam explosion was used to isolate cellulose nanofibres from pineapple leaf (Cherian et al. 2011). Combination of two or more pretreatment methods is recommended for higher micro- or nanofibre yields (Immonen et al. 2013).

3.2.2.2 Drying

Drying of natural fibre and biopolymers is a crucial step in green composite processing. Depending on the industrial setup (in-line vs. batch downstream process), drying may be required prior to pretreatment, compounding, and/or composite fabrication process. Various studies suggested drying of natural fibres to 1–3 % and biopolymers below 1 % humidity prior to processing (Ayrilmis and Jarusombuti 2010; La Mantia and Morreale 2011). La Mantia and Morreale (2011) recommended that small amount of humidity may help to soften the cellulose fibre during processing. Drying time and temperature varied greatly for natural fibre and biopolymers among various studies. Frone et al. (2011) dried PLA pellets and MCC fibres in an oven for 9 h at 80 °C and 24 h at 30 °C, respectively. Other studies dried fibres at 80 °C prior to compounding; however, the time varied between 4 and 48 h (Sawpan et al. 2011; Zampaloni et al. 2007). For drying most natural fibres, recommended drying practice is 80 °C for 4 h.

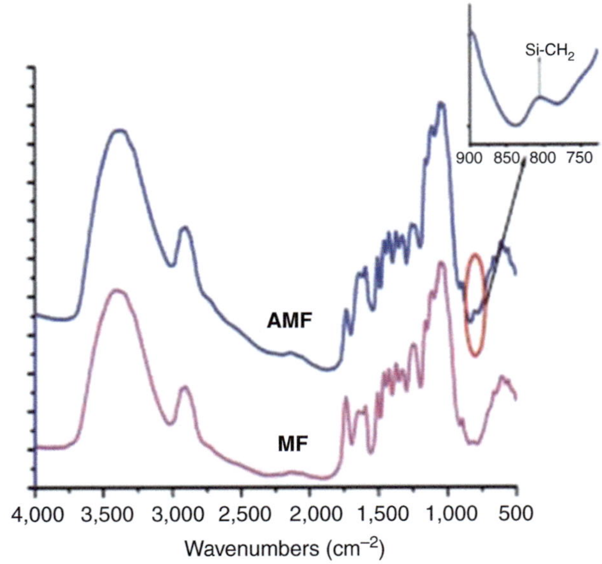

Fig. 3.4 FTIR spectra of untreated (MF) and silane-treated (AMF) cellulose microfibril. Adapted from Frone et al. (2011). Copyright 2014. By permission from John Wiley and Sons

3.2.2.3 Fibre Pretreatment

The basic principle behind pretreatment is to chemically modify the fibre surface by binding the sizing agents or by changing the thermodynamics between fibre and matrix (U.S. Patent No. US20080160567 2011). Various physical and chemical fibre pretreatment methods are used to enhance interfacial bonding (U.S. Patent No. US7208221 B2 2007). Among physical methods, homogenization and vacuum treatment are used; however chemical methods are more common (Takagi and Asano 2008; Zampaloni et al. 2007). Among chemical methods, fibre treatment with surfactants, alkali, acid hydrolysis, silanes, and isocyanates are more common. In a study by Thakur et al. (2011b), cellulosic fibres were mercerized for better surface adhesion with methyl acrylate and reduced water absorption properties. Frone et al. (2011) used 10 % APS (3-aminopropyltriethoxysilane) and acid hydrolysis to remove amorphous regions in cellulose fibres. In Fig. 3.4, FTIR (Fourier Transform Infrared Spectroscopy) spectra of APS-treated cellulose microfibril show the formation of Si-CH_3 bond. In another study, 7.5 % of epoxy and anhydride-based compatibilizers were used to improve adhesion between cellulose fibre and PLA (Immonen et al. 2013).

3.2.2.4 Modification of Biopolymers

Depending on the types of biopolymers used and their application, modification is necessary to achieve the desired processability and performance properties (Frenz et al. 2008). Biopolymers have poor processability due to lower melt strength, sensitivity to water, and lower impact strength (Imre and Pukánszky 2013; Thakur et al. 2011a). For example, PLA and PHA have relatively lower molecular weight and are brittle limiting their rheological properties during CM and TF (Thakur et al. 2011a). Conventional methods of polymer modification such as plasticization and physical blending are low-cost alternatives but may not achieve desired miscibility with biopolymers (Imre and Pukánszky 2013). In order to overcome this issue, chemical methods such as copolymerization, grafting, and transesterification are most widely used. Study by Okamoto et al. (2009) used various plasticizers (polyester diols) to enhance the ductility and elongation at break of PLA. Another study used combination of chain extension and interfacial modification in PLA/TPS blends to enhance its melt strength and the rheological properties (Li and Huneault 2011). For thermoforming of pea starch, grafting with PCL was used to enhance the elongation to break mechanical properties and reduce the water absorption (Fig. 3.5) (Chang et al. 2009). Some studies also used copolymers to enhance fibre dispersion and improve mechanical properties of biopolymers. Drzal et al. (2009) used polystyrene (PS) to copolymerize soy protein, while Frenz et al. (2008) used chain extenders (Joncryl® from BASF) for PLA, PHA, and PHB. These studies showed that copolymerization enhanced fibre-matrix entanglement leading higher melt strength and larger processing window during compounding and thermoforming process.

3.2.3 Semifinished Product Manufacturing

Semifinished products in CM and TF can be in the form of granules, sheets, prepregs, or laminates (Table 3.1). Technologies involved in manufacturing semifinished products can be categorized into compounding and fibre mat technology.

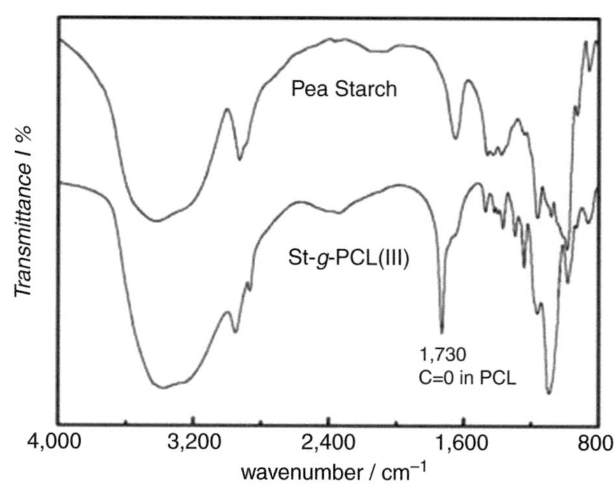

Fig. 3.5 FTIR spectra of pea starch and grafted pea starch with polycaprolactone (St-*g*-PCL). Adapted from Chang et al. (2009). Copyright 2014. By permission from John Wiley and Sons

Typically, fibre mat technology is used with longer fibres (bast fibres of length 10–30 mm) and compounding technology for shorter fibres (wood fibre, seed fibre, core fibre of length < 10 mm) (Faruk et al. 2012; Reddy and Yang 2011). However, some extrusion technologies such as pull drill and pultrusion are capable of compounding longer and continuous fibres (Faruk et al. 2012).

3.2.3.1 Compounding Technology

Compounding of green composites can be done via extrusion or mixing. Extrusion is a continuous process while mixing is a batch process. Among extrusion techniques, corotating twin screw extruders are commonly used for better compounding of green composites (La Mantia and Morreale 2011). Drzal et al. (2009) used this technology for producing composite sheet from soy flour and starch-based biopolymers. Temperature range was 95–130 °C and screw speed was 100 rpm (U.S. Patent No. US20060043629 A1 2009). Author also recommended maintaining processing conditions during extrusion if biopolymers are copolymerized with other biodegradable polymers due to their high susceptibility to denaturing. In another study, melt extrusion was used to make composite sheets from laminated PHA with woven natural fibre (U.S. Patent No. US20080160567 2011). Among mixing technologies, dry blending is the most common method for mixing polymer, additives, and natural fibre (Ayrilmis and Jarusombuti 2010; Benthien and Thoemen 2012). This technique is simple and cost-effective but lacks even distribution of raw materials. Frone et al. (2011) used two-roll mill Brabender mixer to produce composite granules at 165 °C and rotor speed of 27 and 22.5 rpm. Another study used low-intensity mixture such as ribbon blender to compound cellulose and PVC (U.S. Patent No. US6971211 B1 2005). Some study also used pre-blended granules of natural fibre and polymer (Benthien and Thoemen 2012).

3.2.3.2 Fibre Mat Technology

In this process, fibre mat or yarns are impregnated with thermoplastic or thermoset resins to form sheets or prepregs (Du et al. 2010). Unlike conventional glass fibres, natural fibres are not usually supplied in rovings or yarns. For manufacturing micro- or nanocellulose and PLA composite sheets, studies suggested using homogenization technique followed by membrane filtration and drying (Takagi 2011; Wang and Drzal 2012). Study by Kim et al. (2010a) reported use of a carding method to form natural fibre mat for making composite sheets. In the process, PLA and PP fibre (30 mm long) and natural fibre (50–70 mm long) were carded together to form webs, needle punched and pre-pressed at 120 °C to form a prepreg mat. In another method, alternate layers of polypropylene (PP) powder (400 μm) and short kenaf fibre were used to form a composite sheet (Zampaloni et al. 2007). PP powder was sprayed evenly over kenaf fibre and pressed using Carver Laboratory Press. For nonwoven mats with longer fibre (10–50 mm), Nechwatal et al. (2005) reported

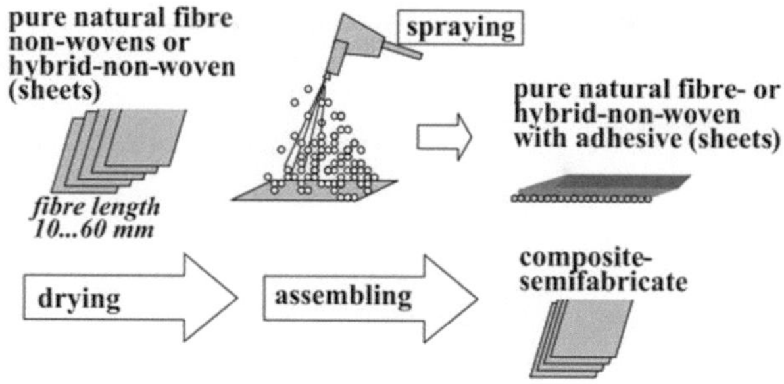

Fig. 3.6 Composite sheet fabrication from nonwoven mats using spraying method. Reprinted from Nechwatal et al. (2005). Copyright 2014. By permission from Wiley-VCH Verlag GMBH & CO

three methods for fabricating composite sheets. Polymers could be sprayed, powdered, or added in fibre spinning process. Method for fabricating composite via spraying polymer into nonwoven natural fibre mat is shown in Fig. 3.6. Khondker et al. (2006) developed a micro-braiding method for continuous yarns of jute and PLA. In this method, composite yarns were wounded in metallic frame to form a mat for compression molding. In a study by Takemura (2010), woven jute fibre sheets were used to make preforms and impregnated with PVA resin to form composite laminate.

In conventional compression molding, SMC (sheet molding compound) and BMC (bulk molding compound) prepregs are formed using thermosetting resin such as unsaturated polyester (UPE), additive and glass-fibre reinforcement, or filler (Du et al. 2010). Recently, SMC technology has been developed for manufacturing green composite sheets with thermoplastic resin. In a patented method developed by Drzal et al. (2009), chopped natural fibres were fed (via vibratory feeder) into the carrier film where thermoplastic resin was sprayed and pressed with compression roller to make continuous sheets. Author also developed another method for making prepregs from natural fibres and UPE (U.S. Patent No. US7208221 B2 2007). In the process, dried layer of natural fibres of moisture content less than 10 % was carried in a first film and mixed with the second film carrying catalyzed polyester resin. Inventor recommended that prepregs must be stored in refrigerated temperatures before compression molding into a final part. In another study, prepregs were made from jute fibre mat that was evenly sprayed with wheat gluten (Reddy and Yang 2011).

3.2.4 *Green Composite Fabrication*

3.2.4.1 Compression Molding

In compression molding (CM), semifinished products such as sheets or prepregs are used to manufacture finished parts. For sheets and prepregs, most studies used alternate stacking method to compression molding of green composites (Du et al. 2010;

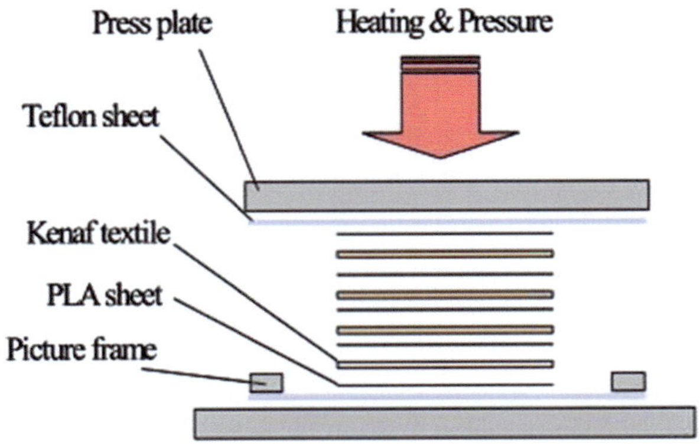

Fig. 3.7 Compression molding of kenaf and PLA via alternate stacking. Reprinted from Ben and Kihara (2007). Copyright 2014. By permission from Trans Tech Publications

Ma and Joo 2010; Reddy and Yang 2011). Study by Ben and Kihara (2007) used alternate stacking of kenaf fibres and PLA sheets for fabricating composite laminates (Fig. 3.7). Similar processing was used by Ma and Joo (2010) for fabricating jute and PLA composites. Processing steps varied on the type of materials and the number of alternate stacking. For kenaf and PLA sheets, processing steps involved melting at 10 MPa pressure for 10 s, holding at 1 MPa for 20 min, and impregnation at 10 MPa for 10 s. Platen temperature throughout the process was maintained at 185 °C (Ben and Kihara 2007). In contrast, fabrication of CM cellulose and PLA sheets involved two steps: pre-pressing and pressing. During pre-pressing, 5 MPa of pressure was applied for 3 min followed by 15 MPa of pressure for 2 min, and the platen temperature was maintained at 165 °C (Frone et al. 2011). Number of alternate sheets or prepregs depends on the thickness of individual sheets or prepregs and the thickness of the finished part. Five mats with alternating UPE resin were stacked prior to molding to achieve 3 mm thickness of the composite laminates (Du et al. 2010). Some study also used prepregs for fabricating green composites. In a study by Du et al. (2010), prepregs of kenaf and polyester resin were compression molded to form composite laminates (dimension = 102 × 178 × 3 mm) at 175 °C. Processing steps involved pressing at 5 MPa for 10 s and turning of heat. It was followed by constant pressure of 5 MPa for 1 h until the temperature of the mold was cooled to 100 °C (Du et al. 2010). Reddy and Yang (2011) also used prepreg of jute fibre and wheat gluten to form composite laminate. A composite press was used at temperature range of 150–180 °C for 5–20 min at 140 MPa of pressure followed by cold-water cooling and laminate removal (Reddy and Yang 2011). In these studies, release agents (silicon) and wax papers were used to avoid contacts between the material and mold platen (Ayrilmis and Jarusombuti 2010; Du et al. 2010).

In CM, curing time, curing temperature, and clamp pressure are the important process parameters that need to be optimized for the desired part (Onal and Adanur 2005). Few studies were found in optimizing processing parameters for green or bio-based composites. Ben and Kihara (2007) investigated in optimizing melt temperature, holding time, and impregnation time of kenaf and PLA composites. Based on the mechanical properties of fabricated composites, combination of melt temperature of 185 °C and holding time of 15 min were optimal for kenaf and PLA composites. For impregnation time, 30 s was recommended time for quasi-isotropic lamination. However, volume fraction of both fibre and PLA was not reported in the study which is an important consideration for optimizing processing. In another study, influence of press temperature on panel properties was investigated (Benthien and Thoemen 2012). Based on physical and mechanical properties of wood flour and polypropylene panels, optimal press temperature was 210 °C. This temperature is high for natural fibres and may result in fibre degradation. Reddy and Yang (2011) also studied effects of molding (holding and impregnation) time and temperature on wheat gluten-jute fibre composites. Results showed that molding time and temperature of 170 °C and 15 min produced the maximum flexural properties. Overall, curing temperature green composites should be in range of 150–180 °C while pressure and time varies with the type of material used and the thickness of the sheets or prepregs.

3.2.4.2 Thermoforming

Prior to thermoforming, drying of semifinished green composites is required to avoid blemish defects on the finished parts (U.S. Patent No. US20130331518 A1 2013). In thermoforming process, composite sheets are roll fed or precut depending on their thickness (Klein 2009). For precut sheets, clamping frame is used to avoid twisting and warping. Next, sheets are heated to its softening temperature (Tg) via convection or radiant heaters located on one or both sides of the sheets (Zampaloni et al. 2007). Prior to forming process, preheating is recommended to eliminate risk of material shear and premature fracture as a result of rapid cooling from ambient temperature (Lim et al. 2008; Zampaloni et al. 2007). Hence, the recommended steps in thermoforming green composites are load/unload, preheat, heat, and forming (Klein 2009). A thermoforming machine with thermoformed part is shown in Fig. 3.8.

Studies in thermoforming green composites used prepregs of cellulose fibre (40–70 % w/w) and thermoplastic (U.S. Patent No. US8012389 B2 2009), sheet laminates of cellulose (17 % w/w), sheets of PLA (Lim et al. 2008), and composite sheets of kenaf and polypropylene (Zampaloni et al. 2007). Forming method and processing parameters of these composites were slightly different. In a patented method developed by Immonen et al. (2013), cellulose and PLA composite sheets (width 10 cm) were dried and pressed at 185 °C for 3 min (U.S. Patent No. US20130331518 A1 2012). Study by Lim et al. (2008) recommended the thermoforming temperature in the range of 80–110 °C and use of aluminum (Al) molds for lower faster thermoforming cycles. Another study on processing parameters optimization of kenaf and polypropylene showed optimal forming temperature, die temperature, heating time, and the draw depth as 190 °C, 165 °C, 15 min, and 50.8 mm, respectively (Zampaloni et al. 2007). Authors also suggested that these composite sheets have better formability due to less wrinkling and distortions.

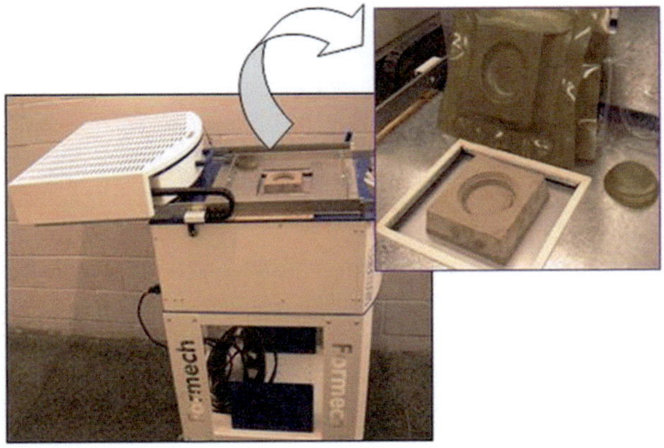

Fig. 3.8 Thermoforming machine with a thermoformed coffee cup lid (courtesy of CBBP, University of Toronto)

3.3 Commercial Green Composite Semifinished Products

Over the past few years, there has been significant growth in compression-molded and thermoformed products using green composites. FuturaMat has developed thermoformable green composites BioCeres® and BioFibra® (FuturaMat© Website: Our Plastics n.d). BioCeres® is made from corn-based thermoplastic starch (TPS) and wheat flour, while BioFibra® is made from TPS and spruce wood flour. Similarly, ©Polyfea has developed Caprowax P™ from cellulose and TPS for thermoforming applications (POLYFEA Polymer 2009). It claims that these compounds do not require pre-drying and recommends processing at lower shear rates and pressures. Some examples of commercial green composites and their processing parameters are summarized in Table 3.2.

Other commercial examples include natural fibres with polyolefins. FlexForm Technologies has developed nonwoven mats and panels for CM and TF based on natural fibres reinforced with polypropylene matrix, while Composites Evolution has developed prepregs using flax fabric and polyolefins for CM (Fig. 3.9). Composites Evolution has recently developed a prepreg from flax fabric and PLA. Their current automotive application includes package trays, door panels, headliners, seat backs, trailer side walls, pillars, etc. These panels are thermoformed at 200 °C and 0.379 MPa pressure using matched metal cooling (Gardiner 2006). Solvay industries also produced Gonaf® and Technogor® sheets using sisal fibres and PP matrix for use in CM and TF application (Gardiner 2006). Recommended processing methods for these sheets were heating at 180–190 °C for 50 s and applying pressure of 0.786–0.979 MPa pressure for 40 s. Example of compression-molded door panel and thermoformed seat back is shown in Fig. 3.10.

Table 3.2 Processing method for some commercial green composites and biopolymers[a]

SN	Trade name/ material composition	Form/ processing method	Properties	Processing parameters
1	Cereplast®6000 PLA	Pellets/sheet extrusion and thermoforming	MFR: 3.0 g/10 min Ten. str @ yield: 49.6 MPa Flex. mod: 3.79 GPa Elong. @ brk: 9.0 % HDT @ 0.45 MPa: 51.1 °C	Drying: 2–4 h @ 71–82 °C Cyl. temp: 154–174 °C Melt temp: 194 °C
2	Mirel™ P3001/ F3002 PHA	Pellet/sheet extrusion and thermoforming	Ten. str @ yield: 19.0 MPa Flex. mod: 1.48 GPa Elong. @ brk: 13 % HDT @ 0.45 MPa: 116 °C	Drying: 4 h @ 80 °C Cyl. temp: 165–175 °C Melt temp: 165–170 °C
3	Biocycle®189C-1/ PHB	Powder/sheet extrusion and thermoforming	MFR: 15 g/1omin Ten. str @ yield: 30 MPa Flex. mod: 2.6 GPa Elong. @ brk: 2.2 % HDT @ 0.45 MPa: 120 °C	Drying: 4 h @ 50 °C Cyl. temp: 140–170 °C Melt temp: 170 °C
4	Caprowax P™6006-11-000 granulat/cellulose fibre and TPS	Granules/ thermoforming	MFR: 2–6 cm^3/10 min Vicat softening temp: 56 °C	No drying required Cyl. temp: 80–150 °C Melt temp: 90–120 °C Thermoforming temp: 75–85 °C
5	BioCeres® BC-LBE01/ TPS and wheat	Pellets/ thermoforming	MFR: 8.1 g/10 min Ten. str @ yield: 32.6 MPa Flex. mod: 3.3 GPa Elong. @ brk: 2.9 % HDT @ 1.8 MPa: 41 °C	N/A
6	Plantic® R1/ TPS	Film/ thermoforming	Ten. str: 42–46 MPa Elong. @ brk: 20–36 % Elmendorf tear str: 3,500–5,500 g Film thickness: 300–550 μm	Glass transition temperature: 40–50 °C Vicat softening temperature: 130–135 °C

[a]All data are available at UL *IDES Prospector* http://www.ides.com/

3.4 Summary

Green composites are made from 100 % renewable and low-cost materials compared to conventional materials. They offer practical solutions to growing environmental and economic concerns in the development of new products. However, fabrication of green composites via compression molding and thermoforming is different to traditional polymer composites mainly due to changes in rheological behavior, thermal instability, and chemical and morphological differences in natural

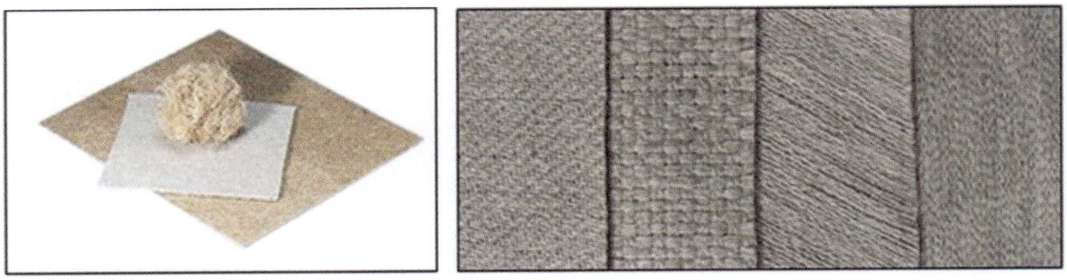

Fig. 3.9 Examples of commercial green composite semifinished products: (**a**) nonwoven bast fibres and polyolefin mat from FlexForm Technologies and (**b**) flax-fibre prepregs. Reprinted with permission from Composite Evolution (*Composite Evolution*: www.compositesevolution.com)

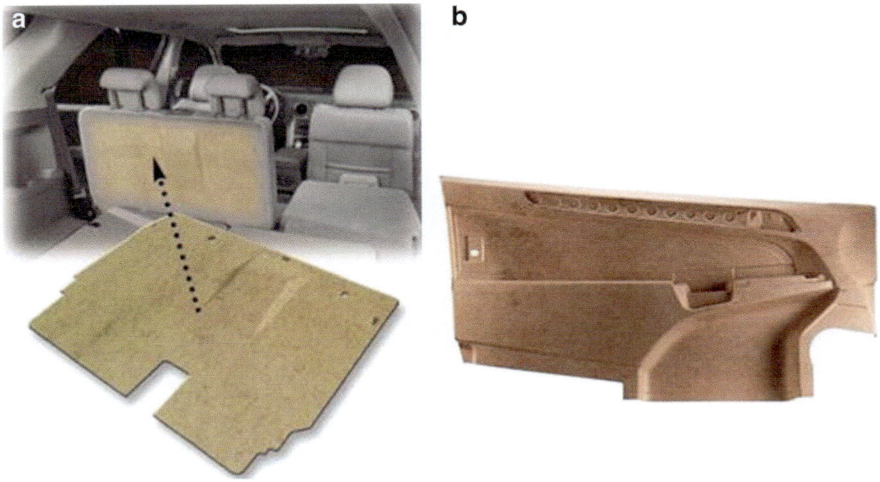

Fig. 3.10 Commercial examples of green composites: (**a**) thermoformed automotive seat back with FlexForm® (*FlexForm Technologies, website*: www.flexformtech.com) and (**b**) compression-molded automotive door panel from Fibrowood® (*Johnson Controls, website*: www.johnsoncontrols.com)

fibres and biopolymers. Hence, specific processing guidelines must be followed for manufacturing green composites. Processing steps involves four main steps: material selection, preprocessing of fibre and/or biopolymers, manufacturing of semifinished products, and part manufacturing. In material selection, raw materials are selected based on functional properties and processability of materials. For processability, properties such as melt viscosity, specific heat capacity, thermal conductivity, and crystallinity of materials. Once the materials are selected, fibre and polymers need to be preprocessed. Preprocessing involved fibre pretreatment, modification of biopolymer, and drying. Next step is to manufacture semifinished products such as granules, prepregs, sheets, or laminates via compounding or fibre mat technology.

In the final step, composite parts are manufactured using semifinished products. In each of these steps, careful attention in regard to moisture and thermal degradation of green composite must be given.

In recent years, there has been increasing number of green composite products in the market. It is expected that green composite market will continue to progress as long as tighter government environmental regulation around the world continues to influence corporate goals toward the use of more sustainable and environmentally friendly materials in their new product designs.

References

Altun Y, Dogan M, Bayramli E (2013) Effect of alkaline treatment and pre-impregnation on mechanical and water absorbtion properties of pine wood flour containing poly (lactic acid) based green-composites. J Polym Environ 21:850–856

Ayrilmis N, Jarusombuti S (2010) Flat-pressed wood plastic composite as an alternative to conventional wood-based panels. J Compos Mater 45:103–112

Ben G, Kihara Y (2007) Development and evaluation of mechanical properties for kenaf fibers/PLA composites. Key Eng Mater 334–335:489–492

Benthien JT, Thoemen H (2012) Effects of raw materials and process parameters on the physical and mechanical properties of flat pressed WPC panels. Compos Part A Appl Sci Manuf 43:570–576

Billington S, Criddle C, Frank C, Morse M, Christian S, Pieja A (2011) US Patent no. US20080160567 A1. Washington: U.S. Patent and Trademark Office

Chang PR, Zhou Z, Xu P, Chen Y, Zhou S, Huang J (2009) Thermoforming starch-graft-polycaprolactone biocomposites via one-pot microwave assisted ring opening polymerization. J Appl Polym Sci 113:2973–2979

Cheng B (2009) The effect of fibre orientation and dispersion on the mechanical properties of natural fibre reinforced polypropylene (Masters thesis). ProQuest dissertations and Theses, p 116

Cherian BM, Leão AL, de Souza SF, Costa LMM, de Olyveira GM, Kottaisamy M, Nagarajan ER, Thomas S (2011) Cellulose nanocomposites with nanofibres isolated from pineapple leaf fibers for medical applications. Carbohydr Polym 86:1790–1798

Drzal L, Mehta G, Misra M, Mohanty A, Thayer K (2007) US Patent no. US7208221 B2. Washington: U.S. Patent and Trademark Office

Drzal L, Mohanty A, Liu W, Thayer K, Misra M (2009) Cellulosic biomass soy flour based biocomposites and process for manufacturing thereof. US Patent no. US20060043629 A1. Washington: U.S. Patent and Trademark Office

Du Y, Zhang J, Toghiani H, Lacy TE, Xue Y, Horstemeyer MF, Pittman CU (2010) Kenaf bast fiber bundle-reinforced unsaturated polyester composites. I: processing techniques for high kenaf fiber loading. For Prod J 60:289–295

Erchiqui F, Godard F, Gakwaya A, Koubaa A, Vincent M, Kaddami H (2009) Engineering investigations on the potentiality of the thermoformability of HDPE charged by wood flours in the thermoforming part. Polym Eng Sci 49:1594–1602

Faruk O, Bledzki AK, Fink HP, Sain M (2012) Biocomposites reinforced with natural fibers: 2000–2010. Prog Polym Sci 37:1552–1596

Frenz V, Scherzer D, Ag B, Germany L, Villalobos M, Awojulu AA, Edison M, Corporation B, Van Der Meer R, Nederland B (2008) Multifunctional polymers as chain extenders and compatibilizers for polycondensates and biopolymers. Chain extension vs. solid state polymerization. Paper presented at the 66th annual technical conference of the society of plastics engineers, Milwaukee

Frone AN, Berlioz S, Chailan J, Panaitescu DM, Donescu D (2011) Cellulose fiber-reinforced polylactic acid. Polym Compos 32:976–985

FuturaMat© Website: our plastics (n.d) Retrieved from http://www.futuramat.com/english/our-products/

Gällstedt M, Mattozzi A, Johansson E, Hedenqvist MS (2004) Transport and tensile properties of compression-molded wheat gluten films. Biomacromolecules 5:2020–2028

Gardiner G (2006) Thermoformable Composite Panels, Part 1: CompositesWorld. In: Composites technology. Retrieved from http://www.compositesworld.com/articles/thermoformable-composite-panels-part-1

Gejo G, Kuruvilla J, Boudenne A, Sabu T (2010) Recent advances in green composites. Key Eng Mater 425:107–166

Hu R-H, Ma Z-G, Zheng S, Li Y-N, Yang G-H, Kim H-K, Lim J-K (2012) A fabrication process of high volume fraction of jute fiber/polylactide composites for truck liner. Int J Precis Eng Manuf 13:1243–1246

Immonen K, Sivonen E, Valta K, Hulkko J, Alto S, Pitkanen P, Salorinne K (2013) US Patent no. US20130331518 A1. Washington: U.S. Patent and Trademark Office

Imre B, Pukánszky B (2013) Compatibilization in bio-based and biodegradable polymer blends. Eur Polym J 49:1215–1233

Khondker OA, Ishiaku US, Nakai A, Hamada H (2006) A novel processing technique for thermoplastic manufacturing of unidirectional composites reinforced with jute yarns. Compos Part A Appl Sci Manuf 37:2274–2284

Kim H-J, Lee B-H, Kim HS (2010a) US Patent no. US20100170649 A1. Washington: U.S. Patent and Trademark Office

Kim S, Xu J, Liu S (2010b) Production of biopolymer composites by particle bonding. Compos Part A Appl Sci Manuf 41:146–153

Klein P (2009) Fundamentals of plastics thermoforming. Morgan and Claypool Publishers, California, USA, pp. 1–97

La Mantia FP, Morreale M (2011) Green composites: a brief review. Compos Part A Appl Sci Manuf 42:579–588

Li H, Huneault MA (2011) Effect of chain extension on the properties of PLA/TPS blends. J Appl Polym Sci 122:134–141

Lim L-T, Auras R, Rubino M (2008) Processing technologies for poly(lactic acid). Prog Polym Sci 33:820–852

Ma H, Joo CW (2010) Structure and mechanical properties of jute—polylactic acid biodegradable composites. J Compos Mater 45:1451–1460

Menzel C, Olsson E, Plivelic TS, Andersson R, Johansson C, Kuktaite R, Järnström L, Koch K (2013) Molecular structure of citric acid cross-linked starch films. Carbohydr Polym 96:270–276

Nechwatal A, Reubmann T, Bohm S, Richer E (2005) The dependence between the process technologies and the effect of MAH-PP adhesives in natural fiber reinforced thermoplastic composites. Adv Eng Mater 7:68–73

Nilsson H, Galland S, Larsson PT, Gamstedt EK, Iversen T (2012) Compression molded wood pulp biocomposites: a study of hemicellulose influence on cellulose supramolecular structure and material properties. Cellulose 19:751–760

Okamoto K, Ichikawa T, Yokohara T, Yamaguchi M (2009) Miscibility, mechanical and thermal properties of poly(lactic acid)/polyester-diol blends. Eur Polym J 45:2304–2312

Onal L, Adanur S (2005) Optimization of compression molding process in laminated woven composites. J Reinf Plast Compos 24:775–780

POLYFEA – a strong partner in compostable plastics (2009) In: BIOPRO Baden-württemb. GmbH. Retrieved from http://www.bio-pro.de/biopolymere/artikelliste_biopolymere/index.html?lang=en&artikelid=/artikel/04261/index.html

Reddy N, Yang Y (2011) Biocomposites developed using water-plasticized wheat gluten as matrix and jute fibers as reinforcement. Polym Int 60:711–716

Sain M, Pervaiz M (2008) Mechanical properties of wood-polymers composites. Wood-polymer composites. Woodhead Publishing, Cambridge, pp 101–116

Sawpan MA, Pickering KL, Fernyhough A (2011) Improvement of mechanical performance of industrial hemp fibre reinforced polylactide biocomposites. Compos Part A Appl Sci Manuf 42:310–319

Takagi H (2011) Strength properties of cellulose nanofiber green composites. Key Eng Mater 462–463:576–581

Takagi H, Asano A (2008) Effects of processing conditions on flexural properties of cellulose nanofiber reinforced "green" composites. Compos Part A Appl Sci Manuf 39:685–689

Takemura K (2010) Molding conditions and mechanical properties of jute fiber reinforced composite. Key Eng Mater 452–453:261–264

Thakur VK, Singha AS, Thakur MK (2011a) Biopolymers based green composites: mechanical, thermal and physico-chemical characterization. J Polym Environ 20:412–421

Thakur VK, Singha AS, Thakur MK (2011b) Graft copolymerization of methyl acrylate onto cellulosic biofibers: synthesis, characterization and applications. J Polym Environ 20:164–174

Thakur VK, Singha AS, Thakur MK (2013) Ecofriendly biocomposites from natural fibers: mechanical and weathering study. Int J Polym Anal Charact 18:64–72

Wang B (2011) Dispersion of cellulose nanofibers in biopolymer based dispersion of cellulose nanofibers in biopolymer based nanocomposites. Doctoral dissertation, T-Space research repository, University of Toronto

Wang T, Drzal LT (2012) Cellulose-nanofiber-reinforced poly(lactic acid) composites prepared by a water-based approach. ACS Appl Mater Interfaces 4:5079–5085

Warnes JM, Fernyhough A, Anderson CR, Lee BJ, Ralph M, Witt J (2009) Method for producing wood fiber composite products. US Patent no. US8012389 B2. Washington: U.S. Patent and Trademark Office

Zampaloni M, Pourboghrat F, Yankovich SA, Rodgers BN, Moore J, Drzal LT, Mohanty AK, Misra M (2007) Kenaf natural fiber reinforced polypropylene composites: a discussion on manufacturing problems and solutions. Compos Part A Appl Sci Manuf 38:1569–1580

Zehner BE (2005) Cellulosic/polymer composite material. US Patent no. US6971211 B1. Washington: U.S. Patent and Trademark Office

Chapter 4
Compaction, Permeability and Flow Simulation for Liquid Composite Moulding of Natural Fibre Composites

Darshil U. Shah and Mike J. Clifford

Abstract With the growing interest in developing high-performance natural fibre composites (NFRPs) for (semi-)-structural applications, researchers are increasingly considering liquid composite moulding (LCM) processes and investigating key manufacturing-related issues. Here, we critically review the literature on LCM of NFRPs. Consequently, we identify key findings concerning the reinforcement-related factors (namely, compaction and permeability) that influence, if not govern, the mould-filling stage during LCM of NFRPs. In particular, the differences in structure (physical and chemical) of natural and synthetic fibres, their semi-products (i.e. yarns and rovings) and their textiles are shown to have a perceptible effect on their processing via LCM.

Keywords Polymer-matrix composites • Liquid composite moulding • Resin transfer moulding • Compaction • Permeability • Flow modelling • Natural fibres • Biocomposites

---

D.U. Shah (✉)
Oxford Silk Group, Department of Zoology, University of Oxford,
South Parks Road, Oxford OX1 3PS, UK

Department of Architecture, University of Cambridge, Cambridge CB2 1PX, UK
e-mail: dus20@cam.ac.uk; darshil.shah@hotmail.co.uk

M.J. Clifford
Polymer Composites Research Group, Division of Materials, Mechanics and Structures,
Faculty of Engineering, The University of Nottingham, Nottingham NG7 2RD, UK

M.S. Salit et al. (eds.), *Manufacturing of Natural Fibre Reinforced Polymer Composites*, DOI 10.1007/978-3-319-07944-8_4

4.1 Introduction

4.1.1 Liquid Composite Moulding

It is estimated that in 2011, liquid composite moulding (LCM) accounted for 4 % (ca. 350 kilotonnes) of the global (Reux 2012) and 11 % (ca. 120 kilotonnes) of the EU (Witten and Jahn 2013) fibre-reinforced plastics' (FRPs) production volume. Notably, the use of LCM has grown consistently over the last few decades, particularly because they require low capital investment and are closed-mould processes offering better working conditions (viz., health and safety) and enhanced part quality (viz., mechanical properties, defects and surface finish) than open-mould processes. In general, LCM is primarily used for the cost-effective production of high-performance components with moderate-to-high complexity geometry and moderate-to-large size (up to 100 m^2) at low-to-medium volumes (e.g. 100–10,000 parts/year): from automotive parts to wind turbine blades (Manson et al. 2000).

With the increasing usage of LCM, numerous process variants have evolved with resin transfer moulding (RTM), vacuum-assisted RTM and light RTM being the most commonly used (Campbell 2003; Long et al. 2005). In general, an LCM process can be divided into four stages (Table 4.1): (1) reinforcement layup, (2) mould filling, (3) post-filling and (4) demoulding. The basic approach in any LCM process is to force a catalysed thermosetting liquid resin to flow through a dry, stationary, porous, compacted reinforcement inside a closed mould by creating a pressure differential between the inlet(s) and outlet(s).

As the primary aim of any LCM process is to ensure complete filling of the mould, successful execution of LCM involves understanding, controlling and optimising the mould-filling stage in particular. This stage dictates the production cycle time, quality (viz., defects including voids and dry spots), geometry (e.g. thickness) and, ultimately, mechanical properties of the final part (Table 4.1). Not surprisingly, computational mould-filling simulations are widely used as a cost-efficient and time-saving tool to optimise the LCM process (Long et al. 2005). However, accurate manufacturing process simulations require accurate input data. Notably, the principal reinforcement-related factors affecting the mould-filling stage are compaction and permeability (Table 4.1).

In this chapter, we examine the processing of natural fibre-reinforced plastics (NFRPs) via LCM. We specifically inspect the through-thickness compaction behaviour and permeability of natural fibre reinforcements and highlight their key differences from synthetic fibre reinforcements. These insights are then used as foundations to discuss the modelling of the mould-filling stage in LCM of natural fibre reinforcements using adapted resin flow models.

Table 4.1 Schematic and description of a generic LCM process illustrating the four principal stages

Stage	1. Reinforcement layup	2. Mould filling	3. Post-filling	4. Demoulding
Schematic	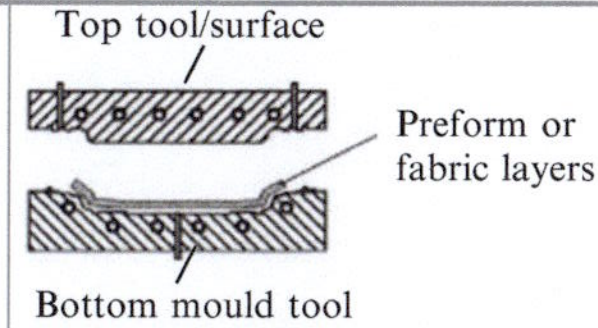	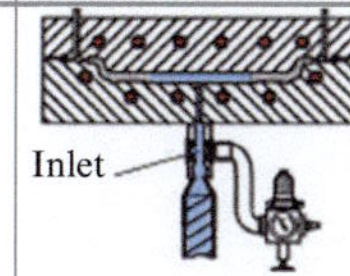	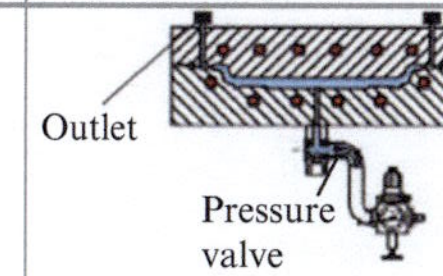	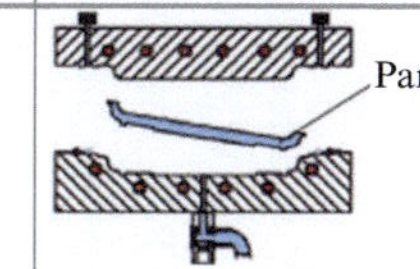
Description	Layup of the dry reinforcement (multiple layers of textiles or a prefabricated 'preform') in a mould with a rigid (metal) or semi-rigid (composite) bottom tool and a rigid (metal), semi-rigid (composite) or flexible (silicone membrane or vacuum bag film) top tool/surface	'Dry compaction' of the reinforcement, followed by resin impregnation via drawing vacuum and/or injecting resin under pressure	'Wet compaction' of the impregnated reinforcement to allow laminate thickness to equilibrate, followed by curing of resin, possibly at elevated temperatures	Demoulding of the cured, stiff composite part
Factors affecting this stage	(a) Tool: design (incl. geometry) and material (b) Reinforcement: layup (incl. type, architecture, number of layers, orientation)	(a) Tool: design (incl. geometry, inlet/outlet gate location) (b) Reinforcement: layup, compaction, permeability (c) Resin: temperature, viscosity, degree of cure (all a function of time) (d) Pressure differential in mould cavity, tooling temperature	(a) Reinforcement: 'wet' compaction and permeability (b) Resin: cure kinetics (c) Heat transfer, pressure differential in mould cavity, tooling temperature	(a) Tool: geometry, surface release properties (b) Resin: cure kinetics
Factors affected by this stage	(a) Cycle time (layup time) (b) Part quality: layup-related defects (e.g. fibre/ply misalignment, incorrect fibre content distribution, mould closure problems) (c) Part mechanical properties (incl. fibre volume fraction)	(a) Cycle time (fill time) (b) Part quality: infusion-related defects (e.g. voids/porosity, dry spots, fibre/ply washing and misalignment), part geometry (incl. thickness) (c) Part mechanical properties (incl. fibre volume fraction)	(a) Cycle time (post-fill time) (b) Part quality: cure-related defects (e.g. incomplete cure, ply cracking, warpage), part geometry (incl. thickness) and tolerances (c) Part mechanical properties (incl. fibre volume fraction)	(a) Part quality: surface finish, defects (e.g. print through, warpage, spring in, delamination), part geometry (incl. thickness) and tolerances (b) Part mechanical properties

Process variations primarily relate to the choice of mould material and the resin injection/infusion mechanism

4.1.2 Natural Fibre Composites in LCM Processes

NFRPs accounted for over 13 % of the 2.4 million tonne EU FRP market in 2010 (Fig. 4.1) (Carus 2011; Carus and Gahle 2008; Shah et al. 2013a, b). While wood and cotton were used as reinforcements in the majority of these NFRPs [most likely due to their wide availability and low cost (Shah 2013b)], even the production of non-wood, non-cotton, plant fibre (e.g. flax, jute, hemp)-reinforced plastics (PFRPs) was comparable to the production of non-glass, synthetic fibre (e.g. carbon, aramid) composites.

The increasing consideration of natural fibres as next-generation, sustainable reinforcements requires tackling the first hurdle, which is composite manufacture [reviewed in (Pickering 2008; Summerscales et al. 2010; Faruk et al. 2012; Ho et al. 2012; Shah 2013b)]. Due to the commercial applications of NFRPs in predominantly small-sized, high-volume, low-cycle time, nonstructural components (e.g. decking for construction industry and interior panels for automotive industry), injection/extrusion moulding and compression moulding are the widely used manufacturing techniques (Fig. 4.1) (Shah 2013b; Carus et al. 2014). The reinforcement form has typically been pellets/granules for the former and random fibre mats for the latter. While wood and cotton reinforced NFRPs largely (>80 %) employ thermosetting matrices, PFRPs are primarily (>70 %) based on thermoplastic matrices (Fig. 4.1) (Carus and Gahle 2008; Shah 2013b; Carus et al. 2014).

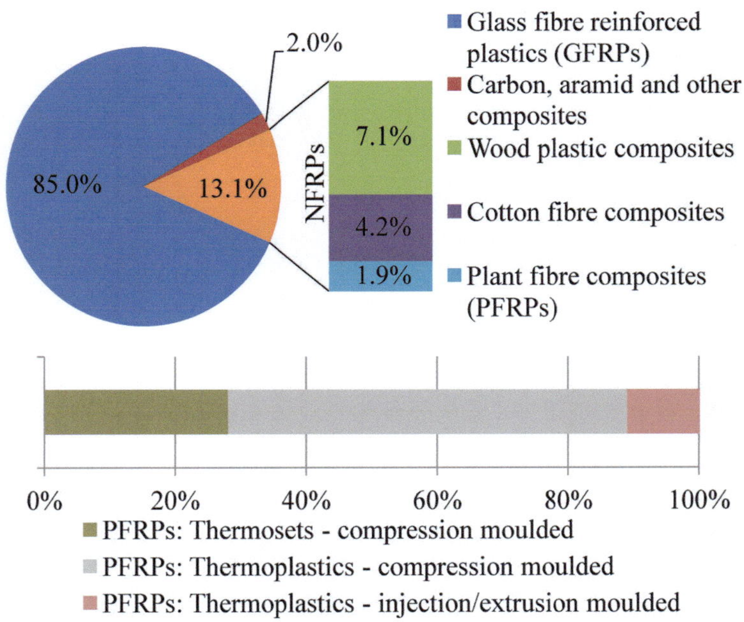

Fig. 4.1 The production of NFRPs in the 2.4 million tonne EU FRP market in 2010 (Carus 2011; Carus and Gahle 2008; Shah et al. 2013a, b)

LCM, on the other hand, is particularly suitable for (semi-)structural components utilising textile reinforcements (comprising of aligned, continuous yarns/tows that are knitted/woven/stitched/braided) in thermosetting matrices at high fibre fractions. Other than the potential to produce high-performance composites, LCM processes are specifically well suited to natural fibre reinforcements for a variety of reasons (Francucci et al. 2012a; Shah 2013b; Shah et al. 2013c):

1. Low processing temperatures (often <120 °C) during composite fabrication, thereby avoiding thermal degradation of natural fibres (which decompose above ~220 °C)
2. Minimal fibre damage during composite processing (as opposed to injection/ extrusion moulding), thereby allowing retention of high reinforcement length, alignment and mechanical properties
3. Use of liquid resins with low viscosities (0.1–1 Pa s), thereby allowing good preform impregnation with low porosity even at low compaction/injection pressures
4. Use of thermosetting resins with polar functional groups that form a better interface with the (typically) polar surface of natural fibres (than polyolefin-based thermoplastics)
5. Relatively low-cost (and often unsophisticated) tooling, making the process compatible with low-cost plant fibres, particularly when manufacturing in less economically developed countries with an abundance of indigenous natural fibres
6. Closed-mould LCM processes reduce exposure to harmful emissions, therefore offering worker-friendly conditions, much like nonhazardous plant and animal fibres

Consequently, the LCM of NFRPs has received much recent attention in scientific research, where critical aspects such as reinforcement compaction (van Wyk 1946; Madsen 2004; Umer et al. 2011; Xue et al. 2011; Francucci et al. 2012a, 2013; Shah et al. 2014a) and permeability (Rodriguez et al. 2004; Francucci et al. 2010, 2013; Masoodi and Pillai 2011; Umer et al. 2011; Xue et al. 2011) and resin flow behaviour (Richardson and Zhang 2000; Rodríguez et al. 2007; Francucci et al. 2010; Masoodi and Pillai 2011; Xue et al. 2011) have been investigated. More recently, commercially viable, structural products have also been manufactured with NFRPs using LCM. These include the world's first flax composite 11 kW wind turbine blade developed by the University of Nottingham (Shah et al. 2013a) and Samsara's flax/hemp biocomposite surfboard (18/06/2012).

Here, we critically review the available literature on LCM of NFRPs. It is expected that the differences in structure (physical, chemical) of natural and synthetic fibres, their semi-products (i.e. yarns and rovings) and their textiles will have a perceptible effect on their processing via LCM. Consequently, we identify key findings concerning the reinforcement-related factors (namely, compaction and permeability) that influence, if not govern, the mould-filling stage during LCM of NFRPs.

4.2 Compaction Behaviour of Natural Fibre Reinforcements

4.2.1 Introduction

The first studies (Winson 1932; Schiefer 1933; van Wyk 1946) on fabric compaction were indeed based on natural fibres like wool, cotton and silk. The studies were, however, conducted by textile engineers to evaluate 'fabric handle', that is, the softness/smoothness, loftiness/resilience and drape of the material. Resistance to compression is still used as an effective parameter to measure the softness and pliability of garment textiles (Matsudaira and Qin 1995; Matsudaira 2006).

In contrast, in relation to composite manufacture, the compaction behaviour of technical textiles affects the reinforcement permeability and part fill time in the mould-filling process and also determines the thickness and volumetric composition (i.e. fibre volume fraction) of the final part (Long et al. 2005). Tight control of part thickness (and therefore weight) is a requisite for quality assurance in any composite manufacturing process. In addition, in their uncompressed state, textile reinforcements have a low fibre volume fraction [typically between 10 and 25 % (Long et al. 2005)]; for semi-structural applications this must be increased (to up to 70 %) during processing to exploit the mechanical properties of the reinforcement. Studying the relationship between compaction pressure P and fibre volume fraction v_f for a given preform also enables determining the maximum (practical) fibre volume fraction, which sets the upper limit of reinforcement efficiency. Consequently, compaction plays an important role in not only LCM processes but also in the stamping of textile-reinforced thermoplastic composites. Knowledge of the compaction behaviour of the reinforcement form is therefore critical.

While the compaction characteristics of synthetic reinforcements is well studied (Long et al. 2005), the compaction response of natural fibre reinforcements is a relatively new topic (van Wyk 1946; Madsen 2004; Umer et al. 2011; Xue et al. 2011; Francucci et al. 2012a, 2013; Shah et al. 2014a).

4.2.1.1 Compaction Measurement and Modelling

Through-thickness compaction is usually measured by pressing flat preform layers between parallel metallic platens at a constant compaction speed (typically 0.5–5 mm/min) to a predetermined load (or pressure) or to a predetermined thickness (or fibre volume fraction) (Fig. 4.2). Tests can be conducted in dry (i.e. unsaturated) or wet (i.e. saturated in a test fluid) conditions. At the end of a compression cycle, the tester may (1) hold at constant load to measure creep, (2) hold at constant thickness to measure relaxation or (3) unload the sample (at the same rate) to measure hysteresis. Multiple repeat cycles can then be applied on the same specimen to examine the effects of rearrangement and compaction history.

From the compaction tests, plots of preform thickness t against compaction force F are obtained, which can be transformed into the more conventional plots of fibre volume fraction v_f against compaction pressure using Eq. (4.1) (Fig. 4.2).

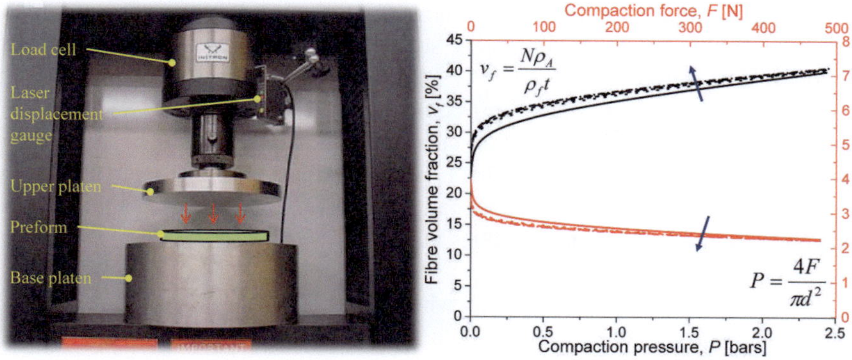

Fig. 4.2 *Left*: testing rig for compaction of a dry reinforcement. *Right*: example compaction curves for a plain-woven flax fabric. The *arrows* indicate the direction in which the curves shift for the successive compaction cycles

$$v_{\mathrm{f}} = \frac{N\rho_{\mathrm{A}}}{\rho_{\mathrm{f}} t}; \quad P = \frac{F}{A}, \tag{4.1}$$

where N is the number of layers, ρ_{A} is the areal density of a single layer, ρ_{f} is the fibre density and A is the sample press area.

At low compaction pressures ($P<0.1$ bar), the fibre volume fraction increases rapidly and the preform thickness decreases rapidly. Subsequent increase in compaction pressure pushes the fibre volume fraction and preform thickness to gradually approach an asymptotic maximum and minimum value, respectively. The experimental data is often approximated by models (assuming no rate dependence), where fitting parameters are representative of reinforcement properties (Robitaille and Gauvin 1998a, b; Correia et al. 2005; Long et al. 2005).

van Wyk (1946) and Gutowski et al. (1987a, b) independently proposed semi-analytical models for the compaction of random fibre mats (Eq. 4.2) and unidirectional plies (Eq. 4.3), respectively, assuming only fibre bending deformation. Both models incorporate fitting terms for the fibre volume fraction of the uncompacted preform v_0 and the (bending) stiffness (spring constant) K of the reinforcement. Notably, van Wyk's work was based on the compaction of natural wool fibres, while Gutowski et al. (1987a, b) studied graphite fibres. For a variety of wool fibre mats, the reinforcement stiffness K in Eq. (4.2) was estimated to range from 10 to 40 MPa.

$$P = K v_0^3 \left(\frac{1}{v_{\mathrm{f}}^3} - \frac{1}{v_0^3} \right), \tag{4.2}$$

$$P = K \frac{\left(\sqrt{\frac{v_{\mathrm{f}}}{v_0}} - 1 \right)}{\left(\sqrt{\frac{1}{v_{\mathrm{f}}}} - \sqrt{\frac{1}{v_{\mathrm{f,max}}}} \right)^4}, \tag{4.3}$$

where $v_{f,max}$ is the maximum fibre volume fraction and K and v_0 are empirical parameters representing reinforcement stiffness and uncompacted fibre volume fraction, respectively.

More commonly, power-law regressions, of the form in Eqs. (4.4) and (4.5), are fitted to experimental compaction curves. Various researchers have employed Eqs. (4.4) (Madsen 2004; Xue et al. 2011) and (4.5) (Shah et al. 2014a) to accurately reproduce compaction data of various natural fibre reinforcements, obtaining non-linear least squares regression R^2-values >0.99.

$$v_f = a_1 \cdot P^{b_1}, \quad \text{where } P \text{ is in} [\text{Pa}], \tag{4.4}$$

where a_1 is the fibre volume fraction at 1 Pa and b_1 is the stiffening index. a_1 and b_1 typically range between 0.02–0.12 and 0.1–0.2 for unidirectional plant fibre reinforcements and between 0.002–0.010 and 0.25–0.35 for random-mat plant fibre reinforcements (Madsen 2004).

$$v_f = v_0 + a_2 \cdot P^{b_2}, \quad \text{where } P \text{ is in} [\text{bar}], \tag{4.5}$$

where v_0 is the initial fibre volume fraction at no compaction pressure and a_2 and b_2 are the 'additional' fibre volume fraction (i.e. $v_f - v_0$) at 1 bar and the reinforcement stiffening index, respectively. v_0, a_2 and b_2 typically range between 0.3–0.4, 15–20 and 0.3–0.5 for woven silk textiles and between 0.2–0.3, 5–10 and 0.35–0.45 for woven plant fibre textiles (Shah et al. 2014a).

To compare materials that have been characterised with different models [e.g. Eqs. (4.4) and (4.5)], often the uncompacted fibre volume fraction [indicated by a_1 in Eq. (4.4) and v_0 in Eq. (4.5)] and the fibre volume fraction at a reference compaction pressure are used.

4.2.1.2 Key Compaction Mechanisms

While empirical models are useful for inputs in process simulations, analytical studies provide an understanding of the mechanisms driving compaction (Chen and Chou 1999, 2000; Chen et al. 2001). Table 4.2 identifies the main factors contributing to the compaction of random-mat, nonwoven, and woven preforms. The primary and secondary mechanisms have been indicated. As shown in Fig. 4.3, natural fibre preforms experience all the compaction mechanisms that synthetic fibre preforms experience (Table 4.2), but not vice versa.

In particular, cross-section deformation at the fibre scale (Table 4.2) is limited to plant fibre reinforcements, which, due to their hollow nature and low transverse stiffness and strength (Shah et al. 2012a; Shah 2013b), undergo lumen closure and transverse cell-wall buckling and delamination during compaction (Lundquist et al. 2004; Francucci et al. 2012a, 2013). This is illustrated in Fig. 4.4. A substantial change in fibre cross-sectional shape is not unimportant in the context of preform compaction, as it may alter the potential for fibre relative motion and yarn

Table 4.2 The dominant and secondary compaction mechanisms at the fibre, yarn and fabric scales in random-mat, nonwoven and woven preforms

Scale	Mechanism		Dominant in	Secondary in
Fibre/filament 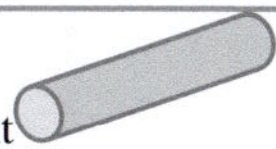	Cell-wall/lumen collapse (fibre cross-section deformation)	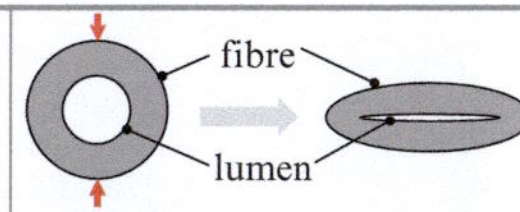	All plant fibre preforms	–
Yarn/tow 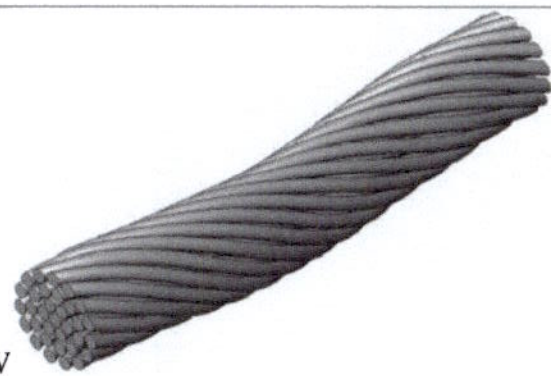	Yarn cross-section deformation	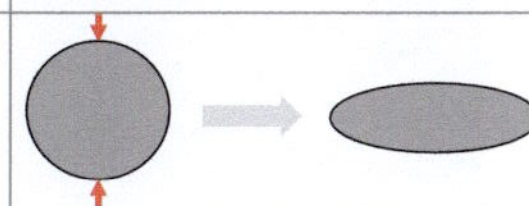	All	–
	Void condensation (i.e. closing gaps between fibres)	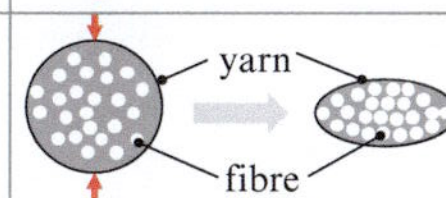	Random mat	Woven and nonwoven
	Yarn flattening	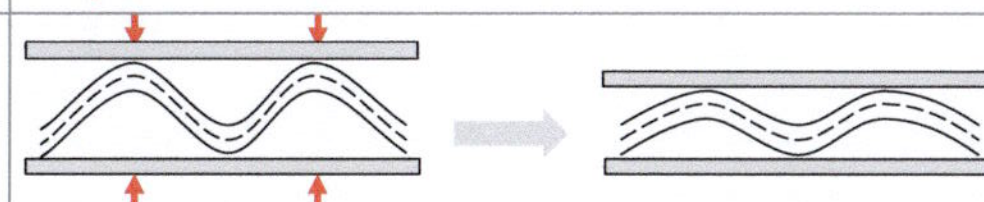	Woven	Random mat
	Fibre/yarn bending deformation	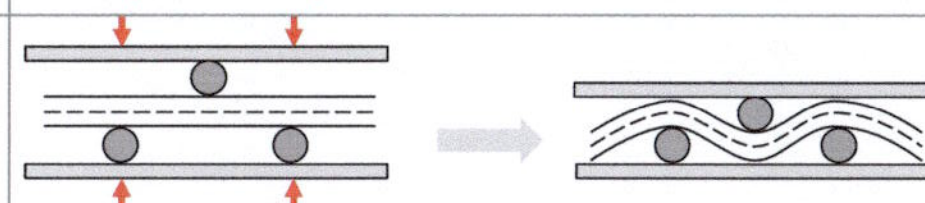	–	All
Fabric/preform 	Nesting and packing	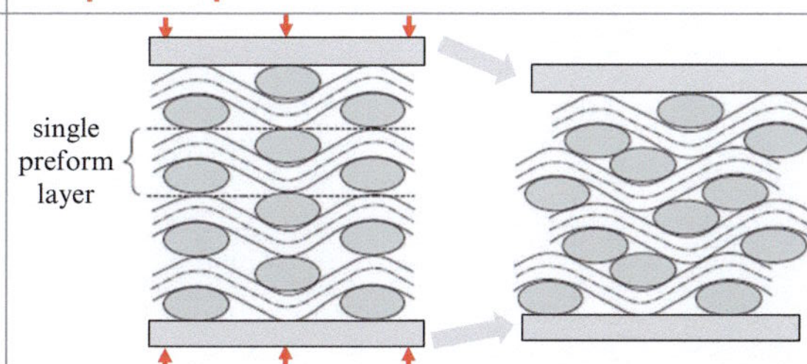	Woven	Nonwoven

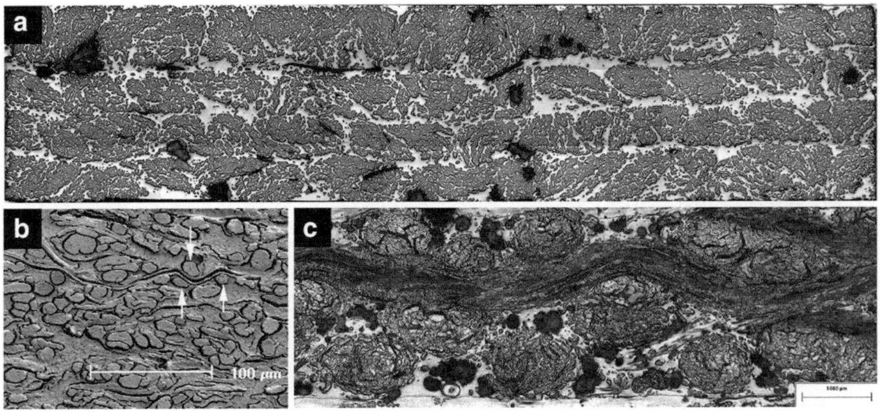

Fig. 4.3 Compaction mechanisms in various natural fibre preforms: (**a**) yarn cross-section deformation, void consolidation and nesting/packing in a nonwoven unidirectional flax composite (Shah et al. 2011, 2014a, b; Shah 2013a, b). (**b**) Fibre bending and flattening and void consolidation in a random-mat wood fibre composite (Lundquist et al. 2004). Yarn cross-section deformation and flattening and nesting/packing in a woven flax composite (Goutianos et al. 2007)

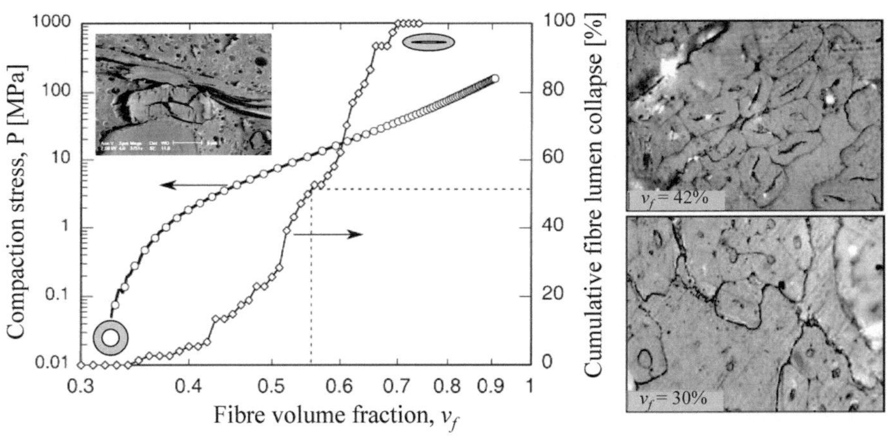

Fig. 4.4 *Left*: increased compaction pressure on a wood pulp random mat increases the fibre volume fraction, with partial contribution from cumulative lumen collapse-associated fibre deformation (Lundquist et al. 2004). *Inset*: SEM depicts a collapsed fibre. *Right*: lumen collapse observed in a woven jute fabric due to increase in compaction pressure at higher fibre volume fractions (Francucci et al. 2012a)

reorganisation, and could possibly lead to hindered impregnation in localised interfibre zones.

Lundquist et al. (2004) found that lumen compression occurred between fibre volume fractions of 34 and 69 % in wood pulp random-mat preforms (Fig. 4.4). Void condensation is a dominant compaction mechanism, particularly at low compaction pressures (e.g. when v_f < 55 %) in such random-mat preforms, with fibre/yarn bending deformation and flattening being additional secondary mechanisms, particularly at high compaction pressures (e.g. when $v_f > 55$ %). Francucci et al. (2012a) have also observed such irreversible transverse cell-wall deformation in compacted woven jute fabrics, and this phenomenon increased as the fibre content increased. They noted that this mechanism would contribute, alongside irreversible yarn cross-section deformation, yarn flattening and yarn nesting, to the compaction of the woven material. However, Francucci et al. opined that such lumen collapse would mostly occur when fibre rearrangement and tow movements are limited, i.e. at high fibre volume fractions. This is in stark contrast with the observation of Lundquist et al. (2004) described previously, where all wood pulp lumen had collapsed by $v_f = 69$ % and the wood pulp random mat was compressed further up to $v_f = 90$ %.

It is worth mentioning, however, that the luminal porosity in plant fibres varies between different species and even evolves with age within the same fibre. For instance, luminal porosities range between 74 and 80 % in kapok fibres, 20 and 70 % in wood fibres and 2 and 11 % in flax fibres (Shah 2013b). The luminal porosity has a significant effect on the transverse stiffness and yield strength in both compression and tension (Gassan et al. 2001; Lundquist et al. 2004; Placet et al. 2012), where fibres with a larger lumen (and smaller second moment of area) would tend to deform more readily. Naturally, therefore, different plant fibres would exhibit different degrees of fibre bending and cross-section deformation during plant fibre preform compaction.

Moreover, not all natural fibres have a central lumen; proteinaceous animal fibres like silk and wool are solid. Nonetheless, natural fibres tend to have low transverse properties due to their hierarchical, anisotropic structure (Ho et al. 2012; Shah et al. 2012a, 2014a). For instance, silks have a low transverse compressive modulus of 0.5–0.7 GPa (Ko et al. 2001) and a high Poisson's ratio of 0.4–0.5 (Zhang et al. 2010). Consequently, they undergo no change in cross-sectional area, but experience substantial deformation (i.e. flattening in cross-sectional shape) with increasing compressive force (Ko et al. 2001). While a 20 % reduction in thickness has been recorded for spider silk for compressive stresses of about 20 MPa (200 bar) (Ko et al. 2001), compaction pressures during LCM rarely exceed 1.5 MPa, below which the expected change in cross section is negligible.

Other micro- and macrostructural features, such as fibre cross-sectional shape and surface roughness (relating to fibre slippage) and degree of fibre alignment and dispersion, affect which (and when) compaction mechanisms will play primary and secondary roles.

4.2.2 *Through-Thickness Compaction of Natural Fibre Preforms*

The compaction response of reinforcement preforms is complex and depends on various elements, such as: type and form of fibre reinforcement, fibre architecture, number of layers in the preform, preform stacking sequence, history of loading, rate

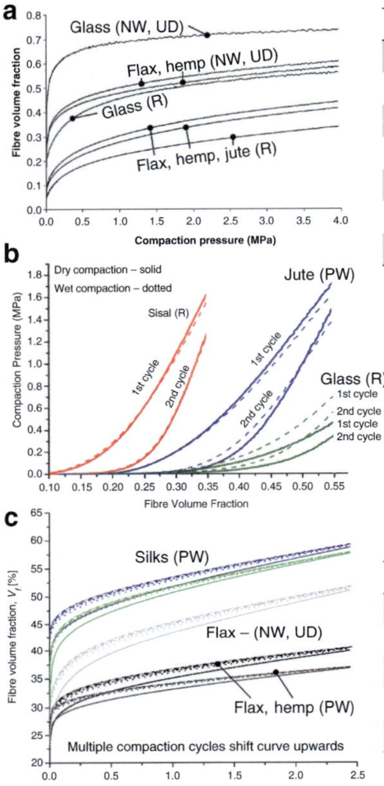

$v_f = a_1 \cdot P^{b1}$	a_1	b_1	v_f at 2 MPa
Glass (NW, UD)	0.297	0.06	71%
Flax (NW, UD)	0.083	0.13	55%
Hemp (NW, UD)	0.080	0.13	53%
Glass (R)	0.044	0.17	52%
Flax (R)	0.006	0.28	38%
Hemp (R)	0.004	0.30	34%
Jute (R)	0.002	0.33	28%

$v_f = v_0 + a_2 \cdot P^{b2}$	v_0	a_2	b_2	v_f at 2 bar
Silks (PW)	33-40	12-18	0.30-0.50	54-57%
Flax (PW)	24	10.9	0.37	39%
Hemp (PW)	26	7.3	0.43	36%
Glass (PW)	44-48	15	0.30	49-54%
Flax (NW, UD)	24	20	0.34	49%

Fig. 4.5 Compaction curves for various reinforcement preforms comparing the effect of fibre type and architecture. Plant fibre preforms are consistently less compactable than glass fibre preforms; natural silk preforms, however, are significantly more compressible than plant fibre preforms (and even more than glass fibre preforms—see text). Generally, nonwoven (NW) preforms tend to be more compactable than woven (W) preforms, which in turn are more compactable than random-mat (R) preforms. PW refers to plain-woven fabrics, while UD refers to unidirectional fabrics. The graphs are adapted from (**a**) Madsen (2004), (**b**) Francucci et al. (2013) and (**c**) Shah et al. (2014a). For reference, fitted model parameters are provided in the table

of compaction, tooling temperature and presence of lubricant (i.e. wet versus dry state) (Kim et al. 1991; Robitaille and Gauvin 1998a; Kelly et al. 2004; Correia et al. 2005; Long et al. 2005; Francucci et al. 2012a).

4.2.2.1 Effect of Fibre Type on Single-Cycle Dry Compaction

The compaction behaviour of various reinforcement preforms is presented in Fig. 4.5. The data is a compilation from Madsen (2004), Francucci et al. (2013) and Shah et al. (2014a), where various natural fibre reinforcements have been characterised. Note that the preform architecture has a substantial effect on the compaction response of a material. In this section, we discuss the sole effect of the fibre type.

It is consistently observed that plant fibre reinforcements are considerably less compactable than synthetic fibre reinforcements, but natural silkworm silk fibres are as compactable as, if not more compactable than, glass fibres. It is notable that both the initial uncompacted fibre volume fraction [a_1 in Eq. (4.4) and v_0 in Eq. (4.5)] and the fibre volume fraction at a reference compaction pressure (e.g. 2 MPa or 2 bar) follow this trend (Fig. 4.5). This suggests that the compaction pressure needed to achieve any fibre volume fraction (within the range studied) is significantly lower in the case of silk and glass reinforcements than for plant fibre reinforcements. Moreover, amongst plant fibres, flax is found to be more compressible than hemp, which in turn is more compressible than jute.

Besides the differences in vertical positions (reflected by differences in fibre volume fractions at zero and reference compaction pressures), the shape of the compaction curves is also different for different fibre reinforcements. One could compare fitting parameters of Eqs. (4.4) (namely, b_1) and (4.5) (namely, a_2 and b_2) to evaluate the differences in fibre type. For instance, glass and silk fibre reinforcements exhibit a steeper increase in fibre volume fraction at low compaction pressures than plant fibre reinforcements; this is indicated by small b_1 values in Eq. (4.4) (Madsen 2004) and large a_2 values in Eq. (4.5) (Shah et al. 2014a) (Fig. 4.5). For reference, fitted model parameters are presented in Fig. 4.5.

Analysing micro- to macrostructural features (i.e. fibre to yarn to fabric) in the silk, plant fibre and glass fibre reinforcements may explain these observations.

At the fibre scale, the (1) degree of fibre separation, (2) fibre surface roughness, (3) fibre cross-sectional shape and (4) fibre mechanical properties are notable factors. Firstly, preforms with well-separated fibres compact more readily due to increased degrees of freedom (Madsen 2004). While plant fibres typically exist as bundles, silk fibres, like synthetic glass fibres, are well separated from one another (Fig. 4.6a, b). Notably, it is the residue of the natural binding agent that dictates whether the fibres are separated or bundled. Pectin, the binding agent between elementary plant fibres to form a technical plant fibre bundle, is usually only partially decomposed during the fibre extraction process. On the other hand, the fibroin strand in silkworm silk is usually well degummed from the sericin binder. Secondly, the friction between fibre and fibre contacts is a relevant factor (Long et al. 2005), as fibre, yarns and fabric layers move (i.e. slide against each other) to dissipate energy during compaction. Plant fibres have a rougher surface than both silk and glass fibres (Fig. 4.6c, d), and consequently the latter may find it easier to adjust to a more efficient packing arrangement due to lower friction forces (Francucci et al. 2013; Shah et al. 2014a). Thirdly, the irregular almost-triangular cross-sectional shape of single silk fibres also permits higher ‘intra-yarn’ packing densities than irregular polygonal cross-sectional plant fibres (Fig. 4.6e, f) (Shah et al. 2014a). Finally, as the longitudinal and transverse elastic tensile moduli of a) silk and b) flax are a) 7–17 GPa and 0.5–0.7 GPa and b) 50–70 GPa and 4–9 GPa (Shah et al. 2012a), respectively, and as the ultimate failure strain of silk and flax is 15–30 % and 2–4 % (Shah et al. 2012a), respectively, silks can undergo much greater bending deformation without fibre breakage to enable efficient nesting and interlayer packing. In fact, fibre breakage and lumen collapse have been reported to be two critical mecha-

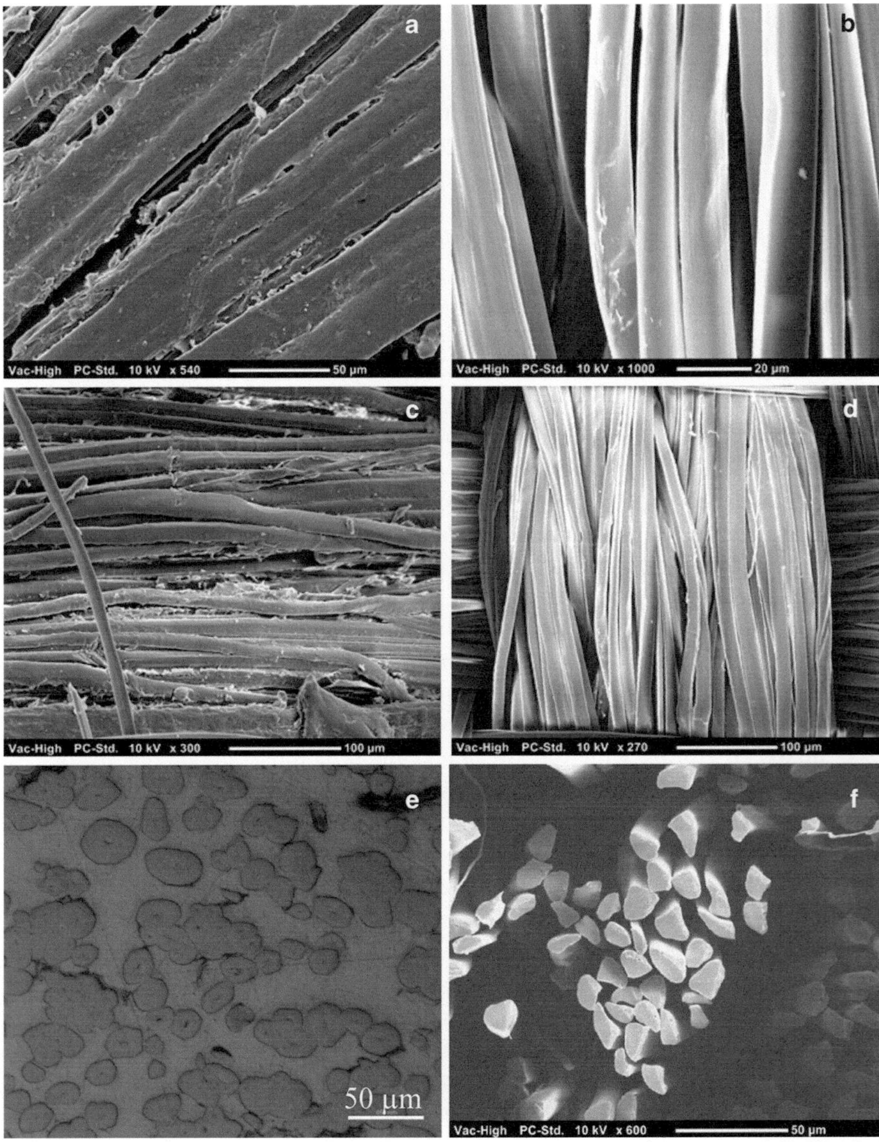

Fig. 4.6 Micrographs depicting: (**a**) the bundled nature of plant fibres and (**b**) the fibrillated nature of silk fibres; (**c**) plant fibres have a rougher surface than (**d**) silk fibres; while (**e**) plant fibres have irregular polygonal cross-sectional shape, (**f**) silk fibres have irregular almost-triangular cross-sectional shape. From Shah et al. (2014a)

nisms responsible for the permanent deformation of plant fibre preforms during compaction (Francucci et al. 2012a). Glass fibres, on the other hand, are stiff, isotropic fibres. It would be expected that fibre bending would be a more dominant compaction mechanism in silk and plant fibre reinforcements than glass fibre reinforcements.

Fig. 4.7 Micrographs depicting woven textiles of flax and hemp (**a**, **b**) and silk (**c**, **d**). From Shah et al. (2014a)

At the yarn and fabric scale, (1) yarn/fabric geometry and (2) fibre alignment are notable factors. Regarding fabric geometry, the tightness of a weave and inter-yarn porosity, for instance, will affect preform compactibility. Silk textiles, like glass fibre textiles, are more tightly woven than plant fibre textiles (Fig. 4.7), thereby imparting high packing density, particularly at low compaction forces. This would explain the high values of v_0 and a for silk textiles in comparison to plant fibre textiles (Shah et al. 2014a). Regarding yarn geometry, while plant fibre yarns tend to have a circular cross section due to high twist factors, silk fibre assemblies, much like glass fibre assemblies, have large width-to-thickness ratios (of ~5). The lenticular cross-sectional shape of silk rovings would enable higher 'inter-yarn' compaction than twisted plant fibre yarns (Shah et al. 2014a). Moreover, this also reduces crimp. High crimp in woven fabrics would lead to higher 'hills' and lower 'valleys'. It would be expected, therefore, that yarn bending deformation and nesting would be more important compaction mechanisms in plant fibre textiles than silk textiles.

Crimp is one source of fibre misorientation that has a detrimental effect on compaction. In terms of orientation, the arrangement of yarns in the fabrics is clearly more ordered and uniform in the silk textiles (Fig. 4.7). Studies on the compaction of both E-glass and plant fibre reinforcements have reported that rovings are significantly more compactable than twisted yarns (Kim et al. 1991; Madsen 2004). This is because yarn twist induces fibre misorientation and transverse pressure, both of which are detrimental to effective packing (Shah et al. 2013b). Note that twist is a 3D phenomenon and that yarn twist (and yarn packing fraction) are a function of yarn radius; the twist angle and packing fraction are highest and lowest, respectively, at the yarn surface (Shah et al. 2013b). As observed in Fig. 4.7, while the plant fibre textiles employ yarns that are twisted (surface twist angle of 20–30°), the silk textiles employ rovings of well-aligned filaments.

Studies have shown that not only yarn twist but yarn hairiness (or fluffiness) is also an important source of misorientation (Kim et al. 1991). Elementary plant fibres have non-uniform width in the range of 10–100 μm and a short length in the range of 4–100 mm (Lewin 2007). Due to the short length of staple plant fibres, during the spinning process not all fibre ends are integrated into the yarn structure. The distribution (length and frequency) of fibre ends protruding from the fibre surface is referred to as yarn 'hairiness' by textile engineers. These protruding fibres have a negative impact on the packing of a fabric preform. Kim et al. (1991) report that the compaction response of a fluffy roving may be as poor as that of a twisted spun yarn, relative to a straight roving; for instance, straight roving preforms have fibre volume fractions of 70 % at compaction pressures of 2 MPa, compared to fibre volume fractions of only 50 % for both fluffy roving and spun yarn preforms. As observed in Fig. 4.7, numerous fibres are protruding from the structure of the yarns in the flax textiles, while filaments in the silk fibre textiles are well integrated into the roving. It is noteworthy that the sources of misorientation (yarn twist and yarn hairiness) are lacking in silk textiles because silks exist as long fibres (i.e. filaments), unlike staple plant fibres. In fact, silks, such as those industrially processed from the silkworm cocoons, exist as single continuous filaments/strands of fibroin with lengths of up to 1,500 m (Lewin 2007). Put it simply, it is easier to align longer fibres than shorter fibres.

4.2.2.2 Effect of Fabric Architecture

As shown in Table 4.2, the fabric architecture plays a key role in determining the mechanisms that drive compaction. The effect of fabric architecture on the compaction of synthetic fibres has been studied in some detail (Kim et al. 1991; Robitaille and Gauvin 1998a; Yang et al. 2012) and similar trends have been reported in the limited articles investigating natural fibre reinforcements (Madsen 2004; Francucci et al. 2013; Shah et al. 2014a).

As observed in Fig. 4.5, unidirectional nonwovens tend to be more compactable than woven fabrics, which in turn are more compactable than random-mat reinforcements. Notably, the compaction curves shift upwards [i.e. higher fibre fractions at zero [a_1 in Eq. (4.4) and v_0 in Eq. (4.5)] and reference compaction pressures], exhibit a steeper rise at low compaction pressures [i.e. smaller b_1 in Eq. (4.4) and larger a_2 in Eq. (4.5)] and become more flat at high compaction pressures.

Fibre alignment/orientation is a mechanism-governing factor in fabric architecture. For instance, in unidirectional fabrics, fibres in adjacent layers are parallel to each other and can therefore fill gaps by dislodging other fibres to produce a compact structure. In woven fabrics, however, fibres/yarns leave voids during crossover (weave density) that cannot be filled by crossover fibres in other layers. Essentially, fibre packing between layers with the same fibre/yarn orientation is easier than packing between layers with different fibre/yarn orientations. Of course, in aligned reinforcements (nonwoven and woven), the stacking sequence also affects the compaction response, but this has not been studied on natural fibre reinforcements so far.

Alongside fibre orientation, fibre/yarn length and diameter may also affect preform compaction. Umer et al. (2011) and Madsen (2004) investigated the compaction behaviour of random mats constructed with yarns of varying length and diameter. Madsen (2004) found that compaction curves of hemp yarn random mats, with mean yarn lengths of 2, 10 and 50 mm, were nearly identical. In contrast, Umer et al. (2011) observed that the required compaction pressure for 50 mm flax yarn random mat was about 40 % higher in comparison to a 15 mm flax yarn random mat. They attributed the poorer compactibility of longer yarn mats to the larger number of bundle–bundle crossover points. Umer et al. (2011) also found that flax random mats were significantly more compactable when larger diameter yarns were employed. They suggested that there were more fibres in larger diameter yarns that were more efficiently packed within the available space (than they would be as loose fibres in the mat), and consequently there were less fibre crossover points within the mat.

4.2.2.3 Effect of Multiple Compaction Cycles

Multiple compaction cycles have a noticeable and well-documented effect on preform compaction (Kim et al. 1991; Robitaille and Gauvin 1998a; Long et al. 2005). Natural fibre reinforcements, like synthetic fibre reinforcements, follow the general trend (Madsen 2004; Francucci et al. 2012a; Shah et al. 2014a): (1) in comparison to the first cycle, the second compaction cycle requires a relatively lower compaction pressure to achieve a given fibre volume fraction, (2) subsequent compaction cycles (i.e. third, fourth and fifth) overlay with the second compaction cycle, and (3) all compaction curves share the same asymptotic fibre volume fraction (at very high compaction pressures) (Fig. 4.5). With regard to the fitting model parameters, in comparison to the first compaction cycle, the uncompressed fibre volume fraction [a_1 in Eq. (4.4) and v_0 in Eq. (4.5)] tends to increase and a steeper rise at low compaction pressures is observed [i.e. smaller b_1 in Eq. (4.4) and larger a_2 in Eq. (4.5)] for the second compaction cycle. Subsequent compaction cycles have fairly identical fitting parameters to the second cycle.

Maximising the fibre content is appealing for NFRPs, not only to improve composite mechanical performance (Shah et al. 2012b; Shah 2013b) but also to increase the content of bio-based material and reduce the amount of polymer.

The latter improves sustainability credentials as natural fibres require significantly less energy for production than both synthetic fibres and polymer matrices (Duflou et al. 2012; Shah 2013b). The aforementioned findings provide useful insights in developing high fibre content NFRPs. If high compaction pressures, such as those realised using an autoclave for LCM (up to 15 bar) or a mechanical press for compression moulding (up to 100 bar), are used, a single compaction cycle is sufficient. That is, pre-compaction is not necessary. However, if low compaction processes are employed, such as in vacuum bagging, pre-compaction of the reinforcement is an attractive technique to improve preform compaction and increase the part fibre volume fraction.

Notably, multiple compaction cycles generally lead to the reorganisation and permanent deformation of fibres/yarns within the textiles, driven through both elastic and irreversible mechanisms, such as those illustrated in Table 4.2. Compaction mechanisms such as fibre/yarn cross-section deformation, yarn flattening and nesting tend to become more significant at higher compaction pressures (and thus fibre volume fractions) (Francucci et al. 2012a); therefore, the maximum compaction pressure of the previous cycle has an effect on the compaction in the next cycle. In addition, stacked textiles have a time-dependent viscoelastic behaviour (Francucci et al. 2012a); therefore, studying preform relaxation is as important as studying preform compaction. In fact, relaxation studies post-multiple compaction cycles have revealed that plant fibre reinforcements undergo significantly more irreversible, permanent deformation than glass fibre reinforcements (Francucci et al. 2013).

4.2.2.4 Wet Compaction

In LCM, while the preform is first compacted in the dry state (in the mould-filling stage), once the preform has been impregnated, the saturated preform undergoes 'wet compaction' (in the post-filling stage) (Table 4.1). During and post-impregnation, the compaction pressure exerted on the partially or fully saturated preform is only partially felt by the reinforcement; the remaining compaction pressure is managed by the liquid resin (Long et al. 2005). However, lubricating effects of the liquid tend to reduce the compaction pressure required to achieve a given fibre volume fraction. This is because lubrication of fibre-to-fibre contact points means that individual fibres find it easier to move and slide and realign and reorganise. As a corollary, it is also expected that as the fibres are likely to attain a higher degree of stability for a given compaction pressure, the preform would exhibit less relaxation (when held at a constant thickness). These assertions have been found to be true for synthetic fibre preforms (Kim et al. 1991; Long et al. 2005).

Natural fibres, including plant and animal fibres, are different from synthetic fibres like glass, in that natural fibres tend to be polar and hydrophilic. That is, natural fibres absorb polar liquids; this not only includes liquids used during compaction and permeability testing like water and glycerine solution but also includes polar thermosetting resins such as vinylester and phenolics (Francucci et al. 2010). Moreover, natural fibres swell upon liquid absorption (10–25 % moisture regain) (Francucci et al. 2010).

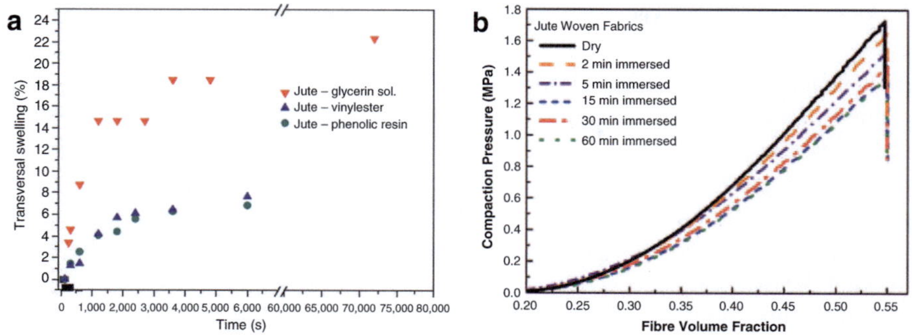

Fig. 4.8 (**a**) Transverse swelling (in terms of % diameter change) of jute fibres in different fluids. Adapted from Francucci et al. (2010). (**b**) Effect of immersion time on the compaction of jute woven fabrics. From Francucci et al. (2013)

For instance, over an immersion time of 6,000 s, jute fibres exhibited equilibrium transverse swelling (i.e. change in diameter) by over 18 % in glycerine solution and between 6 and 8 % in vinylester and phenolic resin (Fig. 4.8a). In comparison, glass fibres do not absorb liquid (<2 % moisture regain), nor do they exhibit swelling (Francucci et al. 2010).

As natural fibres absorb liquids and consequently swell and soften (Francucci et al. 2010, 2013), their wet compaction behaviour requires attention. Since natural fibres will behave differently with different liquids (say, nonpolar oils as test fluids and polar epoxy resin as the matrix), to understand their true compaction behaviour, the natural fibre reinforcement should be studied with the particular resin that is to be used during LCM as the test fluid.

Investigations on plant fibre reinforcements have found that the immersion time is an important test parameter (Francucci et al. 2013). In comparison to the compaction curve of a dry reinforcement, the compaction curve of plant fibre reinforcements shifts to higher fibre volume fractions (i.e. enhanced compactibility) with increasing immersion time of the saturated preform (Fig. 4.8b). This is similar to synthetic fibre performs. However, it is found that the wet relaxation behaviour of plant and synthetic fibre reinforcements stands in contrast, with plant fibre reinforcements showing greater relaxation in the wet state in comparison to the dry state (rather than the expected less relaxation). Francucci et al. (2013) propose that the observed result is due to fibre softening upon swelling during wet compaction.

4.2.2.5 Effects of Other Parameters

The effects of other parameters, such as number of layers and compaction rate, on the compaction behaviour of natural fibre reinforcements have been studied in literature (Francucci et al. 2012a, 2013; Shah et al. 2014a). Generally, their effects are relatively weak, and notably conflicting trends have been reported for synthetic fibre reinforcements.

Shah et al. (2014a) observed that reducing the number of layers of woven silk fabric shifted the compaction curves marginally to higher fibre volume fractions (i.e. enhanced compactibility). Moreover, the curves became steeper at higher compaction pressures. These are indicative of the fact that stacks made of less fabric layers are easier to compact to a given fibre volume fraction, particularly at higher compaction pressures, because nesting, the primary compaction mechanism of woven textiles, becomes progressively difficult with more layers (Long et al. 2005), potentially due to the non-uniform transmission of the applied compressive force through the textile layers.

Conflicting, though weak, trends have been observed on the effect of compaction speed on natural fibre reinforcement compaction. Shah et al. (2014a) reported that woven silk textiles were more compactable at higher compaction rates. For instance, a 100-fold increase in the compaction speed from 0.5 to 50 mm/min increased v_0 from 30.5 to 32.2 % and increased the representative fibre volume fraction v_f at $P=2.0$ bar from 53.7 to 54.1 %. In comparison, Francucci et al. (2012a) found that increasing the compaction speed led to a larger compaction pressure for the same fibre volume fraction of woven jute fabrics.

4.2.3 Conclusions

The compaction response of natural fibre reinforcements can be characterised by the same power-law regression curves as synthetic fibre reinforcements. Good reproducibility of test measurements and strong model curve fits are found. Natural fibre reinforcements also show evidence of similar compaction mechanisms as synthetic fibre reinforcements, with the addition of fibre cross-section deformation through cell-wall and lumen collapse in the case of hollow plant fibres.

In general, plant fibre reinforcements are significantly less compactable than glass fibre reinforcements. However, animal silk fibre reinforcements demonstrate comparable compactibility to the latter. The difference in structure (i.e. geometry, alignment and dispersion) and technical properties of the fibres, their semi-products (i.e. yarns and rovings) and their textiles accounted for the differences in fabric compaction behaviour. Unlike plant fibres, silk, the only natural fibre to exist as long filaments (rather than short fibres), and its semi-products and textiles have much resemblance with synthetic fibres and their semi-products and textiles.

The effects of fabric architecture, multiple compaction cycles, compaction rate and number of layers of fabric on preform compaction follow the same generic trends in natural and synthetic fibre reinforcements. However, irreversible, permanent deformation during compaction is more prevalent in plant fibre reinforcements than glass fibre reinforcements. More studies on the creep and relaxation response of natural fibre reinforcements are therefore required. For low-pressure LCM processes (such as vacuum bagging), pre-compaction through an initial single cycle is identified to be an attractive technique to improve compaction and part fibre volume fraction.

The wet compaction of natural fibre reinforcements requires specific attention, as unlike synthetic fibres, natural fibres absorb the liquid resin and swell and soften. This not only affects the relaxation response of the saturated reinforcement but also its permeability. Moreover, as natural fibres will behave differently with different fluids, the fibre compaction behaviour needs to be studied for the specific resin that is to be used during composite manufacture.

4.3 Modelling the Mould-Filling Process in LCM of Natural Fibre Composites

Controlled and complete filling of the mould with adequate wetting of fibres is a primary objective in LCM. Poor fibre wetting would lead to poor mechanical properties due to micro-void formation and poor interfacial adhesion, while uncontrolled and incomplete filling would lead to defect formation (e.g. dry spots), poor part quality and even part scrappage and material wastage. Optimising mould fill time while avoiding fluid pressure build-up is another important aspect of controlled mould filling. The microscopic and macroscopic flow of the liquid resin through gaps within yarns/tows and the porous preform, respectively, is therefore important to study.

A number of factors affect the complex mould-filling process (Table 4.1), including preform permeability, inlet/outlet gate location, pressure differential in mould cavity, rate of resin injection and resin viscosity (as a function of time). Extensive research has been conducted to measure, predict and simulate the mould-filling process in the LCM of conventional synthetic fibre composites (Long et al. 2005). Often, permeability studies and Darcy's law [which can be derived from the Navier–Stokes equation through averaging methods (Neuman 1977)] are used to model the complicated viscous flow of resin in a porous media. Natural fibre reinforcements, however, require specific considerations.

4.3.1 In-Plane Permeability of Natural Fibre Reinforcements

Permeability is defined as the ease of fluid flow through a preform, and therefore it is an inverse measure of the flow resistance (Long et al. 2005). The greater the preform permeability, the easier it is for the resin to impregnate the reinforcement, and the lesser the time needed to fill the mould. Fabric permeability has a governing effect not only on fill time and flow front shape but also on void and dry-spot formation resulting from the dual-scale flow within preforms (Fig. 4.9) and the anisotropic permeability of anisotropic preforms.

While permeability characterisation of natural fibre reinforcements is a relatively new topic (Rodriguez et al. 2004; Francucci et al. 2010, 2013; Masoodi and Pillai 2011; Umer et al. 2011; Xue et al. 2011), researchers have already found some critical differences between natural and synthetic fibre reinforcements. These will be discussed in this section.

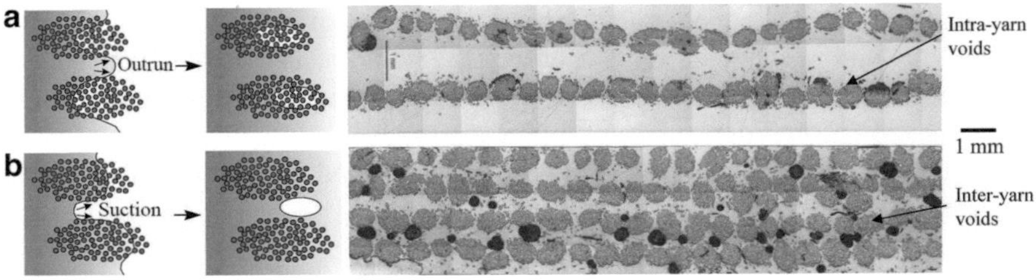

Fig. 4.9 The complex, dual-scale flow of resin in fibrous preforms can generate voids. Macroscale flow relates to the advance of resin between yarns/tows (i.e. inter-yarn flow), while microscale flow relates to the penetration of resin into a yarn (i.e. intra-yarn flow). Note that permeability is also different at the two scales. For instance, (**a**) at low fibre content, due to low yarn permeability but high overall permeability, the yarn is not properly impregnated and thus intra-yarn voids may form, while (**b**) at high fibre content, although yarn and overall permeability are similar, capillary flow in the yarn dominates and therefore inter-yarn voids are formed. From Shah et al. (2012b) and Shah (2013a)

4.3.1.1 Permeability Measurement and Modelling

Typically, preform permeability is experimentally measured by tracking the progression of the fluid flow front and monitoring the pressure field (Fig. 4.10). A planar flow cell is used in which a test fluid (with a similar viscosity as the resin) is injected into the fabric. A 1D unidirectional flow achieved through line-gate injection (Fig. 4.10a, b) is commonly used. However, a 2D radial flow achieved through central injection (Fig. 4.10c, d) may be more appropriate for anisotropic reinforcements to quickly measure anisotropy in permeability. The fluid viscosity is measured using a viscosimeter or rheometer. A flowmeter, typically placed at the inlet or outlet port, may be used to measure fluid velocity. Alternatively, the flow velocity may be estimated by measuring flow front evolution. The flow front can be tracked manually (e.g. with scale bars, as in Fig. 4.10b), with a video camera, or even electrically activated (pressure) sensors. Pressure gradient in the mould may be recorded through manometers or pressure transducers, typically placed at the inlet and outlet ports. Pressure difference is measured relative to the flow front (which is at atmospheric pressure).

Darcy's law, which describes the flow of Newtonian fluids in porous media, is then used to determine permeability. Assuming steady-state 1D flow, Darcy's law (Eq. 4.6) states that the macroscopic volumetric flow rate Q is proportional to the mould cross-sectional area A and the pressure difference over the sample ΔP and inversely proportional to the sample length L (in the flow direction) and fluid viscosity μ. The constant K is termed the permeability.

$$Q = K \frac{A}{L} \frac{\Delta P}{\mu}. \tag{4.6}$$

Saturated (or steady-state) permeability K_{sat} is obtained once the reinforcement is fully saturated, using Eq. (4.7). On the other hand, unsaturated (or transient)

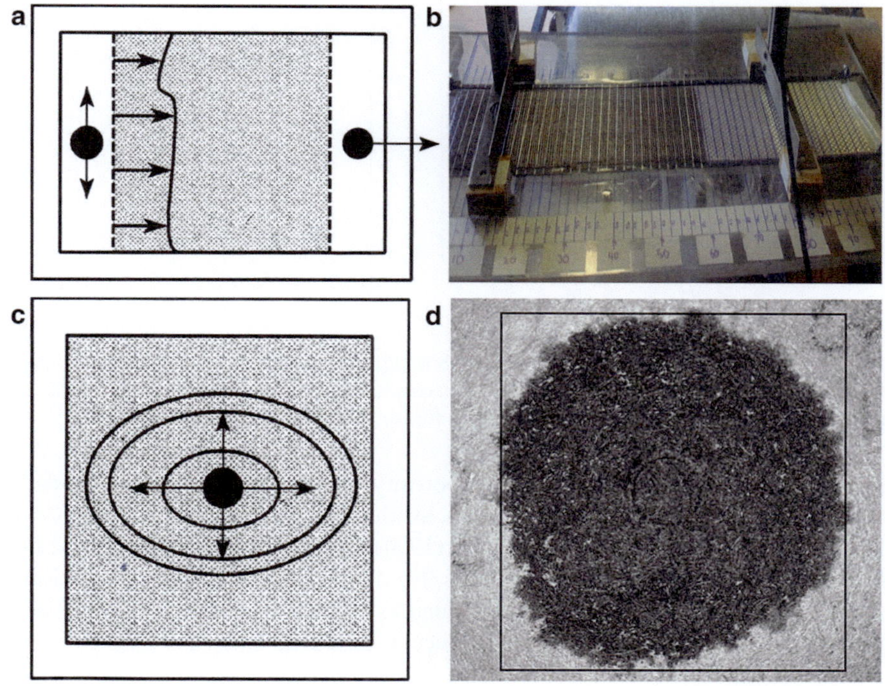

Fig. 4.10 Permeability measurement and flow visualisation set-ups. (**a**, **b**) 1D linear flow through line-gate injection (Parnas et al. 1995). (**c**, **d**) 2D radial flow through central injection (Parnas et al. 1995; Xue et al. 2011)

permeability K_{unsat} is obtained when the flow profile is measured during the injection process. Darcy's law in Eq. (4.6) can be integrated and rearranged to the form in Eq. (4.8) (Cai 1992), where x is the position of the flow front at time t, ϕ is the fabric porosity, and v is the average fluid velocity. Note that fabric porosity is directly related to the fibre volume fraction. For demonstration, Fig. 4.11a shows the linear relationship between the square of the flow front position and fill time, for two separate infusion conditions; the slope of the curve x^2/t is used as an input in Eq. (4.8) to determine K_{unsat}.

$$K_{sat} = Q\frac{\mu}{\Delta P}\frac{L}{A}, \tag{4.7}$$

$$K_{unsat} = \frac{x^2}{t}\frac{\phi\mu}{2\Delta P}, \quad \text{where} \quad \phi = \frac{Q}{vA} \quad \text{and} \quad \phi \equiv 1 - v_f. \tag{4.8}$$

The saturated and unsaturated permeability can be then determined over various conditions, say varying fabric porosity (or fibre volume fraction) as shown in Fig. 4.11b. In fact, reinforcement fibre volume fraction, alongside reinforcement type and orientation, is a key reinforcement-related factor affecting permeability.

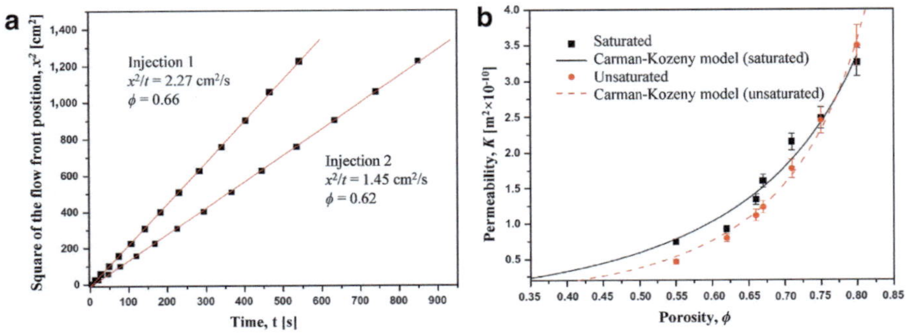

Fig. 4.11 Permeability testing of woven jute fabric (Francucci et al. 2010). (**a**) Plot of square of flow front position with time for different infusion conditions. (**b**) Plot of saturated and unsaturated permeability against porosity (=1—fibre volume fraction)

Increasing the fibre content (decreasing porosity) reduces the number of flow paths and therefore reduces permeability. Often, the three-parameter exponential function [Eq. (4.9); proposed by Gauvin et al. (1994)] or the two-parameter, modified Carman–Kozeny equation (Carman 1937) (Eq. 4.10) is used to model permeability–porosity relationship. Both models are found to be well suited to natural fibre reinforcements (Rodriguez et al. 2004; Francucci et al. 2013).

$$K = a + b \cdot \exp^{c\phi}, \tag{4.9}$$

$$K = \frac{\phi^{n+1}}{C(1-\phi)^n} = \frac{(1-v_\mathrm{f})^{n+1}}{C v_\mathrm{f}^{\,n}}. \tag{4.10}$$

Two problems commonly associated with permeability testing are mould deflection (induced by the pressure gradient on thin moulds) and uncontrolled flow (Parnas et al. 1995). The latter may manifest in the form of fibre washing or edge effects. Fibre washing refers to the displacement of the preform during filling, while edge effects, also known as race-tracking, refer to the faster flow of resin at the edges due to lower permeability resulting from a clearance between the fibre preform and the mould edge. Indeed, such concerns have been voiced in studies on the mould-filling process of natural fibre reinforcements as well (Fig. 4.12) (Richardson and Zhang 2000; Rodriguez et al. 2004; Francucci et al. 2010; Ho et al. 2012). Fibre washing is more prevalent at high injection pressures and low fibre volume fractions (e.g. fewer reinforcement layers), while edge effects are more prevalent at high fibre volume fractions (Richardson and Zhang 2000).

4.3.1.2 Effect of Reinforcement Type

The limited research on permeability studies of natural fibre reinforcements shows that while plant fibre reinforcements exhibit higher permeability than glass fibre reinforcements (Fig. 4.13) (Rodriguez et al. 2004; Francucci et al. 2013), wood fibre

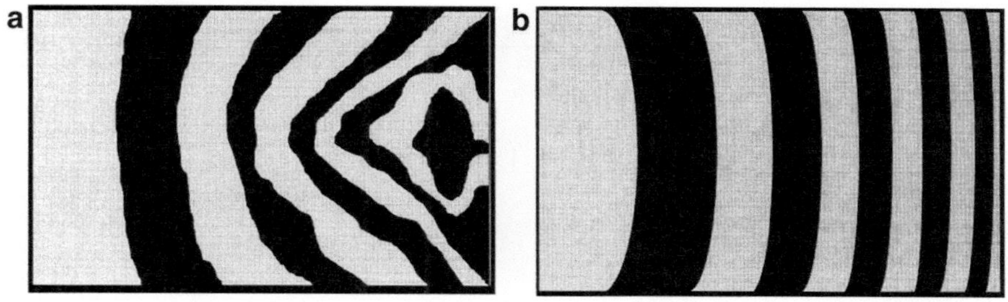

Fig. 4.12 Resin flow in 2-layer random-mat hemp reinforcements before (**a**) and after (**b**) problems of fibre washing and edge flow were resolved. The flow front is smoother, more uniform and quasi-1D in (**b**). Flow front isochrones are shown, where each isochrone represents 20 s. Mould-filling direction is from *left* to *right*. From Richardson and Zhang (2000)

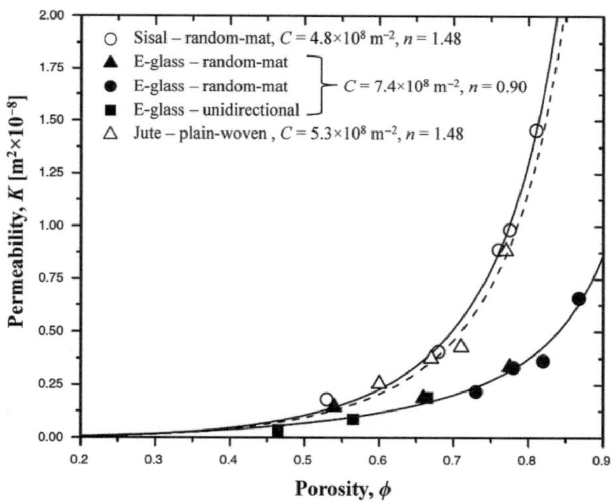

Fig. 4.13 Comparison of the (unsaturated) permeability of sisal, jute and E-glass reinforcements. Carman–Kozeny constants are also provided. Adapted from Rodriguez et al. (2004)

mats exhibit significantly lower (by about two orders of magnitude) permeability than glass fibre mats (Umer et al. 2008). Table 4.3 describes permeability data of various natural fibre reinforcements studied in literature, by showing Carman–Kozeny constants and permeability values at specific fibre volume fractions. Notably, n is close to 2 in most cases, implying that the fluid flow behaviour in natural fibre reinforcements is close to the original Carman–Kozeny model, where fibre arrangement can be described as parallel tubes with low tortuosity (Rodriguez et al. 2004).

Researchers have hypothesised that the higher flexibility of plant fibres in comparison to glass fibres may make the former more permeable (Rodriguez et al. 2004). On the other hand, the lower permeability of wood fibre mats (in comparison to glass fibre mats) has been attributed to the more tortuous flow paths in the less-efficiently packed wood fibre mats, owing to the very short length of wood fibres (Masoodi and Pillai 2011). The argument may have merit as flax yarn random mats composing of

Table 4.3 Permeability data of various natural fibre reinforcements

Reinforcement	C [$\times 10^8$ m^{-2}]	n	K at v_f = 0.2 (or φ = 0.8) [$\times 10^8$ m^{-2}]	K at v_f = 0.5 (or φ = 0.5) [$\times 10^8$ m^{-2}]	Unsaturated or saturated, test fluid, viscosity	Source
Wood fibre – random-mat	2460	1.80	0.00394	0.000203	Saturated Mineral oil, 0.066-0.095 Pa·s	(Umer 2008)
Wood fibre – random-mat	4000	1.76	0.00229	0.000125	Saturated Mineral oil, 0.066-0.095 Pa·s	(Umer 2008)
Sisal – random-mat	4.8	1.48	1.30	0.104	Unsaturated, Glycerine solution, 1.2 Pa·s	(Rodriguez 2004)
Jute – plain-woven	5.3	1.48	1.17	0.0943	Unsaturated, Glycerine solution, 1.2 Pa·s	(Rodriguez 2004)
Sisal – plain-woven	22.5	2.00	0.569	0.0222	Unsaturated, Vinylester resin, 0.5-0.9 Pa·s	(Li 2006)
Jute – plain-woven	81.0	0.88	0.0335	0.00617	Saturated, Glycerine solution, 0.13 Pa·s	(Francucci 2013)
Jute – plain woven	133.8	1.29	0.0357	0.00373	Unsaturated, Glycerine solution, 0.15 Pa·s	(Francucci 2010)
Jute – plain woven	84.6	0.91	0.0334	0.00591	Saturated, Glycerine solution, 0.15 Pa·s	(Francucci 2010)

longer fibres have notably and consistently higher (saturated) permeability across a range of fibre volume fractions (Umer et al. 2011). For instance, the permeability of random mats with 50 mm fibre lengths was found to be 22 and 25 % higher at the lowest (v_f=0.2) and highest (v_f=0.4) fibre volume fractions, in comparison to random mats with 15 mm fibre lengths (Umer et al. 2011).

Yarn diameter (or linear density) is also known to affect the permeability of flax yarn random mats (Umer et al. 2011). Umer et al. (2011) found that the permeability of medium yarn diameter (0.56 mm) mats was consistently (i.e. over a range of porosity levels) 27 % higher than small yarn diameter (0.35 mm) mats, but the permeability of small and medium yarn diameter mats was consistently 68–77 % higher than large yarn diameter (0.81 mm) mats. Given that the permeability of preforms is dominated by the characteristics of open channels and that preform geometric parameters that describe the characteristics of the open channels include the number of fibres in the bundle, twist angle and orientation of bundle, dimension and cross-sectional shape of bundles and single fibres, Umer et al. suggested that large diameter yarns were less compact and had lower twist levels. The loose fibres on the surface (previously referred to as hairiness) of large diameter yarns restrict fluid flow through the open channels, thereby decreasing permeability. The twist level of yarns may also affect permeability and impregnability by altering competition between micro-flow and macro-flow (Goutianos and Peijs 2003; Umer et al. 2011; Shah 2013a); this is discussed further in Sect. 4.3.1.5.

It is clear from Table 4.3 that experimental conditions have a notable effect on the permeability data obtained for even the same natural fibre reinforcement (e.g. jute plain-woven fabric). In particular, it is important to clarify (1) whether it is the saturated or unsaturated permeability that has been measured, (2) the test fluid viscosity and (3) the polarity of the test fluid. While the first two issues apply to reinforcement preforms in general, the latter point applies specifically to natural fibre reinforcements. For natural preforms, saturated permeability tends to be higher than unsaturated permeability (Francucci et al. 2010) [discussed in Sect. 4.3.1.4), and permeability tends to be lower when measured in less viscous fluids (which is also in agreement with Eqs. (4.7) and (4.8)]. The more interesting issue is of the polarity of the test fluid, as natural fibres, unlike synthetic fibres, absorb polar fluids and consequently swell and soften (Umer et al. 2007; Francucci et al. 2010). This was discussed previously in Sect. 4.2.2.4. Conceivably, fluid absorption and swelling are important mechanisms in natural fibre preforms, due to which both saturated and unsaturated permeability are reduced. Francucci et al. (2010) demonstrated through absorption, swelling and permeability tests on untreated and treated jute fibre random mats that untreated mats had up to five times higher absorption levels, up to 20 times higher transverse swelling and an order of magnitude lower permeability than treated mats. Umer et al. (2007) show that permeability levels of wood fibre mats measured in (polar) glucose syrup were lower than that measured in (nonpolar) mineral oil. Fluid absorption removes fluid from the main stream, acting as a sink component and thus decreasing flow velocity during the unsaturated flow. The swelling of the natural fibres constricts the open flow path, reduces porosity and increases flow resistance during saturated flow. Since natural fibres will behave differently with different liquids, to accurately measure their permeability, the natural fibre reinforcement should be studied with the particular resin that is to be used during LCM as the test fluid.

4.3.1.3 Permeability Anisotropy

Random mats are quasi-isotropic reinforcements and would therefore produce quasi-isotropic flow due to quasi-isotropic permeability. Xue et al. (2011) measured the anisotropy ratio in permeability (K_{max}/K_{min}) to range between 1.08 and 1.46 for flax random mats; cross-laid mats were more isotropic than parallel-laid mats. Orientation distribution analysis showed that fibres were uniformly distributed in cross-laid mats, but fibres in parallel-laid mats were oriented primarily along the machine direction. As the preferential resin flow path (path of least resistance) would be along the principal fibre orientation direction, permeability was much higher in this direction, for parallel-laid mats. In cross-laid mats, due to fairly equal paths of least resistance, flow was more isotropic. Note that interestingly parallel-laid mats exhibited lower overall permeability than cross-laid mats compacted under the same pressure, due to fibre nesting and consequently lower porosity in parallel-laid mats.

Aligned reinforcements, such as nonwoven and woven fabrics, tend to produce anisotropic flow due to anisotropic permeability. The permeability anisotropy ratio of plain-woven jute fabric has been estimated to be 1.24 (Francucci et al. 2013).

4.3.1.4 Saturated and Unsaturated Permeability

For synthetic reinforcements, ratios of saturated to unsaturated permeability (K_{sat}/K_{unsat}) have been found to be greater than one, less than one or varying slightly above one (Dungan and Sastry 2002). Various reasons are proposed in literature for the observed result, including (1) flow channelling or fluid flow in preferential high-porosity regions (leading to $K_{sat} > K_{unsat}$), (2) capillary and wicking effects (leading to $K_{unsat} > K_{sat}$) and (3) transverse micro-flow into tows, which occurs after macro-flow between tows and increases penetration/impregnation times (leading to $K_{sat} > K_{unsat}$) (Dungan and Sastry 2002).

It has been observed for natural fibre reinforcements that saturated permeability is higher than the unsaturated permeability. K_{sat}/K_{unsat} ranges between 1.2 and 1.8 for plain-woven jute fabrics (Dungan and Sastry 2002). Like Pillai and Advani (1998) and Dungan and Sastry (2002) argue that the jute yarns act as a 'sink', leading to a delayed impregnation of the tows through transverse micro-flow, which reduces the average macro-flow velocity, thereby reducing permeability.

An alternate explanation was provided for the high saturated permeability related to the absorption of fluid by the natural reinforcements (Dungan and Sastry 2002). Again, as the jute fibres and yarns act as sink components (as fibres absorb liquid, and yarns have micropores), the unsaturated permeability is reduced. Once infusion is complete (i.e. no more porosity within tows need to be filled and fibres cannot absorb more fluid), the fibres and yarns are no longer sink components, and the permeability upon saturation is increased. Note that the swelling of the fibres upon saturation does restrict flow paths and therefore the increase is only marginal. This argument is further strengthened by the fact that the difference between saturated and unsaturated permeability vanishes with increasing porosity (decreasing fibre volume fraction) (Fig. 4.11). In fact, at very high porosity ($\phi > 0.8$), $K_{sat}/K_{unsat} < 1$ (Fig. 4.11).

Notably, the difference between saturated and unsaturated permeability is higher in untreated jute fabrics ($K_{sat}/K_{unsat} = 1.70$–1.84) than in treated jute fabrics ($K_{sat}/K_{unsat} = 1.21$–1.40) (Dungan and Sastry 2002). This suggests that while transverse micro-flow into tows may play an important role in the difference in saturated and unsaturated permeability of natural fibre reinforcements, the polar nature of untreated natural fibres and their tendency to absorb polar fluids and swell have a more dominant role.

4.3.1.5 Capillary Effects and Micro-Flow Versus Macro-Flow

Preform impregnation usually involves a dual-scale flow. Resin flow between yarns (inter-yarn) is referred to as macro-flow, while resin flow through the yarns (intra-yarn) is referred to as micro-flow. As resin flows at low Reynolds numbers, inertial forces can be neglected. Macro-flow is dominated by viscous flow of the resin, while micro-flow is driven by capillary pressure developed within the tows. Capillary effects are important to study, not least because they play a key role in the mechanism of void formation in LCM. As Fig. 4.9 highlights, at low flow velocities (and high fibre volume fractions), capillary flow dominates leading to inter-yarn voids, while at high flow velocities (and low fibre volume fractions), viscous flow dominates leading to intra-yarn voids. This has been shown for plant fibre composites as well (Fig. 4.9) (Shah et al. 2012b; Shah 2013a).

Capillary effects in natural fibre reinforcements have received some attention (Francucci et al. 2012b; Sun et al. 2014), particularly due to the common (but incorrect) notion that the hollow structure of plant fibres may enhance capillary effects. There is little evidence that resin can impregnate the lumen, and in most cases the lumen has shown to remain unfilled after composite manufacture (Madsen 2004; Shah 2013a). Investigations have revealed that the dynamic capillary pressure in plain-woven jute fabrics was −25 and 36 kPa when measured in the glycerine solution and vinylester resin, respectively (Francucci et al. 2012b). That is, spontaneous infiltration occurs during infusion with glycerine solution, but capillary forces act against flow during infusion with the resin. This has significant implications on experimental mould-filling tests as test fluids like glycerine solution may provide deceivingly higher permeability and lower fill times.

The capillary pressure is found to increase exponentially with fibre volume fraction (Francucci et al. 2012b). Notably, the jute fabrics consistently exhibited capillary pressures two to three times higher than synthetic reinforcements (Francucci et al. 2012b), implying that capillary effects and micro-flow are more dominant in natural fibre reinforcements. Francucci et al. (2012b) also demonstrate that the measured capillary pressure may be used to determine a corrected unsaturated permeability, which is found to be similar for both the test fluid and resin.

4.3.2 Flow Modelling of Natural Fibre Composites

4.3.2.1 Introduction

Darcy's law (Eq. 4.11) and the continuity equation (Eq. 4.12) are commonly and conveniently used to describe the flow of thermosetting resins in porous reinforcements.

$$\overline{v} = -\frac{K}{\mu}\nabla P, \tag{4.11}$$

$$\nabla \cdot \overline{v} = 0, \tag{4.12}$$

where $\overline{v}$ is the volume-averaged fluid velocity and ∇P is the applied pressure gradient.

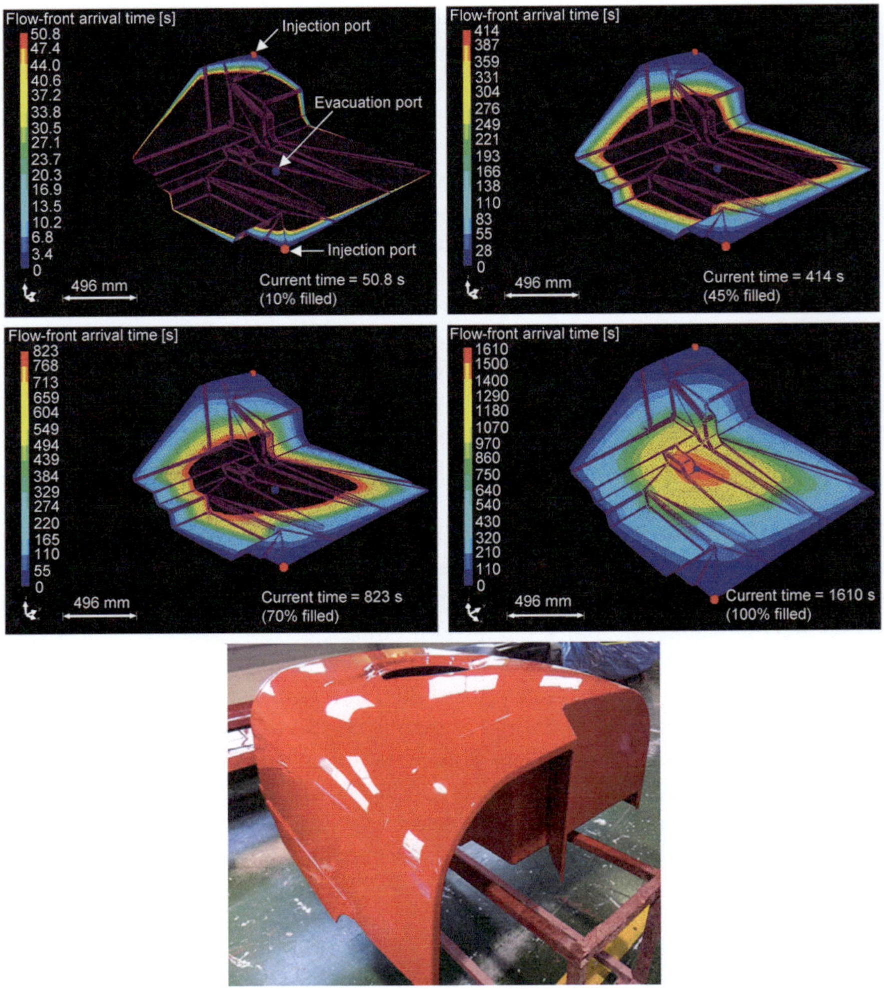

Fig. 4.14 Resin flow simulation conducted in RTM-Worx software for the vacuum-assisted, light-RTM manufacture of the lower part of a flax/vinylester composite agricultural chemical storage tank. Adapted from Kong et al. (2014)

All permeability studies on natural fibre reinforcements so far have been conducted on the presumption of a valid 1D Darcy's law. The mould filling of natural fibre reinforcements has been successfully simulated and experimentally validated using the conventional model by some researchers, including Kong et al. (2014). They performed a resin flow analysis on the upper and lower parts of a flax/vinylester agricultural chemical storage tank, to predict fill times (and select a suitable injection pressure) and ensure complete impregnation. A flow simulation of the lower part is illustrated in Fig. 4.14. The simulations were backed by experiments.

4.3.2.2 Models Suitable for Natural Fibre Reinforcements

During resin impregnation in an LCM process, natural fibre reinforcements, unlike synthetic reinforcements, absorb liquid and subsequently swell. As discussed previously, the percentage absorption and swelling is dependent on the (polarity of the) liquid; for instance, jute swells by 18–22 % in glycerine solution but by 7–8 % in vinylester or polyester resin (Francucci et al. 2010). Due to the absorption of liquid, natural fibre preforms effectively act as a 'sink', soaking up some volume of the resin that is infused. The subsequent fibre volume change due to swelling has a twofold, time-dependent, 'source-like' effect: (1) reducing porosity and thus permeability and (2) increasing flow resistance. The conventional model in Eqs. (4.11) and (4.12) is therefore inappropriate, if not ineffective, in describing the permeability of, and resin flow behaviour in, natural fibre reinforcements as it does not take these sink and source factors into account.

Several researchers have recently attempted to develop flow models specifically for natural fibre reinforcements, including Languri et al. (2010), Nguyen et al. (2014), Masoodi et al. (2009), Masoodi and Pillai (2010, 2011) and Francucci et al. (2014). A modified continuity equation (Eq. 4.13) was initially proposed, which incorporated a sink term $S(t)$ that was related to the rate of fluid absorption by the reinforcement and a porosity term $\phi'(t)$ that was related to the rate of decrease in preform porosity due to increase in fibre volume upon swelling, both of which were functions of time (Languri et al. 2010; Masoodi et al. 2009; Masoodi and Pillai 2010, 2011). However, it was later found that sink and porosity terms had to cancel out to satisfy experimental observations (Masoodi and Pillai 2010, 2011); that is, the rate of fluid absorption (per unit volume) had to be equal to the rate of change of fibre volume (per unit volume). Consequently, Eq. (4.13) simplified to the conventional continuity equation of Eq. (4.12).

$$\nabla \cdot \overline{v} = -S(t) - \phi'(t). \tag{4.13}$$

Following this, Masoodi and Pillai (2011) proposed that the permeability K, alongside porosity ϕ and fibre diameter d, were only functions of time (Eq. 4.14), where the subscript 0 indicates initial values (at $t=0$). Essentially, if the change in fibre diameter over time (due to swelling) was measured, fitted curves could be used to then predict the evolution of porosity with time and the evolution of permeability with time. Then, if the evolution of the pressure gradient as a function of time was known, Eq. (4.15), which is derived by resolving Darcy's law (Eq. 4.11) with the continuity equation (Eq. 4.13), could be used to predict the evolution of the flow front.

$$K(t) = \frac{\phi(t)^{n+1}}{C(1-\phi(t))^n}, \quad \text{where } \phi(t) = 1-(1-\phi_0)\frac{d(t)}{d_0},$$
$$\text{and } d(t) = a\exp\left(\frac{b}{c+t}\right), \tag{4.14}$$

$$x^2(t) = \frac{2}{\phi_0 \mu} \int_0^t \Delta P(t) K(t) dt. \tag{4.15}$$

The above model has been used with much success by Masoodi et al. (2009), Masoodi and Pillai (2011) and Languri et al. (2010) in predicting the flow in natural fibre reinforcements. This particularly demonstrates the importance of tracking the evolution of permeability as a function of time in natural fibre reinforcements.

4.3.3 Conclusions

Permeability studies and data are imperative to understanding and modelling the mould-filling stage in LCM of composites. Plant fibre reinforcements demonstrate higher permeability than glass fibre reinforcements, while wood fibre reinforcements exhibit lower permeability than the latter. Permeability of natural fibre reinforcements increases with porosity and can be modelled by conventional models such as the modified Carman–Kozeny equation. Yarn diameter (linear density), length and twist level are all found to affect preform permeability. Fluid absorption and swelling are key mechanisms in natural fibre preforms, due to which both saturated and unsaturated permeability are reduced.

For natural preforms, saturated permeability tends to be higher than unsaturated permeability, although the difference vanishes with increasing porosity (decreasing fibre volume fraction). Moreover, capillary effects and micro-flow are more dominant in natural fibre reinforcements than in conventional synthetic reinforcements. Importantly, as natural fibres behave differently with different liquids (viz., absorption, swelling and capillary pressure), to accurately measure their permeability, the natural fibre reinforcement should be studied with the particular resin that is to be used during LCM as the test fluid.

To accurately model the mould-filling stage in LCM of natural fibre reinforcements, researchers have proposed experimentally validated Darcy's law-based models, which specifically incorporate effects of liquid absorption and subsequent fibre swelling on preform porosity and permeability as a function of time.

References

(18/06/2012). Samsara 'eco surfboard' features Biotex flax fibre. Retrieved 11/08/2014, from http://www.reinforcedplastics.com/view/26369/samsara-eco-surfboard-features-biotex-flax-fibre/

Cai Z (1992) Analysis of mold filling in RTM process. J Compos Mater 26:1310–1338

Campbell F (2003) Manufacturing processes for advanced composites. Elsevier, Oxford

Carman P (1937) Fluid flow through granular beds. Trans Inst Chem Eng Lond 15:150–166

Carus M (2011) Bio-composites: technologies, applications and markets. In: 4th international conference on sustainable materials, polymers and composites, Birmingham

Carus M, Gahle C (2008) Natural fibre reinforced plastics – material with future. nova-Institut GmbH, Huerth

Carus M, Eder A, Dammer L, Korte H, Scholz L, Essel R, Breitmayer E (2014) Wood-plastic composites (WPC) and natural fibre composites (NFC): European and global Markets 2012 and future trends. WPC/NFC Market Study 2014-03. nova-Institut GmbH, Hürth

Chen B, Chou TW (1999) Compaction of woven-fabric preforms in liquid composite molding processes: single-layer deformation. Compos Sci Technol 59:1519–1526

Chen B, Chou TW (2000) Compaction of woven-fabric preforms: nesting and multi-layer deformation. Compos Sci Technol 60:2223–2231

Chen B, Lang EJ, Chou TW (2001) Experimental and theoretical studies of fabric compaction behavior in resin transfer molding. Mater Sci Eng A 317(1–2):188–196

Correia N, Robitaille F, Long AC, Rudd CD, Simacek P, Advani SG (2005) Analysis of the vacuum infusion moulding process: I. Analytical formulation. Compos A: Appl Sci Manuf 36:1645–1656

Duflou J, Deng Y, Acker KV, Dewulf W (2012) Do fiber-reinforced polymer composites provide environmentally benign alternatives? A life-cycle-assessment-based study. MRS Bull 37:374–382

Dungan F, Sastry AM (2002) Saturated and unsaturated polymer flows: microphenomena and modeling. J Compos Mater 36(13):1581–1603

Faruk O, Bledzki AK, Fink HP, Sain M (2012) Biocomposites reinforced with natural fibres: 2000–2010. Prog Polym Sci 37(11):1552–1596

Francucci G, Rodrıguez ES, Vazquez A (2010) Study of saturated and unsaturated permeability in natural fiber fabrics. Compos A: Appl Sci Manuf 41:16–21

Francucci G, Rodrıguez ES, Vazquez A (2012a) Experimental study of the compaction response of jute fabrics in liquid composite molding processes. J Compos Mater 46(2):155–167

Francucci G, Vazquez A, Ruiz E, Rodrıguez ES (2012b) Capillary effects in vacuum-assisted resin transfer molding with natural fibers. Polym Compos 33:1593–1602

Francucci G, Vazquez A, Rodrıguez ES (2013) Key differences on the compaction response of natural and glass fiber preforms in liquid composite molding. Text Res J 82(17):1774–1785

Francucci G, Rodrıguez ES, Moran J (2014) Novel approach for mold filling simulation of the processing of natural fiber reinforced composites by resin transfer molding. J Compos Mater 48(2):191–200

Gassan J, Chate A, Bledzki AK (2001) Calculation of elastic properties of natural fibers. J Mater Sci 36:3715–3720

Gauvin R, Kerachni A, Fisa B (1994) Variation of mat surface density and its effect on permeability evaluation for RTM modelling. J Reinf Plast Compos 13:371–383

Goutianos S, Peijs T (2003) The optimisation of flax fibre yarns for the development of high-performance natural fibre composites. Adv Compos Lett 12(6):237–241

Goutianos S, Peijs T, Nystrom B, Skrifvars M (2007) Textile reinforcements based on aligned flax fibres for structural composites. In: Composites innovation 2007 – improved sustainability and environmental performance, Barcelona

Gutowski T, Cai Z, Bauer S, Boucher D, Kingery J, Wineman S (1987a) Consolidation experiments for laminate composites. J Compos Mater 21:650–669

Gutowski T, Morigaki T, Cai Z (1987b) The consolidation of laminate composites. J Compos Mater 21:172–188

Ho M, Wang H, Lee J, Hoc C, Lau K, Leng J, Hui D (2012) Critical factors on manufacturing processes of natural fibre composites. Compos Part B 43(8):3549–3562

Kelly P, Umer R, Bickerton S (2004) Compaction of dry and wet fibrous materials during infusion processes. In: 36th international SAMPE technical conference, vol 36. San Diego, pp 785–797

Kim Y, McCarthy SP, Fanucci JP (1991) Compressibility and relaxation of fiber reinforcements during composite processing. Polym Compos 12(1):13–19

Ko F, Kawabata S, Inoue M, Niwa M, Fossey S, Song JW (2001) Engineering properties of spider silk. In: MRS proceedings. vol 702. doi:10.1557/PROC-702-U1.4.1

Kong C, Park H, Lee H, Lee J (2014) Design of natural fiber composites chemical container using resin flow simulation of VARTML process. Int J Mater Mech Manuf 2(3):256–260
Languri E, Moore RD, Masoodi R, Pillai KM, Sabo R (2010) An approach to model resin flow through swelling porous media made of natural fibers. In: 10th international conference on flow processes in composite materials (FPCM10), Monte Verità, Ascona
Lewin M (2007) Handbook of fiber chemistry. CRC Press LLC, Boca Raton
Li Y (2006) Processing of sisal fiber reinforced composites by resin transfer molding. Mater Manuf Process 21(2):181–190
Long A, Boisse P, Robitaille F (2005) Mechanical analysis of textiles. In: Long A (ed) Design and manufacture of textile composites. Woodhead Publishing Limited and CRC Press LLC, Cambridge
Lundquist L, Willi F, Leterrier Y, Manson JAE (2004) Compression behavior of pulp fiber networks. Polym Eng Sci 44(1):45–55
Madsen B (2004) Properties of plant fibre yarn polymer composites – an experimental study. PhD, Technical University of Denmark
Manson J, Wakeman MD, Bernet N (2000) 2.16 – Composite processing and manufacturing – an overview. In: Kelly A, Zweben CH (eds) Comprehensive composite materials. Pergamon & Elsevier Science, Oxford (vol 2, Polymer matrix composites)
Masoodi R, Pillai KM (2010) Darcy's law-based model for wicking in paper-like swelling porous media. Am Inst Chem Eng J 56(9):2257–2267
Masoodi R, Pillai KM (2011) 3 – Modeling the processing of natural fiber composites made using liquid composite molding. In: Pilla S (ed) Handbook of bioplastics and biocomposites engineering applications. Wiley, Hoboken
Masoodi R, Pillai KM, Verhagen MA (2009) Flow modeling in natural-fiber preforms used in liquid composite molding. In: Joint American-Canadian international conference on composites, Newark
Matsudaira M (2006) Fabric handle and its basic mechanical properties. J Text Eng 52(1):1–8
Matsudaira M, Qin H (1995) Features and mechanical parameters of a fabric's compressional property. J Text Inst 86(3):476–486
Neuman S (1977) Theoretical derivation of Darcy's law. Acta Mech 25:153–170
Nguyen V, Lagardère M, Cosson B, Park CH (2014) Experimental analysis of flow behavior in the flax fiber reinforcement with double scale porosity. In: 12th international conference on flow processing in composite materials (FPCM 12), Enschede
Parnas R, Howard JG, Luce TL, Advani SG (1995) Permeability characterization. Part 1: a proposed standard reference fabric for permeability. Polym Compos 16(6):429–445
Pickering K (ed) (2008) Properties and performance of natural-fibre composites. CRC Press LLC, Boca Raton
Pillai K, Advani G (1998) A model for unsaturated flow in woven fiber preforms during mold filling in resin transfer molding. J Compos Mater 32(19):1753–1783
Placet V, Trivaudey F, Cisse O, Gucheret-Retel V, Boubakar ML (2012) Diameter dependence of the apparent tensile modulus of hemp fibres: a morphological, structural or ultrastructural effect? Compos A: Appl Sci Manuf 43(2):275–287
Reux F (2012) Worldwide composites market: main trends of the composites industry. In: 5th innovative composites summit – JEC ASIA 2012, Singapore
Richardson M, Zhang ZY (2000) Experimental investigation and flow visualisation of the resin transfer mould filling process for non-woven hemp reinforced phenolic composites. Compos A: Appl Sci Manuf 31:1303–1310
Robitaille F, Gauvin R (1998a) Compaction of textile reinforcements for composites manufacturing. I: review of experimental results. Polym Compos 19(2):198–216
Robitaille F, Gauvin R (1998b) Compaction of textile reinforcements for composites manufacturing. III: reorganization of the fiber network. Polym Compos 20(1):48–61
Rodriguez E, Giacomelli F, Vazquez A (2004) Permeability-porosity relationship in RTM for different fiberglass and natural reinforcements. J Compos Mater 38:259–268

Rodríguez E, Stefani PM, Vazquez A (2007) Effects of fibers' alkali treatment on the resin transfer molding processing and mechanical properties of jute-vinylester composites. J Compos Mater 41:1729–1741
Schiefer H (1933) The compressometer, an instrument for evaluating the thickness, compressibility, and compressional resilience of textiles and similar materials. Bur Stand J Res 10(6):705–713
Shah D (2013a) Characterisation and optimisation of the mechanical performance of plant fibre composites for structural applications. PhD, University of Nottingham
Shah D (2013b) Developing plant fibre composites for structural applications by optimising composite parameters: a critical review. J Mater Sci 48(18):6083–6107
Shah D, Schubel PJ, Clifford MJ, Licence P (2011) Mechanical characterization of vacuum infused thermoset matrix composites reinforced with aligned hydroxyethylcellulose sized plant bast fibre yarns. In: 4th international conference on sustainable materials, polymers and composites, Birmingham
Shah D, Schubel PJ, Clifford MJ, Licence P (2012a) The tensile behavior of off-axis loaded plant fiber composites: an insight on the non-linear stress–strain response. Polym Compos 33(9):1494–1504
Shah D, Schubel PJ, Licence P, Clifford MJ (2012b) Determining the minimum, critical and maximum fibre content for twisted yarn reinforced plant fibre composites. Compos Sci Technol 72:1909–1917
Shah D, Schubel PJ, Clifford MJ (2013a) Can flax replace E-glass in structural composites? A small wind turbine blade case study. Compos Part B 52:172–181
Shah D, Schubel PJ, Clifford MJ (2013b) Modelling the effect of yarn twist on the tensile strength of unidirectional plant fibre yarn composites. J Compos Mater 47(4):425–436
Shah D, Schubel PJ, Clifford MJ, Licence P (2013c) Mechanical property characterization of aligned plant yarn reinforced thermoset matrix composites manufactured via vacuum infusion. Polym Plast Technol Eng 53(3):239–253. doi:10.1080/03602559.2013.843710
Shah D, Porter D, Vollrath F (2014a) Opportunities for silk textiles in reinforced biocomposites: studying through-thickness compaction behaviour. Compos A: Appl Sci Manuf 62:1–10
Shah D, Schubel PJ, Clifford MJ, Licence P (2014b) Mechanical property characterization of aligned plant yarn reinforced thermoset matrix composites manufactured via vacuum infusion. Polym Plast Technol Eng 53:239–253
Summerscales J, Dissanayake N, Virk AS, Hall W (2010) A review of bast fibres and their composites. Part 2 – composites. Compos A: Appl Sci Manuf 41(10):1336–1344
Sun Z, Zhao X, Ma J (2014) Capillary effect in the impregnation of jute fiber mat reinforced polypropylene composites. J Compos Mater 48:447–453
Umer R, Bickerton S, Fernyhough A (2007) Characterising wood fibre mats as reinforcements for liquid composite moulding processes. Compos Part A 38:43–448
Umer R, Bickerton S, Fernyhough A (2008) Modelling the application of wood fibre reinforcements within liquid composite moulding processes. Compos Part A 39:624–639
Umer R, Bickerton S, Fernyhough A (2011) The effect of yarn length and diameter on permeability and compaction response of flax fibre mats. Compos A: Appl Sci Manuf 42:723–732
van Wyk C (1946) 20 – Note on the compressibility of wool. J Text Inst Trans 37(12): T285–T292
Winson C (1932) 27 – Report on a method for measuring the resilience of wool. J Text Inst Trans 23(12):T386–T393
Witten E, Jahn B (2013) Composites market report 2013: market developments, trends, challenges and opportunities. Industrievereinigung Verstärkte Kunststoffe (Federation of Reinforced Plastics) and Carbon Composites eV, Frankfurt
Xue D, Miao M, Hu H (2011) Permeability anisotropy of flax nonwoven mats in vacuum-assisted resin transfer molding. J Text Inst 102(7):612–620
Yang J, Xiao J, Zeng J, Jiang D, Peng C (2012) Compaction behavior and part thickness variation in vacuum infusion molding process. Appl Compos Mater 19:443–458
Zhang K, Si FW, Duan HL, Wang J (2010) Microstructures and mechanical properties of silks of silkworm and honeybee. Acta Biomater 6:2165–2171

Chapter 5
Manufacturing and Processing of Kenaf Fibre-Reinforced Epoxy Composites via Different Methods

N. Saba, M.T. Paridah, M. Jawaid, K. Abdan, and N.A. Ibrahim

Abstract The twenty-first century is the century of inconsistency regarding transformation, industrial development, environmental crisis, and technological growth. All these are associated with the attempts to minimise the dependence on petroleum-based fuels and products which leads to an increasing need to investigate more sustainable materials and environmentally friendly products to replace existing materials. Lignocellulosic fibrous polymeric composite materials are an attractive replacement for heavier metals due to their comparative superior fatigue and corrosion properties in many applications with respect to other synthetic and ceramics materials. This chapter considers the manufacturing methods by a variety of possible and diverse methods that permit the conversion of kenaf bast (stem) fibres to reinforced epoxy composites into a wide variety of applications and industrial products with properties that are comparable to existing synthetic composites. Among the various method compression moulding and resin transfer moulding processes are the most commonly used for engineered composites involving both thermoplastics and thermosets; but resin infusion is prior for thermosets. The properties of natural fibre-reinforced polymer composites are generally governed by the

N. Saba (✉) • M.T. Paridah
Laboratory of Biocomposite Technology, Institute of Tropical Forestry and Forest Products (INTROP), Universiti Putra Malaysia, 43400 UPM, Serdang, Selangor, Malaysia
e-mail: naheedchem@gmail.com

M. Jawaid
Laboratory of Biocomposite Technology, Institute of Tropical Forestry and Forest Products (INTROP), Universiti Putra Malaysia, 43400 UPM, Serdang, Selangor, Malaysia

Chemical Engineering Department, College of Engineering, King Saud University, Riyadh, Saudi Arabia

K. Abdan
Faculty of Engineering, Department of Biological and Agricultural Engineering, Universiti Putra Malaysia, 43400 UPM, Serdang, Selangor, Malaysia

N.A. Ibrahim
Faculty of Science, Department of Chemistry, Universiti Putra Malaysia, 43400 UPM, Serdang, Selangor, Malaysia

M.S. Salit et al. (eds.), *Manufacturing of Natural Fibre Reinforced Polymer Composites*, DOI 10.1007/978-3-319-07944-8_5

pretreated process of fibre and the manufacturing process of the composites. The performance of a composite material varies greatly with the choice of suitable processing techniques along with the fibre–matrix bond strength.

Keywords Natural fibre • Kenaf fibre • Polymer • Manufacturing methods • Composites

5.1 Introduction

Environmental legislation as well as consumer pressure impelled the probing of novel supernumerary for traditional non-renewable material quite exclusively (Deka et al. 2013). Natural fibre, a distinguished type of new renewable reinforcement sources, is used to supplement polymer-based materials (Ho et al. 2012). All biofibres are cellulosic in nature and exclusively consisted of cellulose, hemicellulose, lignin, and pectin. The amount of cellulose can vary depending on the species and age of the plant (Mahjoub et al. 2014). However, many innumerable demerits such as hydrophilic nature, poor fibre/matrix interfacial adhesion, and poor thermal stability of natural fibres can be overwhelmed by chemical treatment or compatibiliser which amended the adhesion between matrix and fibre face (Saba et al. 2015). Kenaf fibre (*Hibiscus cannabinus* L.) is a type of natural fibre offering many advantages and possesses high potential as reinforcement in composite materials, especially polymer composites. Thermosets resin is one of the well-known polymer matrix which shows maximum reinforcement of both natural and synthetic fibres from historical times which includes epoxies, polyester, and phenolics. Among these epoxies are highly recommended and most noteworthy matrices rendering its involvement in fibre-reinforced polymer for industrial exploitations. Polymer matrix composites are being increasingly used in engineering application due to their lightweight, strength, and stiffness compared to unreinforced polymers (Yahaya et al. 2014). The alarming environmental and renewable issues universally force the progress and advancement of high-performance engineering and novel low-cost products from natural resources (Akil et al. 2011; Deka et al. 2013). Thus, providing an approach to getting rid of the widespread dependence on fossil fuel. Hybrid composite in composite field generally used to describe a matrix containing at least two types of reinforcements or two types of polymer blends with one type of fibre reinforcement to develop some unique features in the end product. The three most important configurations usually obtained are imagined in Fig. 5.1. The simplest and cheapest method for producing a hybrid composite is shown in Fig. 5.1a, while Fig. 5.1b, c are quite complicated, entitled intralayer and intrayarn hybrid composite configuration (Swolfs et al. 2014). Many research studies are also made on kenaf fibre-reinforced hybrid composite involving both synthetic fibre such as glass or Kevlar to maintain the advantages of both fibres and alleviate some disadvantages (Yahaya et al. 2015; Atiqah et al. 2014).

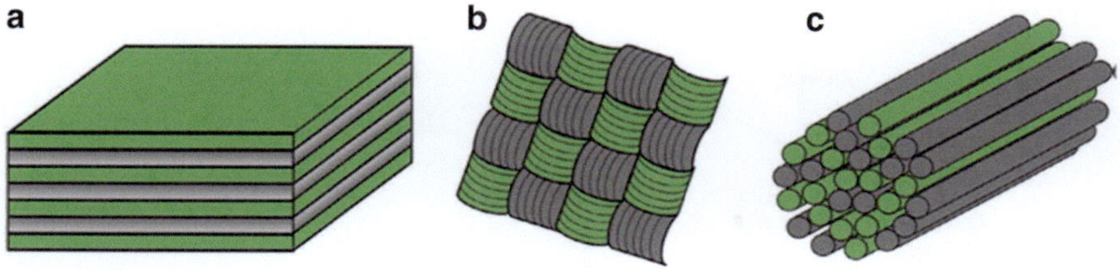

Fig. 5.1 Showing the three main hybrid configurations: (**a**) interlayer, (**b**) intralayer and (**c**) intrayarn (Swolfs et al. 2014)

5.2 Natural Fibre and Kenaf Fibre

Natural fibres are inexpensive, ample, readily available and could be obtained in billions of tons around the world (Karimi et al. 2014). Natural fibres have found to exhibit a number of advantages in terms of specific modulus, cost per weight, and cost per unit length along with the renewable and biodegradable nature (Dittenber and Rao 2012; Nishino et al. 2003; Ochi 2008). Other advantages which include high toughness, low density, reduction in tool wear, comparable specific strength properties, decreased energy of fabrication, ease of separation, CO_2 neutrality or sequestration (Paridah et al. 2011; Mohanty et al. 2000) and renewability are quite attractive (Ishak et al. 2009).

According to researchers, the bast fibres exhibit attractive mechanical properties providing replacements to glass fibres as reinforcing components in polymer composites (Faruk et al. 2012; Liu et al. 2007; Zampaloni et al. 2007). Natural fibres are non-toxic (Zini and Scandola 2011) consequently providing opportunity for developing composites with certainly lesser human health hazards and environmental damage throughout their life cycle during production/processing/use/disposal (Dicker et al. 2014). The life cycle for the natural fibre-reinforced bio-composite is illustrated in Fig. 5.2. Besides this they are safer in handling and processing operational conditions compared to synthetic reinforcements (glass fibres) and are nonabrasive towards moulding and mixing equipment. They can easily subsidise the reductions in substantial equipment maintenance cost, reduction in risk of dermal or respiratory problems and present positive environmental impact (Akil et al. 2011).

They are green composites (environmentally friendly) which reflect appropriate ancillary for traditional fibre (i.e. glass, carbon and aramid fibre) reinforced petroleum-based composites (Low et al. 2007; Marsh 2003). Natural fibres can be highly preferable to synthetic fibres in automotive, construction and furniture production industries (Paukszta and Borysiak 2013) and in another applications where stiffness and weight are primary concerns (Dittenber and Rao 2012). Until now natural fibres are successfully used in different industries to produce environmentally friendly products (Akil et al. 2011; Ochi 2008; Paridah et al. 2011; Xue et al. 2009).

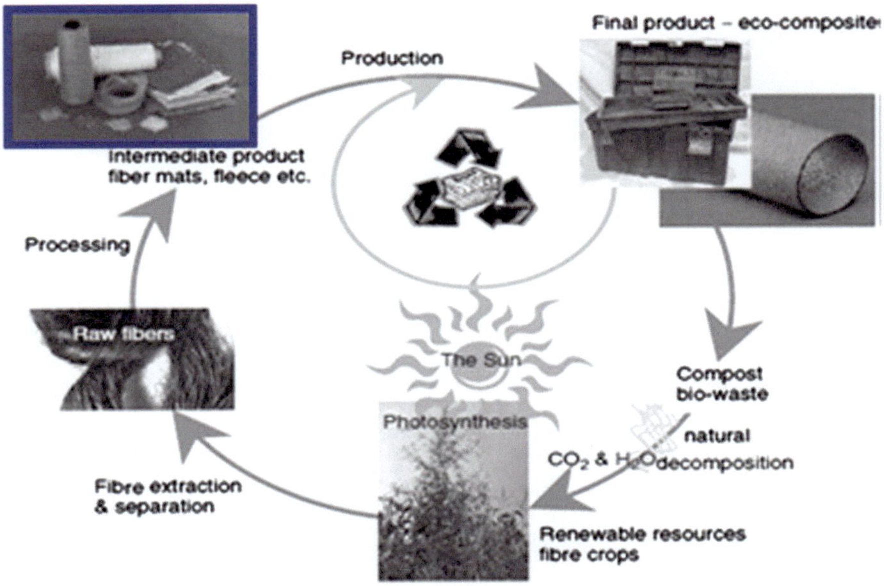

Fig. 5.2 Life cycle of bio-composites (Akil et al. 2011)

Moreover, high moisture absorption tendency, low thermal stability, poor wettability during processing and weaker adhesion with the synthetic counterparts confine their use in certain advanced industrial application (Demir et al. 2006).

5.2.1 Kenaf Fibre

Currently natural fibres are highly appraised, as being cheap, renewable, abundant and simply recyclable (Zampaloni et al. 2007) substitutes of other nonrecyclable fibres replaced by such as oil palm, flax, hemp, kenaf, jute, sisal and roselle fibres. Among them kenaf (*H. cannabinus* L. and family Malvaceae), a viable cellulosic source, economically viable and ecologically friendly fast-growing plant requiring only few treatments during growth and positively withstands harsh and tough climates (Yu and Yu 2010). The kenaf word originated from Persian language in the late nineteenth century. Kenaf plantation and fibre are shown in Fig. 5.3. It can grow up to 5 m in less than 6 months in many parts of the world, including several regions of the USA (Parikh et al. 2002) with maximum dry yields of 4 ton/ha under suitable conditions. Kenaf plant has a particular and fascinating character with long and unbranched stem possessing external layer (bark) and a core providing long stiff fibre (Mahjoub et al. 2014). Kenaf is a multipurpose plant with a diversity of harvestable plant components (seed, leaves, stalk bark and stalk core Yang et al. 2014).

Fig. 5.3 Images showing the kenaf plantation and fibre (Saba et al. 2015)

It yields comparably higher than other plant fibres such as hemp, flax and jute, therefore providing cheaper raw material (Paridah et al. 2011). However, the actual yield and composition of its components are heavily dependent on many factors such as soil type, plant populations, planting date and month, photosensitivity, variety, planting density, length of growing season and plant maturity (Batouli et al. 2014). Furthermore, kenaf has been actively cultivated in recent years owing to some important causes (Rouison et al. 2004):

- Kenaf average absorption rate for nitrogen (0.81 g/m^2/day) and phosphorus (0.11 g/m^2/day) from the soil is several times higher.
- It can accumulate carbon dioxide at a considerably higher rate.
- It has a higher photosynthesis rate than orthodox trees (Nishino et al. 2003; Nishino 2004).

Kenaf fibre compared to glass fibre exhibits high specific mechanical properties, low cost and low density and is easily recycled like most other natural fibres (Mohanty et al. 2000). It is broadly recognised for its stiff and strong agro-based fibres (bast and core fibre) devouring sufficient potential to reinforce polymers. Kenaf fibre consists of mainly cellulose (45–57 %), hemicelluloses (21.5–35 wt.%), lignin (8–14 wt.%), pectin (3–5 wt.%), waxy substances (Akil et al. 2011) and small amounts of extractives and ash (Mazuki et al. 2011; Davoodi et al. 2010). Customarily kenaf plant fibre is denoted as industrial kenaf due to the keen interest in the production of diverse industrial raw and polymer composite materials (Saba et al. 2015). Kenaf fibres are gratifying increasingly widespread throughout the world including Bangladesh, the USA, South Africa, India, Thailand, Indonesia. Beside this, kenaf also extending in south-east Europe and Malaysia as the significant natural material source contributing towards the development of eco-friendly assets for the automotive, sports industries, food packaging and furniture (Anuar and Zuraida 2011), textiles, paper pulp and fibre-board-based industries (Nishino et al. 2003).

5.3 Thermoset Polymer

Recently, the involvement of natural fibres such as hemp, flax, sisal, banana, kenaf and jute as reinforcements in the polymeric matrices has extended a pronounced attention in engineering applications (Alamri et al. 2012). It has unique combinations of properties such as high modulus, excellent chemical, low shrinkage, heat resistance and relatively high strength (Alamri et al. 2012). Thermoset polymer matrices offer the following advantages in the fabrications of natural fibre-reinforced composites (Rosa et al. 2013; Liu et al. 2012; Shahzad 2012; Eichhorn et al. 2010) as shown in Fig. 5.4. Instead of the advantages, these thermoset polymer matrix composites also suffer from few shortcomings such as high curing time and non-recycling process (Akil et al. 2011; Bogoeva-Gaceva et al. 2007; George et al. 2001).

5.4 Natural Fibre Composite

Composite materials are promptly developed into a material of choice as an alternative to other traditional materials such as metals with profound applications in many industrial sectors (Kafi et al. 2011). Over the past few decades, several attempts and needs are breaking out to investigate more environmentally friendly, sustainable materials to replace the existing glass fibre and carbon fibre-reinforced materials in composite solicitations, for minimising the reliance on petroleum fuels and its products (Akil et al. 2011). Besides this, both aerospace and automotive industries for reducing vehicle weight established an awareness and attention towards fabrication and applications of more natural fibre-reinforced composites by shifting from steel

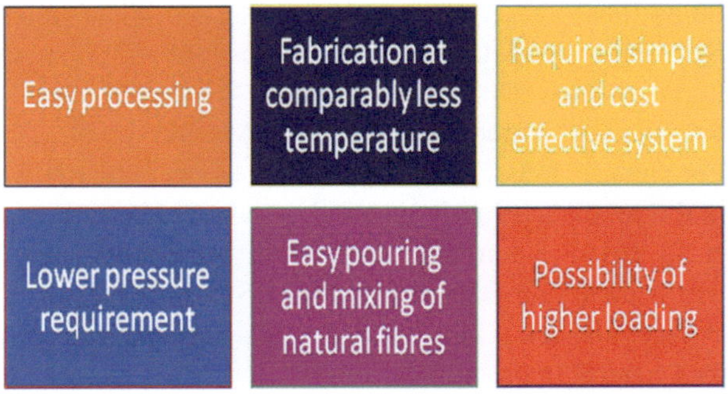

Fig. 5.4 Advantages of thermoset polymer for natural fibre-reinforced composites

to aluminium and from aluminium to composites (Zampaloni et al. 2007). Natural fibre composites are abundant, renewable and cost-effective; possess tool-wearing rates, better formability, thermal insulation, and acoustic properties hence are comparably safer towards health and environment (Davoodi et al. 2010). Many innumerable demerits are associated with plant–fibre-reinforced polymeric composites such as hydrophilic nature, poor wettability, poor fibre/matrix interfacial adhesion, poor thermal stability of natural fibres and incompatibility with some polymeric matrices. This can be overcome by chemical treatment or by adding compatibiliser which amended the adhesion between the matrix and fibre (Saba et al. 2015; Rosa et al. 2010; Dhakal et al. 2007).

5.5 Manufacturing Techniques

Customarily traditional manufacturing techniques are applied for the natural fibre polymer composites, which are designed for conventional fibre-reinforced polymer composites with thermoplastics and thermosets. These techniques include direct extrusion, vacuum infusion, compression moulding, resin transfer moulding (RTM), compounding and injection moulding (Ho et al. 2012). Usually methods such as extrusion, compression and injection moulding are used to introduce fibres into the thermoplastic matrix (Salleh et al. 2014). Manufacturing of natural fibre polymer composites is classified in two types as open and closed moulding techniques that can be further classified as shown in Table 5.1. Closed moulding processes are ideal for efficient fabrication of consistent, quality parts at medium and high volume with relatively little waste and harmful emissions. Open mould methods allow for a rapid product development cycle as the tooling fabrication process

Table 5.1 Classification of composite fabrication techniques on the basis of moulding type

Composites manufacturing processes	
Open moulding	Closed moulding
Hand lay-up	Compression moulding
Filament winding	Pultrusion
Spray up	Resin transfer moulding (RTM)
	Centrifugal casting
	Vacuum infusion processing
	Vacuum bag moulding Vacuum bagging Injection moulding
	Continuous lamination Vacuum-assisted RTM (Va-RTM) Co-injection RTM (CIRTM)
	Reinforced reaction injection moulding (RRIM) Structural reaction injection moulding (S-RIM)

is simple and relatively low in cost. Each of the fabrication technique of reinforced natural fibre polymer matrix composite has its unique importance and running features. Figure 5.5a–n illustrates most common manufacturing techniques used for natural fibre polymer composite both in small and large scale. However owing to some drawbacks of natural fibres with polymeric matrix and their manufacturing techniques (Ho et al. 2012), highly advanced and sophisticated technical processes are required with manageable feature. Consequently, the consequential features of final composites in relation to the choice of suitable materials, preprocessing methods and (fabricating) manufacturing process are intricately entangled (Ho et al. 2012).

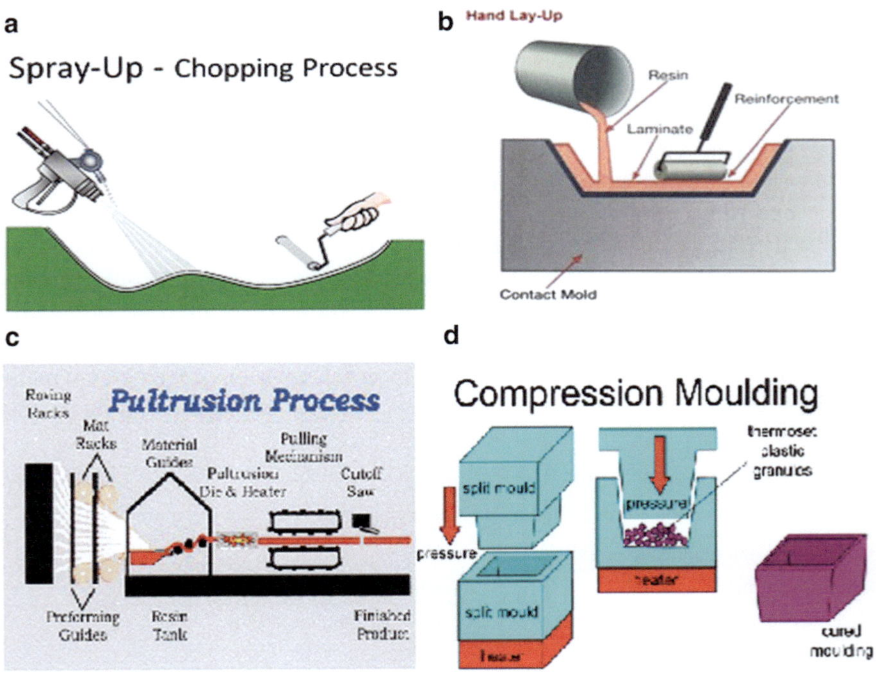

Fig. 5.5 (**a**) Spray-up process (www.cyccomposites.net), (**b**) hand lay-up process (www.flexidynamic.com), (**c**) pultrusion process (www.sparecomposites.com), (**d**) compression moulding process (www.the-warren.org), (**e**) extrusion moulding process (www.the-warren.org), (**f**) vacuum infusion process (www.gopixpic.com), (**g**) vacuum bagging process (www.bertram31.com), (**h**) light resin transfer moulding (LRTM) (www.moldedfiberglass.com), (**i**) filament winding (www.frpmachining.com), (**j**) extrusion process (www.wintechmold.com), (**k**) centrifugal casting (www.custompartnet.com), (**l**) injection moulding (www.imgarcade.com), (**m**) resin transfer moulding (www.coremt.com) and (**n**) reaction injection moulding (www.coremt.com)

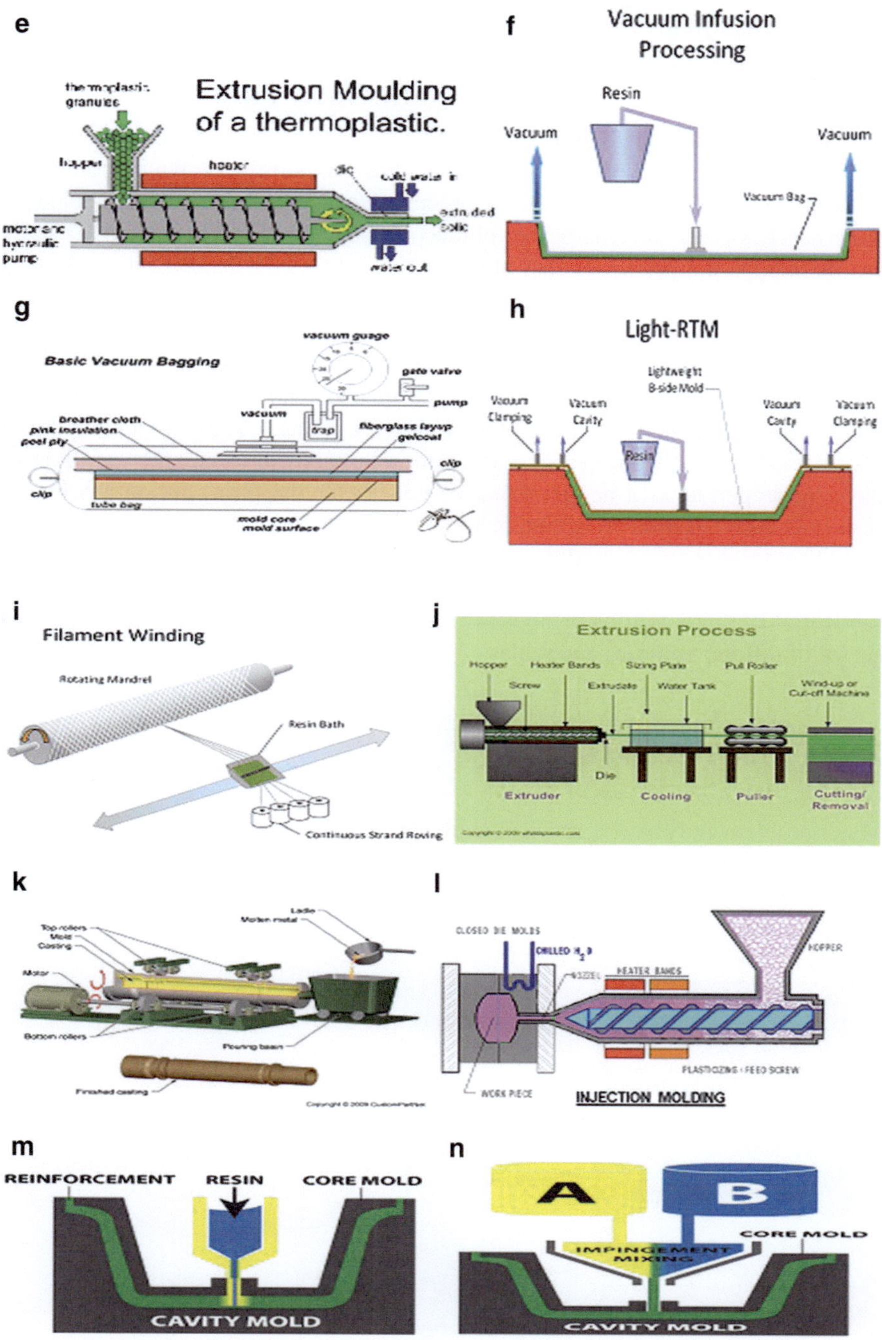

Fig. 5.5 (continued)

5.5.1 Classification and Applications of Manufacturing Techniques

All these techniques can be further divided according to the volume of production produced and the types of structural applications in industry shown by composite products. Selection of manufacturing process depends on many factors including materials, cost, size and chiefly the volume of production. Figures 5.6 and 5.7 show the manufacturing processes that are suitable and greatly depend on production amounts and the possible types of applications in the different sectors for structural components, respectively. The compression and injection moulding processes are general methods used in manufacturing natural fibre-reinforced composites (Liu et al. 2007). Injection moulding upsurges tensile and flexural properties by improving the fibre dispersion (Mohanty et al. 2004). Disparately, compression moulding conserves the isotropic properties of the composites and diminishes the variations in physical properties as it neither damages nor orients the fibres during processing (Liu et al. 2007). Extrusion moulding affects the properties of the composite materials with natural as well as traditional fibres by resulting changes in length and diameter distribution of the fibres consequently (Carneiro and Maia 2000). Noteworthy fabrication of thermosetting composites is a very complex procedure, involving processes of different physics and scales.

5.5.2 Problems and Challenges

Major problem and challenges encountered during manufacturing of composites by using natural fibre such as kenaf, through most frequently used techniques like compression and injection moulding are shown in Fig. 5.8 crafting complication in the

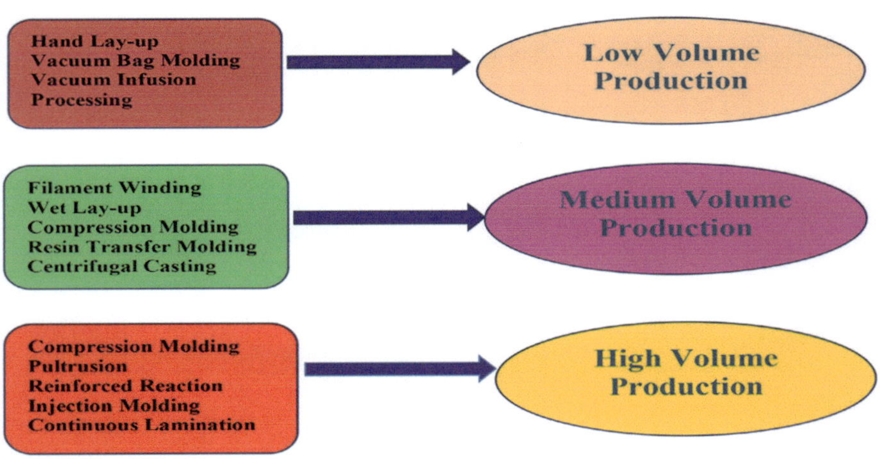

Fig. 5.6 The volume production dependency on different fabrication techniques of composite

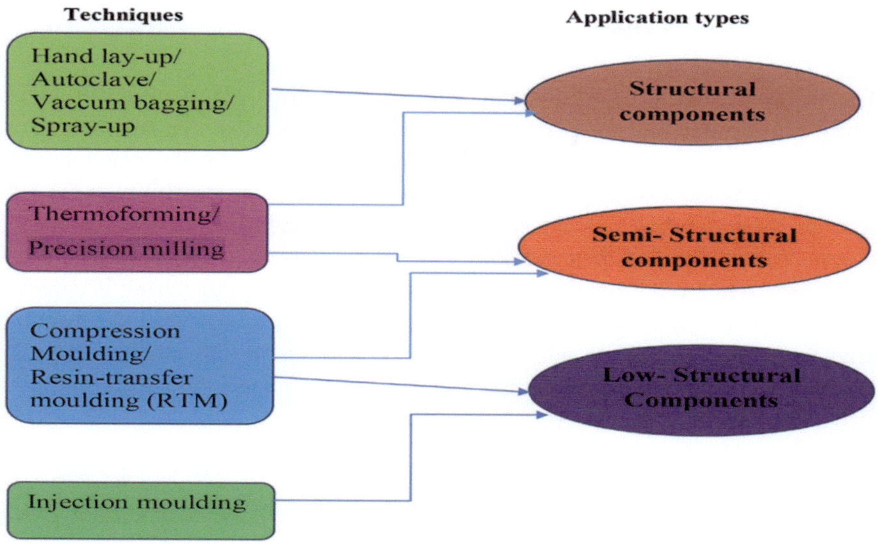

Fig. 5.7 The applications of different manufacturing techniques in fabrication of structural components

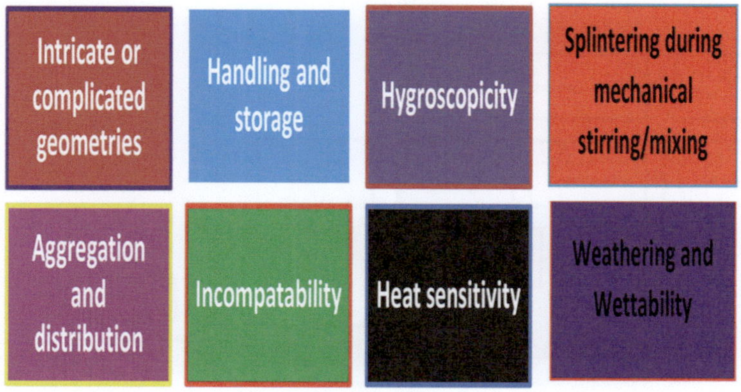

Fig. 5.8 Representing the challenges and technical problems faced during manufacturing of natural fibre composite

overall process. These technical problems restrict the usage of natural fibres and biodegradable polymers for innovative composite development (Ho et al. 2012). As the mechanical, thermal and structural properties of the natural fibres and biodegradable polymers are quite different compared to synthetic fibres and petroleum-based plastics, respectively, verifying their unsuitability for natural fibre-reinforced polymer composites. To overcome these complications, several emerging technologies—such as micro-braiding, long fibre pellet and textile insert moulding—are being considered for fabrication of composite (Leong et al. 2014).

5.6 Manufacturing Uncertainty

The manufacturing process of composite materials involves many uncertainties which can result in a considerable amount of scrap associated with significant cost and environmental implications. Each of the steps of composite manufacturing (forming, consolidation/impregnation and curing) introduces variability to the subsequent manufacturing processes, creating strong interdependencies between the process parameters and their variability and properties of the product. The schematic representation of interdependencies in composite manufacturing is depicted in Fig. 5.9 (Mesogitis et al. 2014). Architecture variability arises during the forming/draping stage, edge effects and void formation and due to different handling and storage conditions or shear deformations.

5.6.1 Dry Textiles and Pre-preg Variability

This type of variability is usually present in all pre-preg and dry textiles and is related with tow waviness shown in Fig. 5.10, tow size, shape variations, distribution of fibres inside the tows and resin content variations. It is generated

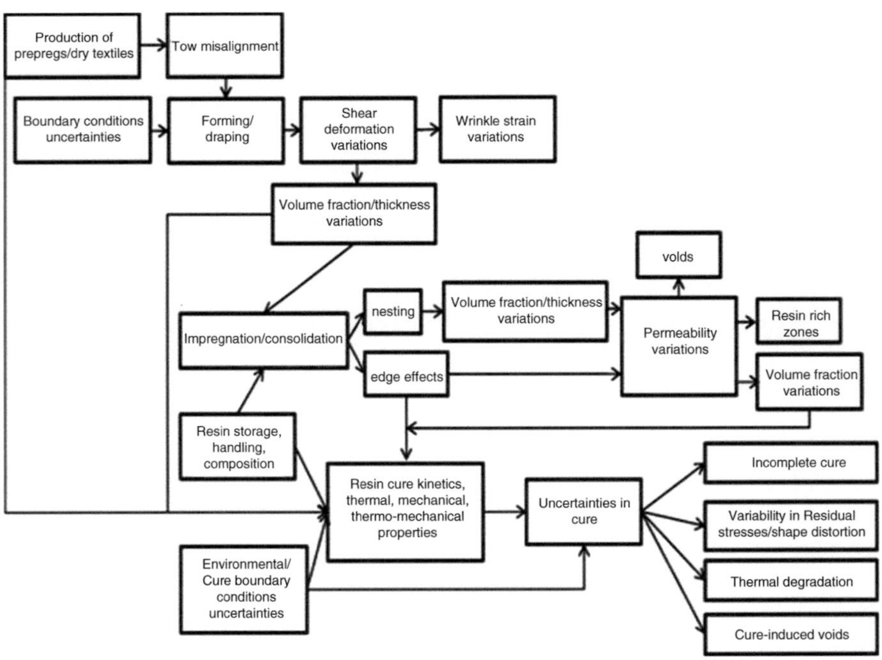

Fig. 5.9 Schematic representation of interdependencies in composites manufacturing (Mesogitis et al. 2014)

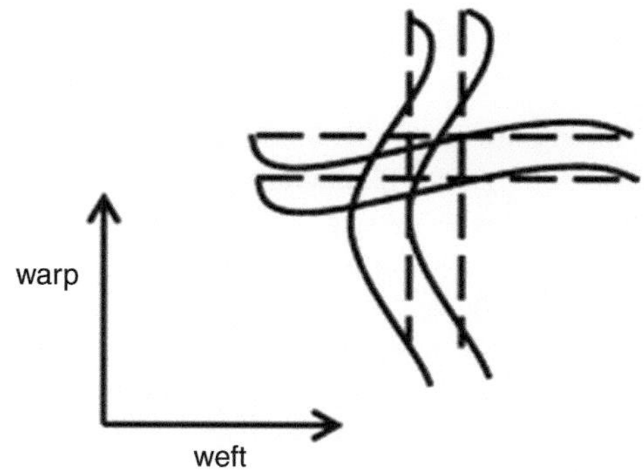

Fig. 5.10 Tow waviness (Mesogitis et al. 2014)

during production, handling or storage stages (Skordos and Sutcliffe 2008; Verleye et al. 2010; Potter et al. 2008).

5.6.2 Forming/Draping Stages

Substantial shear deformation in the case of doubly curved components (Cao et al. 2008) results wrinkles and is eradicated using localised stitching (Duhovic et al. 2011) during the forming/draping stage of the manufacturing process.

5.6.3 Impregnation/Consolidation

The major sources of uncertainty in the impregnation stage involve resin viscosity variations, foreign material inclusion and preform volume fraction variations, as well as accidental misplacement of the preform in the mould (Padmanabhan and Pitchumani 1999; Opperer et al. 2004). These cause macroscopic and microscopic voids and random experimental errors. However, permeability of dry reinforcement is the key parameter controlling the impregnation step in liquid composite moulding processes.

5.6.4 Composites Curing Stage

The cure process is a complex thermo-mechanical phenomenon involving several sources of uncertainty such as material and tooling characteristics variation as well as environmental/boundary condition uncertainties (Mesogitis et al. 2014). These uncertainties result in under-curing, over-curing, cure-induced voids and severe temperature overshoots which may cause thermal degradation and can affect the formation of residual stresses (Potter 2009; Guo et al. 2005).

5.7 Kenaf Fibre Composite

Among the many different types of natural resources such as jute, hemp, coir, ramie sisal, etc., kenaf plants have been extensively exploited over the past few years for the composite fabrication (Akil et al. 2011). Kenaf bast fibre is unique, multipurpose and potentially reliable among the other natural and synthetic fibres; besides this it exhibits low-density, high specific mechanical properties and interesting physical and non-abrasiveness properties, during processing leading to attraction towards composite industry (Fiore et al. 2015). Comparative mechanical properties of some important natural fibres and synthetic fibres are tabulated in Table 5.2. Table 5.2 clearly depicts that properties of kenaf fibre are comparable to conventional fibre, and thus it can be successfully used to reinforce the polymer for fabricating composite through the conventional fibre composite manufacturing process.

Currently, kenaf fibre-reinforced composite displays bright future and great attention among other natural fibre-reinforced composites owing to its specific ecological excellent properties (Akil et al. 2011). Researchers utilise its availability, cost-effectiveness, lightweight and readiness through various manufacturing processes and certify that kenaf fibre-reinforced composite would emerge as a typical substitute composite material, exclusively in construction (building) and automotive industries (Akil et al. 2011). The lightweight of kenaf fibre composites causes improvement in the fuel consumption and emission particularly in automotive industries (Mahjoub et al. 2014). Nowadays kenaf fibre successively been used as

Table 5.2 Mechanical properties of some natural and synthetic fibres

Fibre	Density (g/cm^3)	Tensile strength (MPa)	Elastic or Young modulus (GPa)	Elongation at break (%)
Jute	1.3	393–773	26.5	1.5–1.8
Coir	1.15–1.46	131–220	4–6	15–40
PALF	–	413–1,627	34.5–82.5	1.6
Abaca	–	430–760	–	–
OPEFB	0.7–1.55	248	3.2	25
Ramie	1.55	400–938	61.4–128	1.2–3.8
Nettle	–	650	38	1.7
Sisal	1.5	511–635	9.4–22	2.0–2.5
Flax	1.5	500–1,500	27.6	2.7–3.2
Hemp	1.47	690	70	2.0–4.0
Pineapple	1.56	170–1,627	60–82	2.4
Cotton	1.5–1.6	400	5.5–12.6	7.0–8.0
Kenaf	1.45	930	53	1.6
E-glass	2.55	3,400	71	3.4
Carbon	1.4	4,000	230–240	1.4–1.8

Source: Mahjoub et al. (2014) and Rouison et al. (2004)

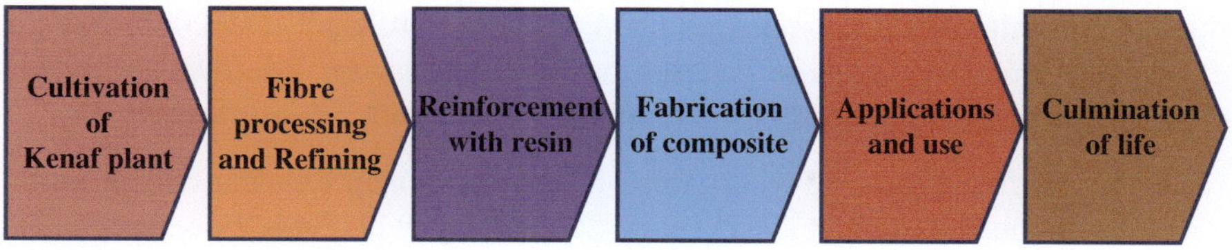

Fig. 5.11 Typical life cycle of kenaf-based polymer reinforced composite

reinforcement for plastic and synthetic product, cosmetic product, organic filler and medicine (Salleh et al. 2014). The most general or typical pathway for the production of kenaf fibre-reinforced polymer composite that can be alienated into six stages initiating from the cultivation of plant and its use through its fibre involvement in the polymer matrix to its termination as shown in Fig. 5.11.

5.7.1 Processing of Kenaf Fibre-Reinforced Composites

Mostly kenaf fibre-reinforced composites are processed by well-known following techniques such as compression moulding, slit jig, sifting process, pultrusion and solution blending. Furthermore, the most important and extensively accepted manufacturing process for fabricating polymeric composite is the compression moulding owing to its simplicity (Liu et al. 2007; Zampaloni et al. 2007; Serizawa et al. 2006; Avella et al. 2008). However, in the development of kenaf fibre-reinforced composites, this technique appeared to be the most effective and appropriate for producing high-quality composite products. The uneven fibre distribution by compression moulding in the kenaf fibre composite system offered major obstacles that need to be addressed during fabrication, as it creates difficulties during manual separation and is found to be discrete evenly while manufacturing (Zampaloni et al. 2007).

Pultruded profiles of kenaf and other natural fibre composite are found to be proven over an extensive range of high-performance and structural constituent requirements in the industrial and engineered products (Velde and Kiekens 2001). Another important novel processing method is solution blending which is considered as an alternative and promising process. However, it but required extra precautions during manufacturing process as it gets extremely entrapped with emulsifiers, contaminants, stabilisers, initiators residues, chain transfer, etc. in open air conditions, which might be altering the final products properties.

However, these techniques are most often being used for fabricating kenaf fibre-reinforced composites, exclusively in the form of thin film (Julkapli and Akil 2008, 2010; Julkapli et al. 2008), and also in the production of higher-quality (uniformity) thinner-film chitosan nanocomposites filled with kenaf (Julkapli et al. 2008) that are exclusively free from pinhole patterns or dull finishes, etc. Morever important demerits for kenaf fibre (or filler) incorporation in polymer composites are:

(a) Poor interfacial bonding within fibre–matrix components results weaker properties in the end product (Tserki et al. 2006; Vilay et al. 2008; Edeerozey et al. 2007).
(b) Humidity aging or water absorption usually originates undesirable effects to the mechanical properties and affects the natural fibre composite durability underneath water (Kim and Seo 2006) and dimensional stability of composites. The only one solution to these limitations and challenges is the chemical treatments or modification of (natural fibres) kenaf by reagents having functional groups adept of bonding with the hydroxyl group of kenaf fibres. The isocyanate treatment, silane treatment, acetylation and alkaline treatment (George et al. 2001) reported by researchers are the most important and effective methods. However, alkaline treatment (or mercerisation), carried out by engrossing the fibres within NaOH solution for a certain interval (Li et al. 2007; Bogoeva-Gaceva et al. 2007; Kulkarni and Ochoa 2006), are most effective especially when reinforcing thermoplastics and thermosets (Han et al. 2007; Huda et al. 2008; Edeerozey et al. 2007; Li et al. 2007).

5.8 Different Fabrication Techniques of Kenaf Polymer Composite

Kenaf fibre-reinforced polymer composite is being fabricated from a variety of techniques. The Table 5.3–5.4 summarised the list of research work done on the Kenaf fibre-reinforced polymer composite with different thermosets and thermoplastics. Some reputed work is also carried out with the biodegradable polymer like PLA, starch, etc.

5.8.1 Fibre/Thermoplastic Polymer Composite

Research studies are extensively investigating on kenaf, and its modified fibres with a wide range of polymer matrices such as polyolefins, polyethylene, and polypropylene showed improvements in mechanical properties. Within the area of biopolymers, polylactides have been investigated (Yang et al. 2014). Table 5.3 tabulates the kenaf fibre-reinforced thermoplastic polymer composite fabricated by different methods.

Table 5.3 List of reported work on kenaf fibre-reinforced thermoplastic polymer composite fabricated by different methods

Reinforcement/polymer	Fabrication technique	References
Kenaf fibre/(PVC), (TPU) poly-blend	Compressing moulding machine	El-Shekeil et al. (2014)
Kenaf fibre/(TPU)	Compression moulding	El-Shekeil et al. (2012)
Soil-buried kenaf fibre/TPU	Melt-mixing method using Haake Polydrive R600 internal mixer	Sapuan et al. (2013)
Kenaf fibre/(PP)	Extrusion	Sallih et al. (2014)
Treated-raw kenaf fibre/ thermoplastic cassava starch	Solution casting	Zainuddin et al. (2013)
Treated kenaf fibre/(HDPE) and (PP)	Hot press machine	Meon et al. (2012)
Nonwoven kenaf fibre/(PP)	Compression moulding	Asumani et al. (2012)
Grafted kenaf/(PVC)/ethylene vinyl acetate	Hot hydraulic press method using Haake Rheomix Polydrive R600/610 internal mixer	Bakar et al. (2015)
Kenaf fibre/polypropylene (PP)	Compression moulding	Zampaloni et al. (2007)
Kenaf fibres/starch-grafted (PP)	Compression moulding	Hamma et al. (2014)
Kenaf fibre thymol/PLA	Melt blending and compression moulding	Tawakkal et al. (2014)
Kenaf fibre/poly (hydroxybutyrate-co-valerate)/poly(butylene adipate-co-terephthalate)	Compression moulding	Yang et al. (2014)
Kenaf–lyocell fibres/(PLA) and (PHB)	Hot pressing followed by compression moulding	Graupner and Müssig (2011)
Kenaf glass/(UP)	Sheet moulding compound	Atiqah et al. (2014)
Grafted and ungrafted kenaf fibres/(PVC)/ethylene vinyl acetate	Haake Rheomix Polydrive R600/610 internal mixer followed by electrically heated hydraulic press	Bakar et al. (2014)
Kenaf fibre/(PP)	Compression moulding	Islam et al. (2011)
Kenaf fibre/tributyl citrate plasticized cellulose acetate	Compression moulding	Pang et al. (2015)
Kenaf mat/(UP)	Hand lay-up and compression moulding methods	Hojo et al. (2014)
Kenaf fibre/(PP)	Press forming	Shibata et al. (2006)
Treated kenaf fibre/PLA	Twin screw extrusion and compression moulding	Shukor et al. (2014)
Kenaf fibre–corn husk flour/(PLA)	Extrusion process	Kwon et al. (2014)
Kenaf/poly(furfuryl alcohol) bioresin	Hand lay-up technique followed by compression moulding	Deka et al. (2013)
Kenaf/polylactide	Hot pressing	Lee et al. (2009)

(continued)

Table 5.3 (continued)

Reinforcement/polymer	Fabrication technique	References
Kenaf fibre/HDPE	Extrusion and compression moulding	Salleh et al. (2014)
Kenaf fibre/(PP)	Compression moulding	Bernard et al. (2011)
Wound kenaf yarn fibre/(UP)	Filament winding	Misri et al. (2015)
Nonwoven kenaf/(PP)	Carding and needle-punching techniques, followed by compression moulding	Hao et al. (2013)
Kenaf fibre/soy	Extrusion, followed by injection or compression moulding	Liu et al. (2007)
Kenaf fibres/(PLA)	Compression moulding using the film-stacking method	Huda et al. (2008)

PLA poly (lactic acid), *PP* polypropylene, *UP* unsaturated polyester, *PHB* polyhydroxybutyrate, *TPU* thermoplastic polyurethane, *HDPE* high-density polyethylene, *PVC* poly vinyl chloride

5.8.2 Fibre/Thermosetting Polymer Composite

Kenaf has been largely used as reinforcement both in thermoplastic polymers (Tajeddin et al. 2009; Avella et al. 2008) and in thermoset polymers (Rassmann et al. 2011; Mutasher et al. 2011). Moreover, kenaf fibres have been used as nonwoven mats in the automotive, textiles, fibreboard and civil and electronic industries (Magurno 1999; Davoodi et al. 2010; Elsaid et al. 2011; Serizawa et al. 2006). Some of the important research studies on the kenaf fibre-reinforced thermosetting polymeric composite by different techniques are summarised in Table 5.4.

5.9 Conclusion

Natural cellulosic fibres most importantly kenaf and their respective polymer composites offer a number of advantages over conventional materials such as considerable toughness, flexibility, easy processing, recyclability, ecofriendliness, etc. Kenaf fibre has appropriate properties for use in engineering as filler in fibre-reinforced polymer composite. The manufacturing method for a composite material affecting the mechanical properties is strongly related to matrix type, size, length and type and direction of fibre reinforcement. The fabrication of thermosetting composites is a very complex procedure, involving processes of different physics and scales at each step of composite manufacturing (forming, consolidation/impregnation and curing). The fabrication with compression moulding machine, hand lay and pultrusion methods is quite critical in order to maintain optimal thermal and structural characteristics of composites without scarifying their environmental performance when using long kenaf fibres. Kenaf fibre-reinforced composite is

Table 5.4 List of reported work on kenaf fibre-reinforced thermosets polymer composite

Reinforcement/polymer	Fabrication technique	References
Kenaf bast long fibres/epoxy	Hand lay-up process followed by hot-press compression moulding	Suriani et al. (2012)
Woven kenaf–aramid/epoxy	Hand lay-up	Yahaya et al. (2015)
Kenaf glass/epoxy	Modified SMC (sheet moulding compound)	Davoodi et al. (2012)
Unidirectional long kenaf fibre/epoxy	Hand lay-up	Yousif et al. (2012)
Kenaf/polyester	Vacuum infusion and hand lay-up method	Yuhazri et al. (2010)
Kenaf glass/epoxy	Modified sheet moulding compound	Davoodi et al. (2010)
Nonwoven kenaf fibres-Kevlar/epoxy laminates	Hand lay-up	Yahaya et al. (2014)
Treated kenaf fibre/epoxy	Initial hand lay-up followed by vacuum bagging	Fiore et al. (2015)
Nonwoven kenaf fibre/epoxy	Hand lay-up	Alkbir et al. (2014)
Kenaf fibreglass/(UP)	Hand lay-up and cold press method	Ghani et al. (2012)

fabricated through different manufacturing processes such as pultrusion, and extrusion. Moreover potentially compression moulding and filament winding. Thus, generating increased attention which conveys an optimistic future compared to other natural fibre-reinforced composites.

5.10 Future Perspective

Currently, the need towards greener and safer environment is growing in the future. So some supplementary fundamental studies and comprehensive researches are required on some of the issues:

- Product commercialisation and manufacturing processing challenges including interfacial attachment and pressure transfer properties to broaden the submissions of the kenaf fibres in resolving ecological complications, especially for large-scale end products in engineering/bioengineering industries
- Investigation on development and integration of stochastic replication strategy within the present viable traditional simulation tools and their design for providing both tremendous benefits in terms of costs and to overcome uncertainty during manufacturing stages
- Adoption of automatic lay-up process instead of existing lay-up process required to be highlighted for increasing the economical production rate of composite along with safety environments to human

- Further highlights needed towards determining the statistical characterisation, relevant sources of uncertainty and variability in composites manufacturing (forming/draping and cure step)

Acknowledgements The first author, N. Saba acknowledges the *International Graduate Research Fellowship* (IGRF) *UPM-Malaysia* to support this work. The authors also thankful to the Universiti Putra Malaysia for supporting this research through Research Grant: GP-IBT/2013/9420700.

References

Akil HM, Omar MF, Mazuki AAM, Safiee S, Ishak ZAM, Bakar AA (2011) Kenaf fiber reinforced composites: a review. Mater Des 32:4107–4121

Alamri H, Low IM, Alothman Z (2012) Mechanical, thermal and microstructural characteristics of cellulose fibre reinforced epoxy/organoclay nanocomposites. Compos B Eng 43:2762–2771

Alkbir MFM, Sapuan SM, Nuraini AA, Ishak MR (2014) Effect of geometry on crashworthiness parameters of natural kenaf fibre reinforced composite hexagonal tubes. Mater Des 60:85–93

Anuar H, Zuraida A (2011) Improvement in mechanical properties of reinforced thermoplastic elastomer composite with kenaf bast fibre. Compos B Eng 42:462–465

Asumani OML, Reid RG, Paskaramoorthy R (2012) The effects of alkali–silane treatment on the tensile and flexural properties of short fibre non-woven kenaf reinforced polypropylene composites. Compos A Appl Sci Manuf 43:1431–1440

Atiqah A, Maleque MA, Jawaid M, Iqbal M (2014) Development of kenaf-glass reinforced unsaturated polyester hybrid composite for structural applications. Compos B Eng 56:68–73

Avella M, Bogoeva-Gaceva G, Bužarovska A, Errico ME, Gentile G, Grozdanov A (2008) Poly (lactic acid) based biocomposites reinforced with kenaf fibers. J Appl Polym Sci 108(3542):51

Bakar N, Chee CY, Abdullah LC, Ratnam CT, Azowa N (2014) Effect of methyl methacrylate grafted kenaf on mechanical properties of polyvinyl chloride/ethylene vinyl acetate composites. Compos A Appl Sci Manuf 63:45–50

Bakar NA, Chee CY, Abdullah LC, Ratnam CT, Ibrahim NA (2015) Thermal and dynamic mechanical properties of grafted kenaf filled poly (vinyl chloride)/ethylene vinyl acetate composites. Mater Des 65:204–211

Batouli SM, Zhu Y, Nar M, D'Souza NA (2014) Environmental performance of kenaf-fiber reinforced polyurethane: a life cycle assessment approach. J Clean Prod 66:164–173

Bernard M, Khalina A, Ali A, Janius R, Faizal M, Hasnah KS, Sanuddin AB (2011) The effect of processing parameters on the mechanical properties of kenaf fibre plastic composite. Mater Des 32:1039–1043

Bogoeva-Gaceva G, Avella M, Malinconico M, Buzarovska A, Grozdanov A, Gen-tile G (2007) Natural fiber eco-composites. Polym Compos 28(1):98–107

Cao J, Akkerman R, Boisse P, Chen J, Cheng HS, de Graaf EF (2008) Characterization of mechanical behavior of woven fabrics: experimental methods and benchmark results. Compos A Appl Sci Manuf 39(6):1037–1053

Carneiro OS, Maia JM (2000) Rheological behavior of carbon fiber/thermoplastic composites. Part I: The influence of fiber type, processing conditions and level of incorporation. Polym Compos 21(6):960–969

Davoodi MM, Sapuan SM, Ahmad D, Ali A, Khalina A, Jonoobi M (2010) Mechanical properties of hybrid kenaf/glass reinforced epoxy composite for passenger car bumper beam. Mater Des 31:4927–4932

Davoodi MM, Sapuan SM, Ahmad D, Aidy A, Khalina A, Jonoobi M (2012) Effect of polybutylene terephthalate (PBT) on impact property improvement of hybrid kenaf/glass epoxy composite. Mater Lett 67:5–7

Deka H, Misra M, Mohanty A (2013) Renewable resource based "all green composites" from kenaf biofibre and poly (furfuryl alcohol) bioresin. Ind Crop Prod 41:94–101

Demir H, Atikler U, Balköse D, Tıhmınlıoglu F (2006) The effect of fiber surface treatments on the tensile and water sorption properties of polypropylene–luffa fiber composites. Compos A Appl Sci Manuf 37:447–456

Dhakal HN, Zhang ZY, Richardson MOW (2007) Effect of water absorption on the mechanical properties of hemp fibre reinforced unsaturated polyester composites. Compos Sci Technol 67:1674–1683

Dicker MPM, Duckworth PF, Baker AB, Francois G, Hazzard MK, Weaver PM (2014) Review Green composites: a review of material attributes and complementary applications. Compos A Appl Sci Manuf 56:280–289

Dittenber DB, Rao GHVS (2012) Critical review of recent publications on use of natural composites in infrastructure. Compos A Appl Sci Manuf 43(8):1419–1429

Duhovic M, Mitschang P, Bhattacharyya D (2011) Modelling approach for the prediction of stitch influence during woven fabric draping. Compos A Appl Sci Manuf 42(8):968–978

Edeerozey A, Akil HM, Azhar A, Ariffin M (2007) Chemical modification of kenaf fibers. Mater Lett 61:2023–2025

Eichhorn SJ, Dufresne A, Aranguren M, Marcovich NE, Capadona JR, Rowan SJ (2010) Review: current international research into cellulose nanofibres and nanocomposites. J Mater Sci 45(1):1–33

Elsaid A, Dawood M, Seracino R, Bobko C (2011) Mechanical properties of kenaf fiber reinforced concrete. Constr Build Mater 25:1991–2001

El-Shekeil YA, Sapuan SM, Abdan K, Zainudin ES (2012) Influence of fiber content on the mechanical and thermal properties of Kenaf fiber reinforced thermoplastic polyurethane composites. Mater Des 40:299–303

El-Shekeil YA, Sapuan SM, Jawaid M, Al-Shujaa OM (2014) Influence of fiber content on mechanical, morphological and thermal properties of kenaf fibers reinforced poly (vinyl chloride)/thermoplastic polyurethane poly-blend composites. Mater Des 58:130–135

Faruk O, Bledzki AK, Fink HP, Sain M (2012) Biocomposites reinforced with natural fibers: 2000–2010. Prog Polym Sci 37(11):1552–1596

Fiore V, Bella GD, Valenza A (2015) The effect of alkaline treatment on mechanical properties of kenaf fibers and their epoxy composites. Compos B Eng 68:14–21

George J, Sreekala MS, Thomas S (2001) A review on interface modification and characterization of natural fiber reinforced plastic composites. Polym Eng Sci 41(9):1471–1485

Ghani MAA, Salleh Z, Hyie KM, Berhan MN, Taib YMD, Bakri MAI (2012) Mechanical properties of kenaf/fiberglass polyester hybrid composite. Proc Eng 41:1654–1659. International Symposium on Robotics and Intelligent Sensors (IRIS)

Graupner N, Müssig J (2011) Comparison of the mechanical characteristics of kenaf and lyocell fibre reinforced poly (lactic acid) (PLA) and poly (3-hydroxybutyrate) (PHB) composites. Compos A Appl Sci Manuf 42(12):2010–2019

Guo Z, Du S, Zhang B (2005) Temperature field of thick thermoset composite laminates during cure process. Compos Sci Technol 65(3–4):517–523

Hamma A, Kaci M, Ishak ZAM, Pegoretti A (2014) Starch-grafted-polypropylene/kenaf fibres composites. Part 1: Mechanical performances and viscoelastic behaviour. Compos A Appl Sci Manuf 56:328–335

Han YH, Han SO, Cho D, Kim HI (2007) Kenaf/polypropylene biocomposites: effects of electron beam irradiation and alkali treatment on kenaf natural fibers. Compos Interfaces 14(5):559–578

Hao A, Zhao H, Chen JY (2013) Kenaf/polypropylene nonwoven composites: the influence of manufacturing conditions on mechanical, thermal, and acoustical performance. Compos B Eng 54:44–51

Ho MP, Wang H, Lee JH, Ho CK, Lau KT, Leng J, Hui D (2012) Critical factors on manufacturing processes of natural fibre composites. Compos B Eng 43:3549–3562

Hojo T, Xu Z, Yang Y, Hamada H (2014) Tensile properties of bamboo, jute and kenaf mat-reinforced composite. Energy Proc 56:72–79. 11th Eco-Energy and Materials Science and Engineering (11th EMSES)

Huda MS, Drzal LT, Mohanty AK, Misra M (2008) Effect of fiber surface-treatments on the properties of laminated biocomposites from poly (lactic acid) (PLA) and kenaf fibers. Compos Sci Technol 68:424–432
Ishak M, Leman Z, Sapuan SM, Salleh M, Misri S (2009) The effect of sea water treatment on the impact and flexural strength of sugar palm fibre reinforced epoxy composites. Int J Mech Mater Eng (IJMME) 4:316–320
Islam MS, Church JS, Miao M (2011) Effect of removing polypropylene fibre surface finishes on mechanical performance of kenaf/polypropylene composites. Compos A Appl Sci Manuf 42:1687–1693
Julkapli NM, Akil HM (2008) Degradability of kenaf dust-filled chitosan biocomposites. Mater Sci Eng C 28:1100–1111
Julkapli NM, Akil HM (2010) Influence of a plasticizer on the mechanical properties of kenaf-filled chitosan bio-composites. Polym Plast Technol Eng 49:944–951
Julkapli NM, Ahmad Z, Akil M (2008) Preparation and properties of kenaf dust-filled chitosan biocomposites. Compos Interf 15(7):851–866
Kafi AA, Magniez K, Fox BL (2011) Effect of manufacturing process on the flexural, fracture toughness, and thermo-mechanical properties of bio-composites. Compos A Appl Sci Manuf 42:993–999
Karimi S, Tahir PMT, Karimi A, Dufresne A, Abdulkhani A (2014) Kenaf bast cellulosic fibers hierarchy: a comprehensive approach from micro to nano. Carbohydr Polym 101:878–885
Kim HJ, Seo DW (2006) Effect of water absorption fatigue on mechanical properties of sisal textile-reinforced composites. Int J Fatigue 28:1307–1314
Kulkarni R, Ochoa O (2006) Transverse and longitudinal CTE measurements of carbon fibers and their impact on interfacial residual stresses in composites. J Compos Mater 40:733
Kwon HJ, Sunthornvarabhas JS, Park JW, Lee JH, Kim HJ, Piyachomkwan K, Sriroth K, Cho D (2014) Tensile properties of kenaf fiber and corn husk flour reinforced poly (lactic acid) hybrid bio-composites: role of aspect ratio of natural fibers. Compos B Eng 56:232–237
Lee BH, Kim HS, Lee S, Kim HJ, Dorgan JR (2009) Bio-composites of kenaf fibers in polylactide: role of improved interfacial adhesion in the carding process. Compos Sci Technol 69: 2573–2579
Leong YW, Thitithanasarn S, Yamada K, Hamada H (2014) Compression and injection molding techniques for natural fiber composites. Nat Fibre Compos 216–232
Li X, Tabil LG, Panigrahi S (2007) Chemical treatments of natural fiber for use in natural fiber-reinforced composites: a review. J Environ Polym Degrad 15:25–33
Liu W, Drzal LT, Mohanty AK, Misra M (2007) Influence of processing methods and fiber length on physical properties of kenaf fiber reinforced soy based biocomposites. Compos B Eng 38(3):352–359
Liu D, Song J, Anderson DP, Chang PR, Hua Y (2012) Bamboo fiber and its reinforced composites: structure and properties. Cellulose 19(5):1449–1480
Low IM, McGrath M, Lawrence D, Schmidt P, Lane J, Latella BA (2007) Mechanical and fracture properties of cellulose fibre reinforced epoxy laminates. Compos A Appl Sci Manuf 38:963–974
Magurno A (1999) Vegetable fibres in automotive interior components. Die Angew Makromol Chem 272:99–107
Mahjoub R, Yatim JM, Sam ARM, Hashemi SH (2014) Construct Build Mater 55:103–113
Marsh G (2003) Next step for automotive materials. Mater Today 6:36–43
Mazuki AAM, Akil HM, Safiee S, Ishak ZAM, Bakr AA (2011) Degradation of dynamic mechanical properties of pultruded kenaf fiber reinforced composites after immersion in various solutions. Compos B Eng 42(1):71–76
Meon MS, Othman MF, Husain H, Fairuz M, Remeli MF, Syahar M, Syaw M (2012) Improving tensile properties of kenaf fibers treated with sodium hydroxide. Proc Eng 41:1587–1592. International Symposium on Robotics and Intelligent Sensors (IRIS)
Mesogitis TS, Skordos AA, Long AC (2014) Uncertainty in the manufacturing of fibrous thermosetting composites: a review. Compos A Appl Sci Manuf 57:67–75

Misri S, Sapuan SM, Leman Z, Ishak MR (2015) Torsional behaviour of filament wound kenaf yarn fibre reinforced unsaturated polyester composite hollow shafts. Mater Des 65:953–960

Mohanty AK, Misra M, Hinrichsen G (2000) Biofibres, biodegradable polymers and biocomposites: an overview. Macromol Mater Eng 1(24):276–277

Mohanty AK, Wibowo A, Misra M, Drzal LT (2004) Effect of process engineering on the performance of natural fiber reinforced cellulose acetate biocomposites. Compos A Appl Sci Manuf 35(3):363–370

Mutasher SA, Poh A, Than AM, Law J (2011) The effect of alkali treatment mechanical properties of kenaf fiber epoxy composite. Key Eng Mater 471:191–196

Nishino T (2004) Green composites: polymer composites and the environment. Elsevier, United Kingdom

Nishino T, Hirao K, Kotera M, Nakamae K, Inagaki H (2003) Kenaf reinforced biodegradable composite. Compos Sci Technol 63:1281–1286

Ochi S (2008) Mechanical properties of kenaf fibers and kenaf/PLA composites. Mech Mater 4(5):446–452

Opperer JG, Kim SK, Daniel IM (2004) Characterization of local preform defects in resin transfer molding by the gas flow method and statistical analysis. Compos Sci Technol 64(13–14): 1921–1935

Padmanabhan SK, Pitchumani R (1999) Stochastic modeling of nonisothermal flow during resin transfer molding. Int J Heat Mass Transfer 42(16):3057–3070

Pang C, Shanks RA, Daver F (2015) Characterization of kenaf fiber composites prepared with tributyl citrate plasticized cellulose acetate. Compos A Appl Sci Manuf 70:52–58

Paridah MT, Basher AB, SaifulAzry S, Ahmed Z (2011) Retting process of some bast plant fibres and its effect on fibre quality: a review. BioResources 4:5260–5281

Parikh DV, Calamari TA, Sawhney APS, Blanchard EJ, Screen FJ, Warnock M, Muller DH, Stryjewski DD (2002) Improved chemical retting of kenaf fibers. Text Res J 7:618–624

Paukszta D, Borysiak S (2013) The influence of processing and the polymorphism of lignocellulosic fillers on the structure and properties of composite materials—a review. Materials (Basel) 6:2747–2767

Potter K (2009) Understanding the origin of defects and variability in composites manufacture. In: Proceedings of the 17th international conference on composite materials, Edinburgh, UK

Potter K, Khan B, Wisnom M, Bell T, Stevens J (2008) Variability, fibre waviness and misalignment in the determination of the properties of composite materials and structures. Compos A Appl Sci Manuf 39(9):1343–1354

Rassmann S, Paskaramoorthy R, Reid RG (2011) Effect of resin system on the mechanical properties and water absorption of kenaf fibre reinforced laminates. Mater Des 32:1399–1405

Rosa DIM, Santulli C, Sarasini F (2010) Mechanical and thermal characterization of epoxy composites reinforced with random and quasi-unidirectional untreated phormium tenax leaf fibres. Mater Des 31:2397–2405

Rosa LAD, Cozzo G, Latteri A, Recca A, Bjorklund A, Parrinello E (2013) Life cycle assessment of a novel hybrid glass-hemp/thermoset composite. J Clean Prod 44:69–76

Rouison D, Sain M, Couturier M (2004) Resin transfer molding of natural fiber reinforced composites: cure simulation. Compos Sci Technol 64:629–644

Saba N, Paridah MT, Jawaid M (2015) Mechanical properties of kenaf fibre reinforced polymer composite: a review. Construct Build Mater 76:87–96

Salleh FM, Hassan A, Yahya R, Azzahari AD (2014) Effects of extrusion temperature on the rheological, dynamic mechanical and tensile properties of kenaf fiber/HDPE composites. Compos B Eng 58:259–266

Sallih N, Lescher P, Bhattacharyya D (2014) Factorial study of material and process parameters on the mechanical properties of extruded kenaf fibre/polypropylene composite sheets. Compos A Appl Sci Manuf 61:91–107

Sapuan SM, Pua FI, El-Shekeil YA, AL-Oqla FM (2013) Mechanical properties of soil buried kenaf fibre reinforced thermoplastic polyurethane composites. Mater Des 50:467–470

Serizawa S, Inoue K, Iji M (2006) Kenaf-fiber-reinforced poly (lactic acid) used for electronic products. J Appl Polym Sci 100:618–624

Shahzad A (2012) Hemp fiber and its composites – a review. J Compos Mater 46(8):973–986

Shibata S, Cao Y, Fukumoto I (2006) Lightweight laminate composites made from kenaf and polypropylene fibres. Polym Test 25:142–148

Shukor F, Hassan A, Islam MS, Mokhtar M, Hasan M (2014) Effect of ammonium polyphosphate on flame retardancy, thermal stability and mechanical properties of alkali treated kenaf fiber filled PLA biocomposites. Mater Des 54:425–429

Skordos AA, Sutcliffe MPF (2008) Stochastic simulation of woven composites forming. Compos Sci Technol 68(1):283–296

Suriani MJ, Alia A, Khalina A, Sapuan SM, Abdullah S (2012) Detection of defects in kenaf/epoxy using infrared thermal imaging technique. Procedia Chem 4:172–178

Swolfs Y, Gorbatikh LG, Verpoest I (2014) Fibre hybridization in polymer composites: a review. Compos A Appl Sci Manuf 67:181–200

Tajeddin B, Rahman RA, Abdulah LC, Ibrahim NA, Yusof YA (2009) Thermal properties of low density polyethylene-filled kenaf cellulose composites. Eur J Sci Res 32:223–230

Tawakkal ISMA, Cranb MJ, Bigger SW (2014) Effect of kenaf fibre loading and thymol concentration on the mechanical and thermal properties of PLA/kenaf/thymol composites. Ind Crop Prod 61:74–83

Tserki V, Matzinos P, Panayiotou C (2006) Novel biodegradable composites based on treated lignocellulosic waste flour as filler. Part II. Development of biodegradable composites using treated and compatibilized waste flour. Compos A Appl Sci Manuf 37:1231–1238

Velde VDK, Kiekens P (2001) Thermoplastic pultrusion of natural fiber reinforced composites. Compos Struct 54:355–360

Verleye B, Lomov SV, Long A, Verpoest I, Roose D (2010) Permeability prediction for the meso–macro coupling in the simulation of the impregnation stage of resin transfer moulding. Compos A Appl Sci Manuf 41(1):29–35

Vilay V, Mariatti M, Mat Taib R, Todo M (2008) Effect of fiber surface treatment and fiber loading on the properties of bagasse fiber-reinforced unsaturated polyester composites. Compos Sci Technol 68:631–638

Xue Y, Du Y, Elder S, Wang K, Zhang J (2009) Temperature and loading rate effects on tensile properties of kenaf bast fiber bundles and composites. Compos B Eng 3:189–196

Yahaya R, Sapuan SM, Jawaid M, Leman Z, Zainudin ES (2014) Quasi-static penetration and ballistic properties of kenaf–aramid hybrid composites. Mater Des 63:775–782

Yahaya R, Sapuan SM, Jawaid M, Leman Z, Zainudin ES (2015) Effect of layering sequence and chemical treatment on the mechanical properties of woven kenaf–aramid hybrid laminated composites. Mater Des 67:173–179

Yang B, Nar M, Visi DK, Allen M, Ayre B, Webber CL III, Lu H, D'Souza NA (2014) Effects of chemical versus enzymatic processing of kenaf fibers on poly (hydroxybutyrate-co-valerate)/poly (butylene adipate-co-terephthalate) composite properties. Compos B Eng 56:926–933

Yousif BF, Shalwan A, Chin CW, Ming KC (2012) Flexural properties of treated and untreated kenaf/epoxy composites. Mater Des 40:378–385

Yu H, Yu C (2010) Influence of various retting methods on properties of kenaf fiber. J Text Inst 5:452–456

Yuhazri MY, Phongsakorn PT, Sihombing H (2010) A comparison process between vacuum infusion and hand lay-up method toward kenaf/polyester composites. Int J Basic Appl Sci IJBAS-IJENS 10

Zainuddin SYZ, Ahmad I, Kargarzadeh H, Abdullah I, Dufresne A (2013) Potential of using multiscale kenaf fibers as reinforcing filler in cassava starch- kenaf biocomposites. Carbohydr Polym 92:2299–2305

Zampaloni M, Pourboghrat F, Yankovich SA, Rodgers BN, Moore J, Drzal LT, Mohanty AK, Misra M (2007) Kenaf natural fiber reinforced polypropylene composites: a discussion on manufacturing problems and solutions. Compos A Appl Sci Manuf 38:1569–1580

Zini E, Scandola M (2011) Green composites: an overview. Polym Compos 32:1905–1915

Chapter 6
Critical Concerns on Manufacturing Processes of Natural Fibre Reinforced Polymer Composites

Mohammad Reza Ketabchi, M. Enamul Hoque, and M. Khalid Siddiqui

Abstract For the last one decade, natural fibre reinforced polymer composites have attracted a considerable attention around the globe due to their inherent performances such as biodegradability, recyclability and accessibility. A suitable manufacturing process is required to develop a promising reinforced polymer composite accordingly. It has been observed that the matrices, materials and chemical, physical, mechanical and thermal properties of both the hosting polymer and the employed natural fibres play the most important role in choosing an appropriate processing technique. Recently, valuable studies have been carried out in this field, as engineers have tried different processing methods and conditions. In this chapter and later a brief introduction to biomass materials, a comprehensive review has been done over the latest biocomposite processing techniques. Beneficial solutions and treatment methods have been addressed. Furthermore, a combination of both traditional and modern processing technique has been recommended for future studies.

Keywords Natural fibre • Biocomposites • Manufacturing • Mechanical properties

6.1 Introduction

Since the last two decades, manufacturers have been progressively looking for higher strength, lower weight, cheaper and more eco-friendly material to produce novel products in order to create entirely new products by taking advantage of these characteristics. Natural fibres as an alternative reinforcement agent in biocomposites present numerous advantages over conventional glass and carbon

---

M.R. Ketabchi • M. Khalid Siddiqui
Faculty of Engineering, Division of Manufacturing and Industrial Processes, University of Nottingham Malaysia Campus, Jalan Broga, 43500 Semenyih, Selangor, Malaysia

M. Enamul Hoque (✉)
Department of Biomedical Engineering,
University of King Faisal, Al-Hofuf, Al-Hassa, Saudi Arabia
e-mail: enamul1973@gmail.com

M.S. Salit et al. (eds.), *Manufacturing of Natural Fibre Reinforced Polymer Composites*, DOI 10.1007/978-3-319-07944-8_6

fibres, low cost, comparable specific tensile properties, low density, nonabrasive to the equipment, nonirritating to the skin, reduced energy intake, fewer health risk, renewability, recyclability and biodegradability.

Since the 1980s, natural fibres are experiencing increased demand as fillers in polymer matrices. The application of natural materials in polymer composites in a broad range of production lines has resulted in decreasing the production of carbon-based products and the employment of petrochemical products. Meanwhile, following the fibre complex structure, various mechanical performances have been observed through the application of the same fibre in different hosting matrices. The product strength can be affected and reduced due to a number of issues, wettability of the natural fibre, presence of degradation at the fibre/matrix interface (following the fibre attractive or repulsive response to water) and fibre destruction during the manufacturing process. A pretreatment process of fibre can prevent this reduction in the product strength and moreover enhance the processing procedure. In this chapter, following a brief introduction to agricultural residues (as a rich source for natural fibres) and the suitability of known manufacturing processes, based on the fibres and matrices materials, mechanical and thermal properties are discussed in detail.

6.2 Biomass Materials

At present, agricultural residue materials are supplying 14 % of the world's energy, while yet a huge amount of bypassed biomass has no particular application (Saxena et al. 2009). Generally this large quantity of biomass is burnt or dumped, producing undesirable contaminants in the air, SO_X and NO_X. Studies have shown that this carbon renewable resource can be transformed to solid, liquid and gaseous product over numerous treatment processes. Natural fibres are derived from lignocellulose materials, and therefore they are hydrophilic. From Fig. 6.1, lignocellulosic biomass contains several important noteworthy natural polymers, cellulose $[C_6H_{10}O_5]_n$ (35–83 % dry weight basis), hemicelluloses $[C_5H_8O_4]_m$ (0–30 % dry weight basis) and lignin $[C_9H_{10}O_3{\cdot}(OCH_3)_{0.9\text{–}1.7}]_x$ (1–43 % dry weight basis) (Balat et al. 2008).

6.3 Manufacturing Techniques and Concerns

6.3.1 *Manufacturing Principles*

Appropriate manufacturing processes are essential to be employed to convert the raw materials to the desired product. For manufacturers, ideal application, performance, size, structure, processing characteristic, speed of production and costs of processing are the major topics to discuss about while designing a production line. In the meantime for initial scheming, through the mentioned principles, the composite finishing size is highlighted and a leading feature. Injection moulding and compression moulding are simple and swift processing techniques that are both

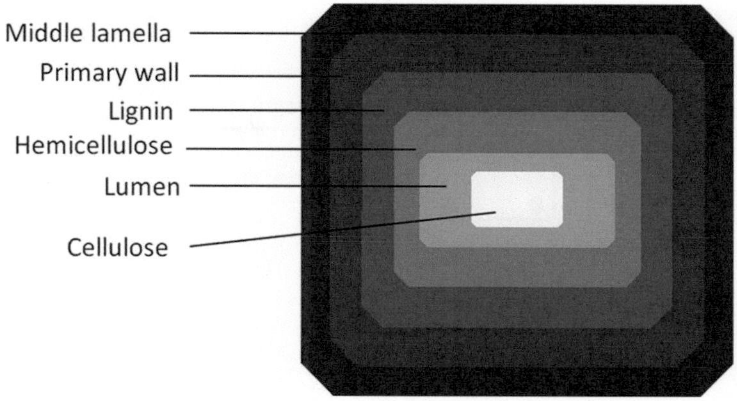

Fig. 6.1 Schematic diagram of lignocellulosic fibre's physical structure

endorsed for small/medium-sized feeding materials. Generally, for a simple form of composite product, compression moulding is recommended. Open moulding and autoclave process are addressed for bulk productions correspondingly. Furthermore, the product shape complexity has a direct impact on the processing technique to be used likewise. Recently a comprehensive review on the effect of different processing conditions and different surface modification techniques on the cellulose fibres and their composites has been carried out (Thakur and Thakur 2014).

For short fibre and nanomaterial-based processing, injection moulding and compression moulding are employed, while pultrusion is mainly used to produce stretched uniform products (Memon and Nakai 2013a, b). Pultrusion is a continuous composite processing method that combines reinforcement impregnation with composite consolidation. In this method, individual prepreg layers are flattened out and a bond is formed between them. The die structure is carefully chosen according to the desired product shape. Continuous controlled steps of heating are required to cure the composite product. Following an appropriate fabrication technique, certain properties of the raw materials require further attention, thermoplastics/thermosets, low/high viscosity and manufacturing temperature. The elementary principle of pultrusion process is the continuous melt impregnation of fibre ropes (roving) in a pultrusion tool (shown in Fig. 6.2). In terms of reducing the energy consumption, the location of the implanted heaters can be more influential than the number or the power of the heaters (Silva et al. 2014). In addition to stretched uniform products, studies have employed pultrusion process to manufacture phenolic foam composites (Yun and Lee 2008).

As mentioned earlier, the aspects in selecting an appropriate processing technique for reinforced polymer composites are totally different with the neat commercial polymers. The performance of a reinforced composite is significantly reliant on the alignment, length, content and diameter of the employed fibre. Additionally, the interface between the fibre and the matrix can be directly influenced by the natural fibre surface properties. To enhance this interface, mercerization is an eminent treatment

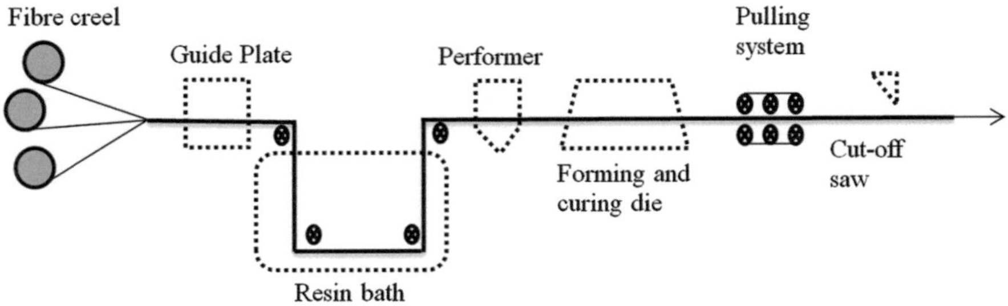

Fig. 6.2 Schematic diagram of pultrusion process

process suitable for surface treatment of fibres. Fibre surface coating removal or pretreatment by using chemical treatment is required to ensure good bonding properties. Fibre treatment enhances the matrix–fibre interface; thus it allows the stress to transfer from the matrix to the fibre which holds the higher flexural modulus of the composites. A heterogeneous system is assembled during the introduction of natural fibres to polymer composites following a weak adhesion at the matrix–fibre border (Matuana et al. 2001; Keener et al. 2004).

Natural fibre pretreatment can play an important role in enhancing the adhesion properties. Chemical and mechanical modifications are employed to increase the fibre resistance to moisture adsorption as well as the surface roughness. Several well-known fibre mechanical and chemical treatment processes are as follows:

– Grinding of fibres
– Mercerization of fibres
– Acetylation treatment
– Electron beam irradiation modification of fibre. (*This method is known economical for mass-produced frameworks*)
– Silanization of fibres
– Steam explosion of fibres
– Graft copolymerization of a monomer/polymer onto fibres. (*This method can be introduced to the applications where degradability of the hosting polymer composite is in priority*)

Following the reviewed studies on various natural fibre treatment processes, fibres with a particle size less than 500 μm are recommended for reinforcing purposes. Moreover, since 2012, to enhance the performance of the polymer composites, while using smaller amount of material at the same time, studies have shifted from using micron-sized particle to nanosized particle. A temperature of about 80–100 °C is initiated to be suitable for drying the fibres. Sodium hydroxide is known to be the most compatible and efficient chemical for an alkali treatment.

In addition to alkali treatment (mercerization), silanization of fibres is also considered to be an appropriate treatment for interface modification; notable improvements in specific tensile and flexural strength properties of fibres are reported following this type of treatment. Also, acetylation is found to bring changes in fibre surface morphology and results in the surface to become much smoother. Acetylation in vapour phase is reported to decrease the hydrophilic properties of the fibres and enhance the thermal stability.

6.3.2 Moulding Processes

6.3.2.1 Injection Moulding Process

For thermoplastic-based processing, injection moulding and compression moulding processes are recommended. Rheological measurements (shear viscosity measurements and extensional viscosity) have been carried for injection moulding machine (Kelly et al. 2009; Bariani et al. 2007) and extruder (Vera-Sorroche et al. 2014). During the injection, as an advantage of using an injection moulding machine, following the screw speed, the volumetric flow rate can be calculated accordingly. Through this determination, minor leakage flows from the valve are negligible, and the speed of the screw is considered constant through the process. The installed screw inside the electric injection moulding machine is empowered and accurately controlled using a servomotor (shown in Fig. 6.3). The precision of rheological measurements is based on the accuracy of this servomotor.

During the injection moulding process, a molten blend of both polymer and fibre is forced into the voids of the mould. Various researches have been carried out to determine the opportunities of reinforcing polymer composites through injection moulding (Serizawa et al. 2006; Huda et al. 2005, 2006a, b). Generally, plastic pellets of the neat thermoplastic polymer are employed for this particular type of processing. At the feeding stage, both plastic pellets and ground fibres are fed, respectively, through a funnel into a preheated compression barrel with installed screw/screws shown in Fig. 6.3. The heat of the barrel melts the plastic pellets and forms them into a viscous liquid. The phase transition of plastic pellets enhances the

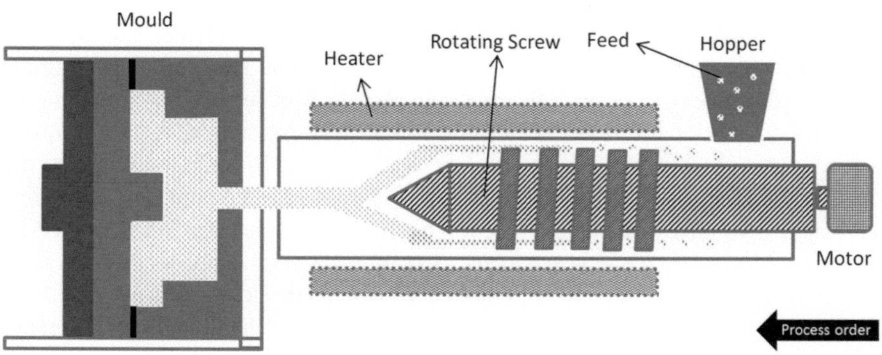

Fig. 6.3 Schematic diagram of an injection moulding machine

injection performance to the designed moulds and later to the sprue nozzle. The heat is applied to melt the plastic pellets fed in between the barrel and screw. Also, the purpose of heating the barrel is to transform the solid pellets into viscous liquid which can be drove through the screw and sprue nozzle and lastly forced into the matched-metal closed mould cavities. Pressurizing is required to drive the blend through sprue nozzle. Following the changes in the process pressure, the moulds are required to be clamped securely. After a proper cooling and solidification stage, the shaped reinforced composite is collected from the moulds.

The screws are employed due to several purposes and benefits:

- To change the phase of the plastic pellets into viscous liquid
- To generate a shear force to enhance the blending and dispersion properties
- To drive the fibre-melted plastic pellet blend through the barrel, the sprue nozzle and into the moulds

For biodegradable polymers such as poly(lactic acid) (PLA), an easier and smoother flow of the polymer is observed following a drop in the shear viscosity which is obtained following an increase in process temperature. Further to an increase in process temperature, an increase in shear rate can decrease the molten polymer viscosity significantly. Generally, this drop in polymer viscosity is caused by following a break in molecule chains influenced by an increase in temperature or shear force.

For a reinforced polymer composite, in the presence of force, the stress is expected to transfer from the polymer matrix to the natural fibre. Furthermore, during the reinforcement, it is desired to load the fibre to its full capacity while having a strong interfacial bonding. One of the common techniques used to reach these valuable goals is chopping the fibre short accordingly. Further studies on fibre size are required while employing the common injection moulding process; in this process, following the great shear rates in the barrel as well as the sprue nuzzle, the application of too small fibres causes notable attrition. Hence, in practice, a strong attrition can result in a shorter fibre length than expected. As a consequence, the fibres in the product are shorter than the critical length; this limits the fibres to carry their determined load effectively and efficiently.

Meanwhile, the fibre can relatively act as a defect in the polymer matrix due to its length and surface properties. It needs to be noted that the fibre could fracture prior to the failure of the matrix formation in case the fibre length drives beyond the expected critical length. Hence, determination of fibre critical length prior to injection moulding is essential. Commonly, the strength as well as the toughness of the finishing product improves following an increase in fibre content. However, this advantage is limited while using injection moulding machine, due to a slim feeding gate, fibre filling, and blend viscosity. Also, there are several concerns that influence the modulus distribution, fibre volume extension and later to the mixing stage with the molten polymer, residual stress and also orientation of the fibres with respect to the depth.

Fast cooling process of the molten composite apart from any external force results in an internal stress called residual stress. The tensile stresses at the surface and the core areas are expressed by the residual stress distribution, and compressive stresses at the intermediate areas are known as the characteristic residual stress distribution in injection-moulded parts (Kim et al. 2009). Since the 1990s, three levels of stress in laminated structure have been recognized: the 'micro-stresses' present among fibres within each ply, the 'macro-stresses' developing in multiaxial laminates at the ply-to-ply scale and a third more dominant level of stress resulting from different thermal histories of different parts of a laminate during the cooling sage (Shokrieh 2014).

Residual stress causes an early fracture in the composites which directly affects the quality of the finishing product (a product based either on a neat thermoplastic polymer or fibre reinforced composites).

The physical and mechanical pressures that the molten polymer goes through from the beginning of the injection moulding process until the end (filling up the mould cavity) have a direct influence on the residual stress distribution along the flow path (White 1985). The residual stress results in shrinkage of the final product. Hence it can be said that the dimensional accuracy of the finishing product is strongly linked to the residual stress distribution in the moulded part.

Furthermore, due to the high process temperature, change in moisture absorption characteristics, in addition to undesired layers and internal void formations, could take place. Therefore studies on residual stress can play a critical role in parameter selection. To avoid any progress of residual stress causing warpage, stress cracking or ageing, optimizations on several parameters can play an important role, processing technique, employed material and linear parameters. In formation of the residual stress, the linear parameters (such as the mould cavity shape and size, the locations of injection gates and the vents that allow the air to escape) play the major role (Batra et al. 2012). This prevention involves controlling the temperature of the molten blend as well as the mould and also the speed of the screw and injection during the process. A notable decrease in stress level and later to a compressive shift of stress onto the surface can be obtained following an increase in mould temperature (Carpenter et al. 2014; Kim et al. 2007). The processing properties and the finished product performance are both influenced by the rheology of the molten polymer as well as the fibre source and content (Park et al. 2013; Garancher and Fernyhough 2014).

Natural fibres are only subjected to a low thermal stress (Nechwatal et al. 2003). The matrix orientation is directly linked to the fibre content adjustments (Folkes and Russell 1980). Following the heat generated during the injection moulding, fibres are therefore oriented. During the solidification of the matrix, the orientation of the fibres is fixed. At low flow rates, the rotating screws affect the parallel alignment of the flow direction. Fibre alignment follows the shearing as well as the injection flow stretching direction; this takes place near the walls of the mould called the skin shown in Fig. 6.4. Beneath the top layer, the liquid viscose mixture remains undergoing the applied shear, and during this time, fibres align along the shear directions accordingly. After the formation of the top layer (skin), the core layer is formed as the fibres are swayed by the bulk deformation of the flow in the mould. The flow structure shown in Fig. 6.4 presents a general 'microstructural' observation.

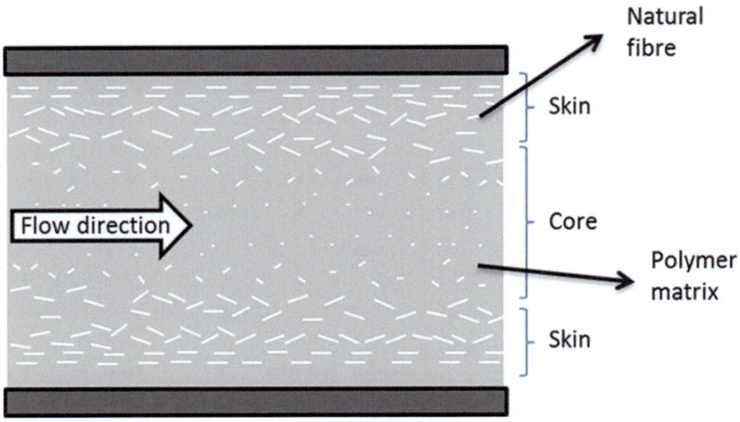

Fig. 6.4 Influence of flow on fibre orientation during injection moulding

The technique to observe the skin–core arrangement in lab-scale preparations with low fibre content is challenging. During 2001, dumbbell-shaped composites have been suggested to enhance the observation (Lee et al. 2001). Recently, studies have suggested the application of computational tomography (CT scan) to identify the amount, dispersion and orientation of components in the hosting matrix (Pujadas et al. 2014).

6.3.2.2 Compression Moulding Process

This process is a high-pressure, high-volume, plastic moulding technique that is appropriate for moulding high-strength products. Many production lines have chosen compression moulding to produce parts due to many advantages of this process, low cost, short cycle time, high-volume production, dimensional accuracy, improved impact strength, uniform shrinkage due to uniform flow and uniform density. To reinforce polymer composites with natural fibres, the application of compression moulding (a mixture of autoclaving and hot-pressing process) has been studied by researchers (Serizawa et al. 2006; Memon and Nakai 2013a, b; Dhakal et al. 2013).

It is mainly followed by two steps: preheating and pressurizing. As shown in Fig. 6.5, the composite is placed in a cavity of the matched mould, and after closing the mould (by bringing the two halves together), pressure and heat are applied in order to squeeze the composite filling the mould cavity. The mould is opened and the part is ejected subsequently. Generally this process operates at a pressure in a range from 14 to 20 MPa and at a temperature of 150–190 °C. There are two traditional types of compression moulding: 'sheet moulding compound' (SMC) and 'bulk moulding compound' (BMC). Composite density, strength and fibre orientation are the material properties affected by the process.

Applying extreme pressure may result in the fracture of fibres. To minimize the presence of fracture and damage on fibres, the fibres are recommended to be

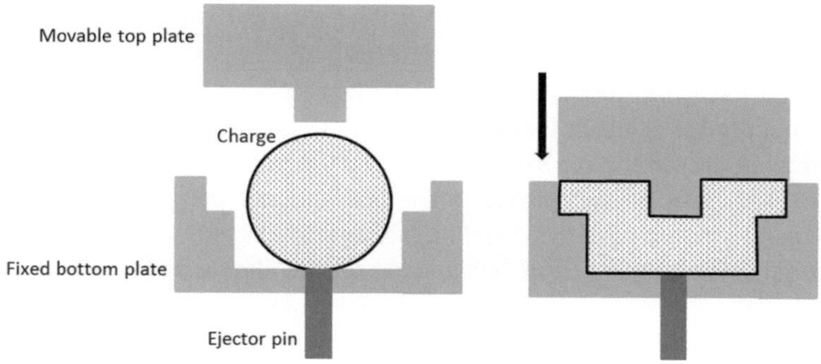

Fig. 6.5 Schematic diagram of compression moulding process

smoothly located inside the mould divest of any shear stress. Additionally, a premix of the natural fibre with the polymer compound is advised. Long fibres could be employed to produce biocomposites with higher volume fracture accordingly.

6.3.2.3 Hot-Pressing Moulding Process

Previously, it was understood that hot-pressing process is a part of compression moulding process. During the hot-pressing stage, two plates are required for compressing and heating the fibre–polymer matrix subsequently. Through this process, the main challenge is to control the matrix viscosity, mainly for thin samples. Minor faults such as warpage, minor voids, fibre breakage, sink marks and residual stresses are able to decrease the mechanical properties notability.

The viscosity level needs to be optimal enough to allow the polymer matrix to fill in the space between fibres, at the same time to maintain the matrix inside the mould, and generally to minimize possible defects to appear. Optimization of viscosity is essential to push out the trapped air in the mould and charge. Formation of voids which cause undesirable effects on the flexural and tensile properties emerges (Chambers et al. 2006; Hagstrand et al. 2005). More recently studies have been able to study the void morphology with 3D μCT images (Lambert et al. 2012). Following the filament structure of fibres, in order to produce quality composites, the wetting procedure requires additional attention (Yan et al. 2014). To enhance the wetting procedure, process parameters need to be controlled adequately, the initial charge shape, process viscosity, temperature, holding time and the composite thickness. The control of process parameters such as temperature becomes much more essential as natural polymers become the hosting matrix (Zhang et al. 2014; Shabanian et al. 2013). Through a study on the influence of various parameters on the mould heating rate, at a mould plate area of 60 mm by 60 mm, it is found that a change in mould material from stainless steel 420 to stainless steel N700 can raise the heating rate from 5.7 to 7.6 °C/s (Chen et al. 2012). Furthermore a range of temperature within 6–8 °C/s has been recommended for practical applications.

Meanwhile, it should be reminded that fibre quantity and quality both play an important role in having a smooth and effective biocomposite process. Optimized pretreatment of natural fibres can avoid many processing damages and additional costs. Pretreatment of fibres involves washing out any undesirable excess component, bringing the fibre into a suitable shape, and increasing the thermal stability of the fibre. Recently through novel treatments and procedures, researchers have observed and characterized the correlation between temperature and the porosity of the materials which could be followed to enhance the fibre–matrix interface (Brewer et al. 2014; Shaaban et al. 2013). Furthermore, the mould configuration itself has a direct influence on the flow behaviour, and appropriate mould shapes may lead to enhance the final product anisotropy. Generally, in the mould, fibre content inside sharp curves is lower than that of other smooth sections.

6.3.2.4 Resin Transfer Moulding Process

For thermo-based processes, resin transfer moulding (RTM) process is advised. RTM is a low-pressure, low-speed (vacuum assisted) moulding process suitable to many well-known applications of biocomposite reinforcements (Laurenzi et al. 2014). During this process a blend of resin-hardener catalyst is added to the mould containing the preform fibre shown in Fig. 6.6. To avoid any resin leakage during injection, the two matching mould are compressed tightly. During this process, a former curing stage is recommended assuring that the resin is completely cured. Normally for polyester or vinyl ester resins where load bearing is not a critical filler, colourants and other resin modifiers are added into the resin to reduce costs, prevent problems such as shrinkage, improve flame resistance properties and also enhance the mechanical properties.

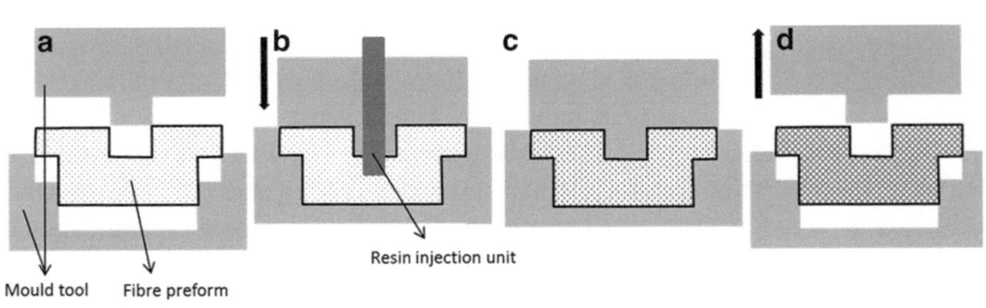

Fig. 6.6 Schematic diagram of resin transfer moulding process stages: (**a**) tools, (**b**) injection, (**c**) curing, (**d**) demould stage

RTM process involves notable advantages and disadvantages as follows:

Advantages	Disadvantages
– Very large and complex shapes can be made efficiently	– Professional tooling design and construction are required
– Dimensional tolerances	– Reinforcement loading may be difficult with complex parts
– Faster production	– Employed tools are costly
– Labour saving	– Mould design is complex
– Surface finish	
– Lower material wastage	
– Design flexibility	
– Significant strength to weight characteristics	

A rich application of natural fibres is in Lamborghini Aventador LP 700-4 Roadster 2014. To fabricate lightweight reinforced composites for this luxury model, the RTM technology has been employed. RTM technology has been applied in many common applications of natural fibre reinforced composites such as automotive, aircraft and construction.

During the RTM process, numerous parameters and variables are required to be controlled and optimized, permeability of the fibre mat, temperature of the mould, the resin injection pressure, the viscosity of the resin, the location and configuration of the gate, vent control and placement techniques of the preform and the preform architecture and permeability (Bickerton and Advani 1997). Following the resin's low viscosity, higher mould temperature and injection pressure can shorten the duration of the process cycle. Interestingly scientists have concluded that neither a predrying process nor any adjustments in the mould temperature has any noteworthy influence on the performance of kenaf fibre reinforced polyester laminates (Rassmann et al. 2010). Nevertheless, there are chances of deformation and damages in the preform fibre at high temperatures and pressures. Hence practical optimizations are required following the correlation between the parameters and variables. Optimizations have a direct influence on the fibre wetting properties, resin injection unit selection and void formations during the process (Park et al. 2011).

Various architectures of preform fibre are being used for RTM process, biaxial weave, triaxial weave, knit, multiaxial multilayer warp knit, 3D cylindrical construction, 3D braiding, 3D orthogonal fabric and angle interlock construction. The design of the preform fibre plays a significant role during the RTM process. It is challenging to assure full wetting inside the preform due to the inconsistent geometry of natural fibre and inhomogeneous fibre architecture. Furthermore, during the filling stage, incomplete wetting process and interruption in uniformity of the flow may occur due to the poor fitting size of the preform. Poor permeability of the preform may worsen this scenario. The inhomogeneity of preform fibre leads to non-uniform permeability.

During the process, the velocity differences progressively decrease due to the gradual increase in flow resistance. Through the flow from point to point, an uneven local velocity is observed due to the non-uniform microstructure of the preform fibre. At the same time, a vacuity between the preform and the mould edge can result in early arrival of edge flow causing an interruption in flow uniformity. Surprisingly, fibre concentration plays much more important role in the propagation of the edge flow compared to the injection pressure. Studies have suggested the application of larger preforms than the mould size to prevent the effect of edge flow. The edge flow results in dry spots and spillage. Furthermore, through different techniques, fast resin flow has been found applicable; however, inappropriate methods may result in the formation of unwanted bubbles affecting the finishing product properties. To prevent any void formation, vacuuming of the mould and injection pot has been recommended prior to the start of the injection process. Simulation studies have been carried out to minimize the void formation during RTM process (Lee et al. 2006). More recently, studies have been able to overcome the challenge of producing a uniform biocomposite by applying ultrasound technology during the RTM process (Planellas et al. 2014).

6.4 Conclusion and Future Perspective

Over the past one decade, much effort has been devoted to the use of natural fibre reinforced polymer composites due to their unique properties, biodegradability, environmental friendliness, notable accessibility, flexibility, easy processing and impressive physicomechanical properties. The material as well as the mechanical characteristics of these biocomposites is unlike the old carbon and glass fibre composite; hence research is still ongoing following the presence of an uncertainty in applying suitable manufacturing process for producing these composites. A combination of both traditional and modern processing techniques has been employed to produce novel reinforced polymer composites. This chapter has briefly addressed critical concerns on various processing techniques.

RTM process is found to be suitable for thermoset-based processes, while injection moulding and compression moulding process are found to be more appropriate for simple thermoplastic-based processing. Compression moulding is found to be a very productive and economical technique for producing reinforced composites. During the production of such composites, increase in yield ratio and decrease in cycle time have been recommended to be cost-effective. Following the limited natural fibre length, processes such as pultrusion are not applicable as high-tensile-strength fibres are necessary during the pultrusion process. Conventional dosing in an extruder is challenging in case of poor flow ability of natural fibres. A homogeneous dispersion of fibres into the polymer matrix is found to have a high impact on the product mechanical performance. Furthermore, the wettability of fibres was found to play an essential role during the process. Pretreatment of fibres is suggested to enhance the performance natural fibres through all processing stages.

Three fields of investigations are recommended for future studies: the influence of polymer viscosity on the processing quality, speed and cycle time; the application of ultrasonic technology during compression moulding process; and finding a suitable processing technique to reinforce polymer composites with recycled old newspaper.

References

Balat M, Balat H, Öz C (2008) Progress in bioethanol processing. Prog Energy Combust Sci 34:551–573

Bariani PF, Salvador M, Lucchetta G (2007) Development of a test method for the rheological characterization of polymers under the injection molding process conditions. J Mater Process Technol 191:119–122

Batra RC, Gopinath G, Zheng JQ (2012) Material parameters for pressure-dependent yielding of unidirectional fiber-reinforced polymeric composites. Compos B Eng 43:2594–2604

Bickerton S, Advani SG (1997) Experimental investigation and flow visualization of the resin-transfer mold-filling process in a non-planar geometry. Compos Sci Technol 57:23–33

Brewer CE, Chuang VJ, Masiello CA, Gonnermann H, Gao X, Dugan B, Driver LE, Panzacchi P, Zygourakis K, Davies CA (2014) New approaches to measuring biochar density and porosity. Biomass Bioenergy 66:176–185

Carpenter HW, Reid RG, Paskaramoorthy R (2014) Extension of the layer removal technique for the measurement of residual stresses in layered anisotropic cylinders. Int J Mech Mater Des 1–12

Chambers AR, Earl JS, Squires CA, Suhot MA (2006) The effect of voids on the flexural fatigue performance of unidirectional carbon fibre composites developed for wind turbine applications. Int J Fatigue 28:1389–1398

Chen SC, Minh PS, Chang JA, Huang SW, Huang CH (2012) Mold temperature control using high-frequency proximity effect induced heating. Int Commun Heat Mass Transfer 39: 216–223

Dhakal HN, Zhang ZY, Guthrie R, Macmullen J, Bennett N (2013) Development of flax/carbon fibre hybrid composites for enhanced properties. Carbohydr Polym 96:1–8

Folkes MJ, Russell DAM (1980) Orientation effects during the flow of short-fibre reinforced thermoplastics. Polymer 21(11):1252–1258

Garancher JP, Fernyhough A (2014) Expansion and dimensional stability of semi-crystalline polylactic acid foams. Polym Degrad Stab 100:21–28

Hagstrand PO, Bonjour F, Månson JAE (2005) The influence of void content on the structural flexural performance of unidirectional glass fibre reinforced polypropylene composites. Compos A Appl Sci Manuf 36:705–714

Huda MS, Mohanty AK, Drzal LT, Schut E, Misra M (2005) "Green" composites from recycled cellulose and poly(lactic acid): physico-mechanical and morphological properties evaluation. J Mater Sci 40:4221–4229

Huda MS, Drzal LT, Misra M, Mohanty AK (2006a) Wood-fiber-reinforced poly(lactic acid) composites: evaluation of the physicomechanical and morphological properties. J Appl Polym Sci 102:4856–4869

Huda MS, Drzal LT, Mohanty AK, Misra M (2006b) Chopped glass and recycled newspaper as reinforcement fibers in injection molded poly(lactic acid) (PLA) composites: a comparative study. Compos Sci Technol 66:1813–1824

Keener TJ, Stuart RK, Brown TK (2004) Maleated coupling agents for natural fibre composites. Compos A Appl Sci Manuf 35:357–362

Kelly AL, Gough T, Whiteside BR, Coates PD (2009) High shear strain rate rheometry of polymer melts. J Appl Polym Sci 114:864–873

Kim CH, Kim S, Oh H, Youn JR (2007) Measurement of residual stresses in injection molded polymeric part by applying layer-removal and incremental hole-drilling methods. Fibers Polym 8:443–446

Kim SY, Kim CH, Kim SH, Oh HJ, Youn JR (2009) Measurement of residual stresses in film insert molded parts with complex geometry. Polym Test 28:500–507

Lambert J, Chambers AR, Sinclair I, Spearing SM (2012) 3D damage characterisation and the role of voids in the fatigue of wind turbine blade materials. Compos Sci Technol 72:337–343

Laurenzi S, Grilli A, Pinna M, Nicola FD, Cattaneo G, Marchetti M (2014) Process simulation for a large composite aeronautic beam by resin transfer molding. Compos B Eng 57:47–55
Lee KS, Lee SW, Youn JR, Kang TJ, Chung K (2001) Confocal microscopy measurement of the fiber orientation in short fiber reinforced plastics. Fibers Polym 2:41–50
Lee DH, Lee WI, Kang MK (2006) Analysis and minimization of void formation during resin transfer molding process. Compos Sci Technol 66:3281–3289
Matuana LM, Balatinecz JJ, Sodhi RNS, Park CB (2001) Surface characterization of esterified cellulosic fibers by XPS and FTIR spectroscopy. Wood Sci Technol 35:191–201
Memon A, Nakai A (2013a) Fabrication and mechanical properties of jute spun yarn/PLA unidirection composite by compression molding. Energy Procedia 34:830–838
Memon A, Nakai A (2013b) Mechanical properties of jute spun yarn/PLA tubular braided composite by pultrusion molding. Energy Procedia 34:818–829
Nechwatal A, Mieck KP, Reußmann T (2003) Developments in the characterization of natural fibre properties and in the use of natural fibres for composites. Compos Sci Technol 63:1273–1279
Park CH, Lebel A, Saouab A, Bréard J, Lee WI (2011) Modeling and simulation of voids and saturation in liquid composite molding processes. Compos A Appl Sci Manuf 42:658–668
Park SH, Lee SG, Kim SH (2013) Isothermal crystallization behavior and mechanical properties of polylactide/carbon nanotube nanocomposites. Compos A Appl Sci Manuf 46:11–18
Planellas M, Sacristán M, Rey L, Olmo C, Aymamí J, Casas MT, del Valle LJ, Franco L, Puiggalí J (2014) Micro-molding with ultrasonic vibration energy: new method to disperse nanoclays in polymer matrices. Ultrason Sonochem 21:1557–1569
Pujadas P, Blanco A, Cavalaro S, Fuente ADL, Aguado A (2014) Fibre distribution in macro-plastic fibre reinforced concrete slab-panels. Construct Build Mater 64:496–503
Rassmann S, Reid RG, Paskaramoorthy R (2010) Effects of processing conditions on the mechanical and water absorption properties of resin transfer moulded kenaf fibre reinforced polyester composite laminates. Compos A Appl Sci Manuf 41:1612–1619
Saxena RC, Adhikari DK, Goyal HB (2009) Biomass-based energy fuel through biochemical routes: a review. Renew Sustain Energy Rev 13:167–178
Serizawa S, Inoue K, Iji M (2006) Kenaf-fiber-reinforced poly(lactic acid) used for electronic products. J Appl Polym Sci 100:618–624
Shaaban A, Se SM, Mitan NMM, Dimin MF (2013) Characterization of biochar derived from rubber wood sawdust through slow pyrolysis on surface porosities and functional groups. Procedia Eng 68:365–371
Shabanian M, Kang NJ, Wang DY, Wagenknecht U, Heinrich G (2013) Synthesis of aromatic–aliphatic polyamide acting as adjuvant in polylactic acid (PLA)/ammonium polyphosphate (APP) system. Polym Degrad Stab 98:1036–1042
Shokrieh M (2014) Residual stresses in composite materials, 1st edn. Woodhead Publishing Limited, Cambridge
Silva FJG, Ferreira F, Ribeiro MCS, Castro ACM, Castro MRA, Dinis ML, Fiúza A (2014) Optimising the energy consumption on pultrusion process. Compos B Eng 57:13–20
Thakur VK, Thakur MK (2014) Processing and characterization of natural cellulose fibers/thermoset polymer composites. Carbohydr Polym 109:102–117
Vera-Sorroche J, et al (2014) The effect of melt viscosity on thermal efficiency for single screw extrusion of HDPE. Chemical Engineering Research and Design 92(11):2404–2412
White JR (1985) On the layer removal analysis of residual stress. J Mater Sci 20:2377–2387
Yan L, Chouw N, Jayaraman K (2014) Flax fibre and its composites – a review. Compos B Eng 56:296–317
Yun MS, Lee WI (2008) Analysis of bubble nucleation and growth in the pultrusion process of phenolic foam composites. Compos Sci Technol 68:202–208
Zhang X, Wub X, Haryonob H, Xia K (2014) Natural polymer biocomposites produced from processing raw wood flour by severe shear deformation. Carbohydr Polym 113:46–52

Chapter 7
Challenges in Machining of Natural Fibre Composites

Piyush P. Gohil, Vijaykumar Chaudhary, and Kundan Patel

Abstract The attention in natural fibre-reinforced polymer composites is growing rapidly for many engineering systems. However, their assorted nature, engineer's lack of experience, a little knowledge of machinability databases, parameter setting and difficulty in manufacturing are barriers to large-scale use of composites. Various conventional and unconventional techniques are used for machining composites to obtain the final shape. Machining of these composites creates challenges that do not usually come across in machining of conventional metallic/monolithic material as the composite materials are heterogeneous and anisotropic. Several manufacturing and machining defects include matrix imperfection, resin-starved area, resin-rich area, voids, cracks, blisters, debond, delamination, fibre pull-out and burning. Different methodologies and tools are aimed to prevail over the above-mentioned issues. This chapter focuses on the prospective use of natural fibres in composite materials and challenges in machining of natural fibre composites.

Keywords Natural fibre composites • Mechanical and physical properties • Damages • Machining

7.1 Composite Material

A mixture of more than two materials, i.e. reinforcement and resin (Schwartz 1992), which differs in structure or composition on a macroscale, is known as composite material. This structure results in a material that increases specific performance properties. The components do not disband completely and as a result normally show an interface between one another. Both reinforcing and resin keep their

P.P. Gohil (✉)
Faculty of Technology and Engineering, Department of Mechanical Engineering, The M S University of Baroda, Vadodara 390 001, Gujarat, India
e-mail: push4679@yahoo.com; piyush.p.gohil-me@msubaroda.ac.in

V. Chaudhary • K. Patel
Faculty of Technology and Engineering, Mechanical Engineering Department, CHARUSAT, Changa 388 421, Gujarat, India

M.S. Salit et al. (eds.), *Manufacturing of Natural Fibre Reinforced Polymer Composites*, DOI 10.1007/978-3-319-07944-8_7

physical and chemical identities; however, constituents make a combination of properties that cannot be attained with either of the constituent acting alone.

Composite materials are widely used in several engineering areas due to their better properties like:

- Composite materials are lighter than woods and metals which are vital for good mileage in aircrafts and automobiles.
- Nowadays, in structural applications, composite materials are having greatest strength to weight ratio. One can design the composite strong as well as light (e.g. bamboo-based composites).
- Composites can be used to store chemicals as they resist corrosion from the weather and from harsh chemicals.
- Composites are flexible to the design as they can be moulded into challenging shapes very easily than other materials.
- The whole assembly of the metal parts can be replaced by making a single piece of composite material.
- Many of the composites are nonconductive and nonmagnetic, so they can be used in electrical/electronics applications.

7.2 Importance of Natural Fibre Composites

There are many advantages of natural fibre over synthetic fibres. E-glass fibres have equivalent or lesser specific modulus and specific strength than natural fibres. In the future, many opportunities are there for replacing E-glass fibres by natural fibres (Drzal et al. 2004).

The production of natural fibres requires less energy and CO_2 is used, whereas oxygen is released to the atmosphere. In the case of natural fibres, there is possibility of thermal recycling, while glass fibres cause difficulties in burning. Natural fibres have worthy thermal and acoustic insulating properties. There are many other advantages like no skin irritations, friendly processing and lower cost.

7.3 Challenges in Preparation of Natural Fibre Composite Materials

Natural fibres absorb definite amount of water which degrade the fibres. The water absorption may be high because the hydroxyl is present all the way through the natural fibre. It signs to reduced wettability and fragile interfacial bond amongst fibres and polymers, by way of matrix (Brouwer 2000). To develop composite with good mechanical properties, it is essential to perform appropriate treatments for improving the compatibility of fibres with matrices (Brouwer 2000; Drzal et al. 2004).

The characteristics of constituents of natural fibres are such that they could not be taken equal to temperature beyond 220–230 °C deprived of concerning degradation.

So, composites based on natural fibres could not be operated at temperature beyond 230 °C.

The problem encountered with the natural fibres is the availability in required form. The lower cost of the natural fibres can be annulled out if they should be conveyed to long distances. The nature of the natural fibre is very complex, and it depends not only on the type of fibre but also on how and where the plants are cultivated. Natural fibre composites always contain porosity which decreases density and mechanical performance of composites.

7.4 Machining of Composites

Composites are created to near-net shape which minimizes the need for machining, though machining processes like trimming for giving finished shape and drilling to facilitate joining of parts in intricate assemblies are common operations in industries that deal with these kinds of materials. Amongst these techniques, turning, milling, grinding and drilling are examples that can be used for machining composites to obtain the final shape or achieve required tolerance. Figure 7.1 shows the various machining methods.

7.4.1 Conventional Machining

Machining processes are conventionally transferred to composite machining after appropriate modifications to process parameters like tool geometry, cutting speeds and feed rates. However, kinematics of the machining process for composites and metals remains the same.

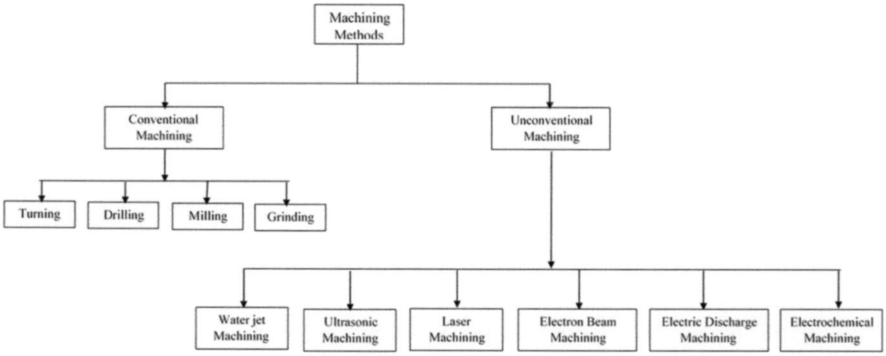

Fig. 7.1 Machining methods

7.4.1.1 Turning

Turning of fibre-reinforced composite materials with HSS or carbide tools has been found to be not cost-effective due to the abrasiveness and inhomogeneity of these materials. Machined parts have poor surface finish and high tool wear (Ramulu 1998; Ramulu et al. 1989). In composites reinforced with carbon fibres, the use of high-speed tools is rare, due to the great abrasiveness of these fibres. Carbide tools are more normally used. Ceramic tooling, due to its low thermal conductivity, is not adequate for this type of materials. Ramulu (1998) investigated machining by turning graphite/epoxy composite material with the purpose of comparing different PCD (synthetic polycrystalline diamond) grain sizes and the identification of dominant parameters in the tool edge geometry. Graphite fibre-reinforced materials are extremely abrasive, leading to the need to use diamond abrasive cutters. The finish of machined surface depends largely on fibre orientation, but roughness values show some variation according to the orientation of the fibres to the cutting edge.

7.4.1.2 Milling

The main characteristic of milling operations on parts made of fibre composite is a low ratio of removed volume to total part volume. Milling is used as a corrective operation in order to produce well-defined and quality surfaces. The fibre has the greater influence in the selection of tools and parameters (Klocke et al. 1998). Optimization of milling process is influenced on one side by the nature of the fibres which must be cut in and near the top layers and on the other side by the thermal stress on the contact surface. The latter is caused by material abrasiveness and the friction between the material and tool, affecting the quality of the cut surface (Klocke and Wurtz 1998a, b). The combination of high cutting speeds with low feeds can lead to matrix melting and stuck to laminate surface or to the chips. Under some conditions, the matrix could even burn. A significant reduction in cutting velocity and an increase in feed can limit this problem. However, these cutting parameters lead to the presence of inadequately cut fibres that remain attached to the surface layers (Klocke and Wurtz 1998a, b). Thermal stress on tools becomes critical when small-diameter tooling is used. Fibre-reinforced composites are characterized by their low thermal conductivity, so the tool has to absorb a considerable amount of generated heat, in opposition to what happens to steels. If the capacity of the tool to absorb the heat is also low, an increase in friction leads to higher thermal stress, more tool wear and poorer operational safety (Klocke and Wurtz 1998a, b).

7.4.1.3 Grinding

Grinding of carbon fibre-reinforced plastics poses considerable problems such as fibre pull-out, delamination, burrs and burning (Hu and Zhang 2001). Surface roughness is a function of fibre orientation, with better results when they are perpendicular to grinding direction. Greater depths of grinding cause more damage to workpiece. When compared with drilling, grinding causes less burrs and edge effect problems.

7.4.1.4 Drilling

In maximum circumstances composite parts need to be assembled with other parts. Composite materials could not be assembled by welding and gluing is somewhat difficult; therefore, mechanical joining by means of fasteners and rivets is the option usually taken on to join composite parts with other parts. The holes required by mechanical joining are generally drilled in the semi-finished composite part. Only bigger-size diameter or intricate contour holes can be created by placing core in the mould while curing. As the composite part must perfectly match with other parts during the joining process, the holes must be located at the exact required position and should have precise diameter. Additionally, because of load transfer when in use, the holes usually undertake concentrated localized stress. Drilling includes the removal of material from a workpiece such that a hole is attained. The holes created are used mainly for joining one component with another, to provide passage for coolants, and for wiring purposes. It is a common process for removing unwanted material. Drilling is generally used to create holes in metals, but because of its availability, it is now used for composites too. Drilling is extensively used as it is the most economical process than other methods and because there are not many other methods that can create deep circular hole. Drilling is repeatedly used for machining composites, because of freely existing machinery. Composites are anisotropic materials, so drilling increases particular problems that will affect the strength of the parts (Hamdoun et al. 2004).

7.4.2 Unconventional Machining

Conventional machining might cause impractical operating conditions because of lower material removal rates (MRRs), tool wear and undesirable responses. Hence, unconventional machining processes may become practical and cost-effective methods for composite machining.

7.4.2.1 Waterjet Machining

To generate thin waterjet by means of high pressure and high velocity is the working principle of waterjet machining. Material removal from workpiece is possible by localized fracture. The performance can be improved by adding abrasive particles. Various methods like kerf cutting, piercing and milling can be used to create holes and slots using waterjet. In piercing, material from the workpiece is removed by means of shock loading (Friend et al. 1973; Adams 1986). The analytical models available for predicting delamination using conventional drilling can be used to determine the required pressure for introducing the delamination (Hocheng 1990). Delamination could be reduced by using lower jet speed, but it affects the piercing capabilities (Hashish 1989). Cutting beside a circle can create better quality holes deprived of delamination. This method can be efficiently used for cutting narrow slits, small holes and non-through holes. The main advantage of using this method is no burning of fibre and no environmental problem, and the main disadvantage is generation of great noise. If the internal pressure generates during cutting, then it will cause layer delamination for composite materials.

7.4.2.2 Ultrasonic Machining

Ultrasonic machining is the method in which a slurry containing abrasives is driven at high velocity beside the workpiece by a vibrating tool. Studies are available on the effect of various process parameters like abrasive concentration, abrasive grain size and feed rate on machining ability with respect to material removal rate (Pentland and Ektermanis 1965; Kremer 1981). The main advantage of ultrasonic machining is that it does not generate thermal or chemical damage on workpiece. With the use of this method, good quality hole can be created in composites, but the disadvantage is that it is a very slow process as it takes time to create a hole with slower speed. Ultrasonic machining can create good quality hole than conventional drilling. This method is appropriate for drilling and slitting some of the composites.

7.4.2.3 Laser Machining

Laser machining uses the light energy from a laser to remove material by vaporization and ablation. Laser beam machining has previously been used in industrial applications for several times; for cutting composites, the general types of lasers used are Nd:YAG laser and CO_2 laser. Nd:YAG laser can be used to cut metal matrix composites which do not contain organic resin because these materials do not absorb laser and decompose easily (Sadat 1991). CO_2 laser can be used for cutting glass fibre-reinforced composites. It is also used for cutting graphite/epoxy composites, but as graphite is a conductive material and temperature is about 3,600 °C, it causes degradation. Laser machining of composites is difficult because the components of composites have varied thermal conductivity (Tagliaferri et al. 1985). Many experiments indicate that good quality of cut surface can be attained, if thermal conductivity of fibres is nearer to that of the matrix. Pan and Hocheng (1998) showed that the heat-affected zone amongst cut could be anticipated by the model of anisotropic heat conduction. Tuersley et al. (1998) studied the effect on the depth of the hole and damage induced in silicon carbide-based composites for various process parameters like gas pressure, focal depth and speed. Rodden et al. (2002) carried out laser drilling on carbon fibre-reinforced composites using long pulse Nd:YAG laser.

7.4.2.4 Electrical Discharge Machining

The process of electric discharge machining is based on the eroding effect of electric spark which is created between electrode (tool) and workpiece in dielectric fluid (Tsao and Hocheng 2004). Electric discharge machining is used for the material which has electrical conductivity, so this method can be used for graphite fibre-based composites (Lau et al. 1990). It is also used for machining metal matrix composites. The main process parameters in this process are current and frequency which affect the surface finish. Better surface finish can be achieved by using higher frequency and lower current. Lau et al. (1990) studied about tool wear in machining of carbon fibre-reinforced composites and showed that copper electrode gives better performance than graphite.

7.4.2.5 Electron Beam Machining

In electron beam machining the electron impinge on the surface of workpiece in vacuumed which generates heat and material vaporizes (Sadat 1991). Vacuum is necessary to eliminate collisions of the electrons with gas molecules. This process is considered to be micromachining. The main advantage of using this method is there is no heat-affected zone and closed tolerance could be achieved. This method is generally used for creating small holes and for slitting some composites.

7.4.2.6 Electrochemical Machining

Electrochemical machining removes material from an electrically conductive workpiece by anodic dissolution. It is the reverse of electroplating. The workpiece is the anode and the tool is the cathode. Material is deplated form the anode (positive pole) and deposited onto the cathode (negative pole) in the presence of an electrolyte bath (Groover 1996). The electrolyte is usually sodium chloride mixed with water, sodium nitrate and other fluids that can chemically react with the workpiece. This process can be used for machining complex cavities and for slitting, drilling and cutting most composites that exhibit continuous and uniform electrical conductivity. One advantage of this process is that it does not cause any thermal damage.

7.5 Research Review on Machining of Natural Fibre Composites

Many researchers worked on machining of polymer matrix composites based on synthetic fibres. Applications of composites based on natural fibres are increasing nowadays. Few of the case studies on machining of natural fibre composites are discussed here.

Jayabal and Natarajan (2010) developed coir–polyester composite and investigated the effect of drill diameter, feed and speed on torque, thrust force and tool wear. Experiments were performed with the help of box–cox method on coir–polyester composites. Various optimization techniques like genetic algorithm and Nelder–Mead technique to optimize parameters and reduce tool wear. Mathematical model for responses was developed to compare significant input parameters. They concluded that if there is an increase in tool wear, then tool life becomes negligible.

Athijayamani et al. (2010) used HSS drill bits to conduct a series of drilling experiments on roselle–sisal hybrid composite. They used artificial neural network and regression model for comparing torque and thrust with speed, feed and tool diameter. They compared experimental values with predicted response values, artificial neural network and regression model. They showed that the artificial neural network model is more valued than the regression model to predict torque and thrust force in the drilling of natural fibre hybrid composites.

Chandramohan and Marimuthu (2011) studied the prediction of torque and thrust force for the drilling of commercially available natural fibre like sisal and roselle–bio epoxy resin composite materials, and the values were compared with the regression model. They used machine vision system to find delamination factor. They found that larger diameter drills and higher feed rates give larger thrust force. They also observed that as volume fraction and feed increase, the thrust force increases, but thrust force was decreased as cutting speed increases at elevated feed values. They also found that with increase in feed and speed, torque increases slightly, but it decreases when increasing the cutting speed.

Jayabal and Natarajan (2011) studied the mechanical and machinability characteristics of coir–polyester composites. They developed regression equations and optimized for studying drilling characteristics using the Taguchi approach. Drill diameter, feed rate and speed were the input parameters to examine the torque, thrust force and tool wear in the drilling process with the greatest hole quality and accuracy. They found the optimum values of machining parameters to reduce tool wear for the drilling of coir–fibre-reinforced composites.

Babu et al. (2012) studied the effect on delamination of hemp fibre-reinforced composite of drilling parameters. They used the Taguchi technique to design the experiments and ANOVA analysis to achieve the conditions for minimum delamination. They examined both the peel-up delamination and push-down delamination. They found that the most critical parameters are the feed rate and cutting speed and should be selected suspiciously to reduce all kinds of damages.

Rakesh et al. (2012) did a survey of the literature published in the last 20 years in this area and found that around 31 % of total papers published are in the last 3 years. Therefore, it can be quantified with sureness that the area of drilling of polymer matrix composites has created wide interest amongst the research group.

Venkateshwaran and ElayaPerumal (2013)studied analysis of delamination at the entrance and exit of the composite plates as a function of the drilling process parameters. By varying the feed and speed, they carried out the drilling on banana fibre epoxy composite. They used machine vision technique to analyse the quality of the hole and also the ultrasonic C-scan imaging method was carried out to find the amount of delamination. They found that with an increase in feed, the delamination increases. They compared experimentally obtained delamination factor with ANOVA technique and observed that the effect of feed rate is more on delamination than speed.

Bajpai and Singh (2013) studied about drilling performance of sisal–polypropylene composite material. They took speed, feed and drill point geometry as input parameters. They used solid and hollow shaped drill geometries for drilling. They used the RSM and ANOVA method to design the experiments for studying the effect of cutting forces and parameters. They found that the cutting forces are considerably affected by the selection of the drill point geometry as the cutting mechanism varies with various drills. They observed that drill point geometry has a significant effect and cutting speed has a minimum effect on the forces during drilling of sisal–polypropylene composites. They concluded that thrust force produced using trepanning tool is significantly small equaled to twist drill through drilling operation and the cutting mechanism of hollow drill is suitable for the drilling of sisal fibre-reinforced PP composites.

7.6 Challenges Observed

Damage in composite materials is caused by several ways during manufacture, assembly and transport or field deployment, besides expected damages during normal use. Composites are used in some critical applications; damages that remain undetected or that are of difficult detection can turn into a serious problem (Strong 1991). One of the possible solutions is to increase the design safety factor, overdesigning the parts, with a penalty in weight and final cost. However, this solution does not prevent damage nor considers its possible extent, which may turn out to be wider than predicted exceeding the safety factor allowance. Unexpected ruptures during service must be avoided. Therefore, the solution most often used is to implement special damage protection methods (Strong 1991). There are a variety of defects that can be caused in composite parts during machining. These operations are normally included in the assembly phase, but one should remind that some damage can be caused prior to assembly, i.e. during parts manufacturing.

Physical properties of composites make them an attractive group of materials; machining poses a certain number of specific issues, when compared with metallic materials. Usually, machined components have poor surface appearance and higher tool wear. The major problem of machining composites is fibre reinforcement which is generally abrasive and causes tool wear and deterioration of machined surfaces (Abrate 1997). Machining could cause some damages like delamination, fibre pull-out and thermal damages (Wern et al. 1994). In drilling of composites, better results depend more on fibre than matrix (Boldt and Chanani 1987).

Manufacturing and machining defects include matrix imperfection, resin-starved area, resin-rich area, voids, cracks, blisters, debond and delamination. If too much resin is used in a part, it is called resin-rich area; for nonstructural applications, this

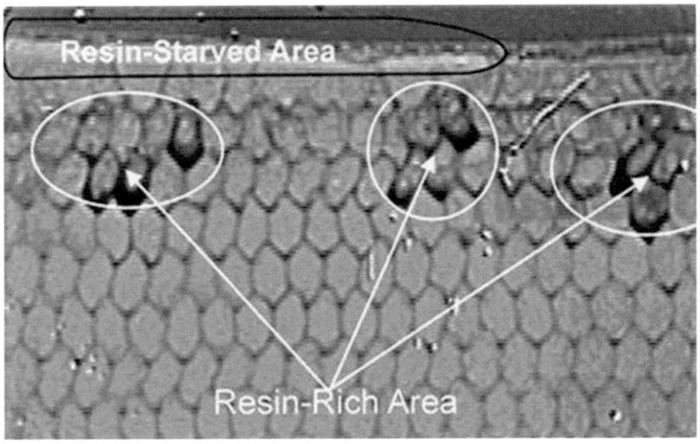

Fig. 7.2 Resin-starved areas and resin-rich areas (Balaskó et al. 2004)

is not necessarily bad, but it adds weight. Apart is called resin starved if too much resin is exploited through the curing or if not enough resin is used during the wet layup process. Resin-starved areas are indicated by fibres that show to the surface. Resin-rich areas and resin-starved areas are shown in Fig. 7.2. The ratio of 60:40 fibre to resin ratio is considered optimum. Matrix imperfections generally arise at the matrix–fibre interface. These imperfections can somewhat decrease certain material properties but are rarely dangerous to the structure, except the matrix degrades extensively.

Damages that are generally initiated by machining composites are burrs, debonding, intralaminar cracks, delamination and thermal damage. The amount of these defects is mainly reliant on selected process parameters. Care should also be paid for circularity of holes after machining. Burrs are small portions of broken material at the corner or at the edge of the surface, but quite attached to a part. Their presence is associated with roundness which does not allow clean cut. Their significance is mostly artistic.

Intralaminar cracking which is shown in Fig. 7.3 begins at the inner plies of the laminate. Generally, it starts at a preference of 60° to the layer plane. The crack spreads along with direction till it extends at the intralaminar plane and turns into delamination (Caprino and Tagliaferri 1995).

Debonds form because there may be no adhesion beside bond line amongst two components and generate delamination in neighbouring laminate layers shown in Fig. 7.4. Under several circumstances debond can raise when it is subjected to frequent loading which causes cataclysmic failure. The acute of debond depends on various parameters like dimension, location in laminate, type of loading and quantity of delamination at specified locality.

Thermal damage occurs because of contact amongst workpiece and tool which generates localized heat, and in machining of composites cooling fluid cannot be used. Higher temperature can cause damage like burning or melting.

Amongst the several defects which are caused by machining, delamination is acknowledged as the most dangerous. Delamination is the separation of layers in

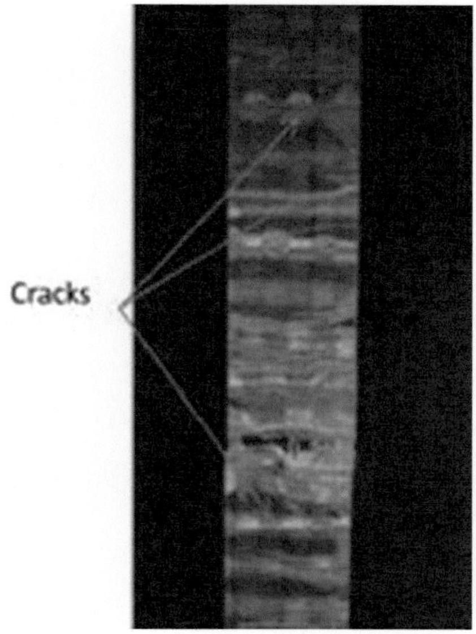

Fig. 7.3 Intralaminar crack (Rios-Soberanis et al. 2012)

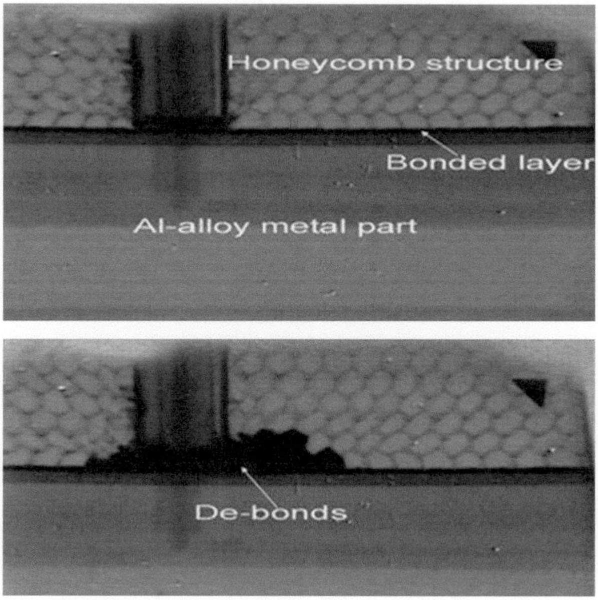

Fig. 7.4 Debonds (Balaskó et al. 2004)

laminate and fibre pull-out which is shown in Fig. 7.5. Delamination factor can be defined by the ratio of damaged area around the hole to actual hole area which is shown in Fig. 7.6. Composite delamination can be divided into two types: peel-up delamination and push-down delamination. If the delamination arises at the

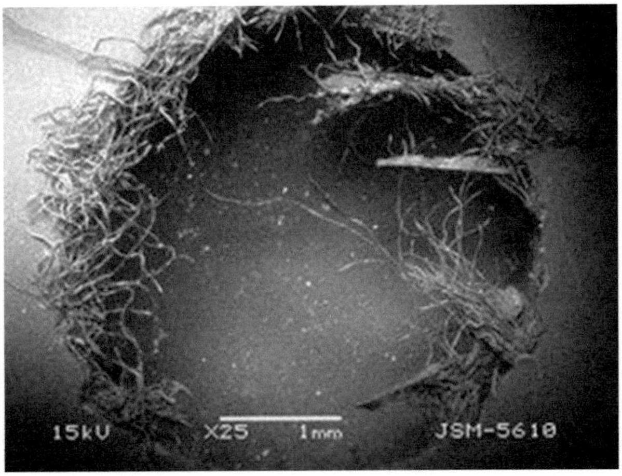

Fig. 7.5 Delamination in drilled hole

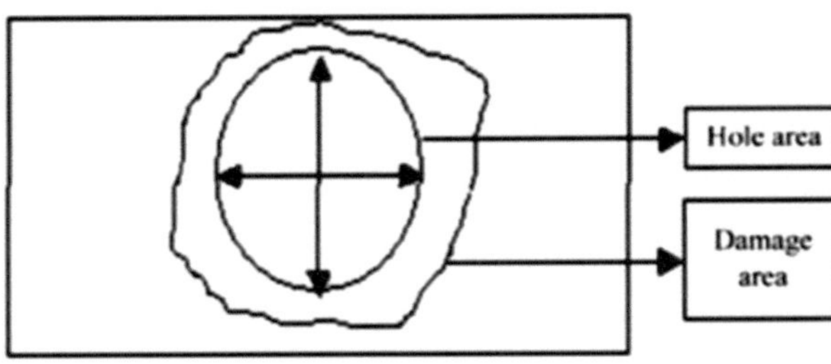

Fig. 7.6 Schematic layout of the damage area and hole area

entrance, then it is called peel-up delamination, and if it occurs at the exit, then it is called push-down delamination which is shown in Fig. 7.7.

Delamination affects the durability of composites and decreases bearing strength which results in performance problems. Other main reasons for delamination are higher thrust force, high feed rate and prompt tool wear. Certain earlier approaches to reduce delamination were to decrease feed rate and thus decrease thrust force with the use of backing plate.

In conventional drilling of composites, circularity shall also be observed, as there is a bouncing-back tendency of the material that causes hole deformation. The return to its initial position causes tightening around the drill, and the drilled diameter is less than the drill diameter. This roundness error is due to the anisotropy of the material (Piquet et al. 2000).These defects are responsible for the rejection of producing parts and contribute to the rise of fabrication costs.

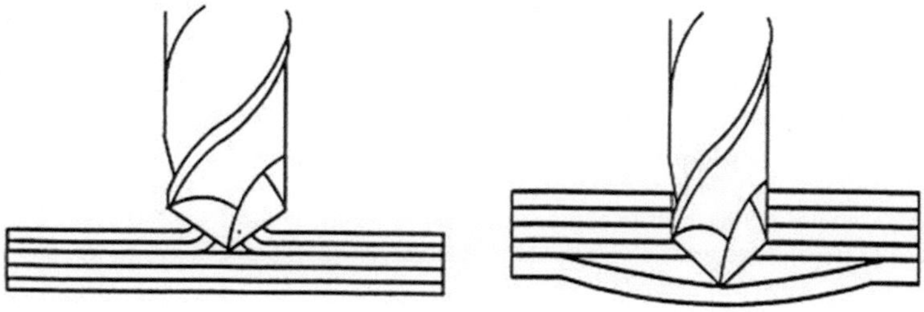

Fig. 7.7 Peel-up delamination at entrance and push-out delamination at exit (Abrate 1997)

7.7 Remarks

The present work was aimed to examine the manufacturing and machining challenges for natural fibre composites. This study also shows the defects observed during manufacturing as well as machining of natural fibre composites. Reduction of defects can be achieved by selecting appropriate cutting parameters, drilling conditions and tool geometry and tool material. It is identified that few of the data are available for machining of natural fibre composites; however, there is a paucity of data for unconventional machining of natural fibre composites. Some important research issues are identified and studies of some critical issues are proposed to meet the challenges in machining of natural fibre composites.

References

Abrate S (1997) Machining of composite materials. In: Mallick PK (ed) Composites engineering handbook. Marcel Dekker, New York, pp 777–809

Adams RB (1986) Water jet machining of composites, composites in manufacturing. SME Paper EM 86-113

Athijayamani A, Natarajan U, Thiruchitrambalam M (2010) Prediction and comparison of thrust force and torque in drilling of natural fibre hybrid composite using regression and artificial neural network modelling. Int J Mach Mach Mater 8:131–145

Babu D, SivajiBabu K, Uma B (2012) Effects of drilling parameters on delamination of hemp fiber reinforced composites. Int J Mech Eng Res Dev 2(1):01–08

Bajpai P, Singh I (2013) Drilling behavior of sisal fiber-reinforced polypropylene composite laminates. J Reinf Plast Compos 32(20):1569–1576

Balaskó M, Veres I, Molnár G, Balaskó Z, Sváb E (2004) Composite structure of helicopter rotor blades studied by neutron- and X-ray radiography. Physica B Condens Matter 350:107–109

Boldt JA, Chanani JP (1987) Solid-tool machining and drilling. Engineered materials handbook, vol 1. ASM International Handbook Committee, Metals Park, pp 667–672

Brouwer W (2000) Natural fibre composites in structural components: alternative application for sisal? Seminar, Commond fund for commodities- alternative applications for sisal and henecuen. Food and Agriculture Organization of the UN (FAO) and the Common Funds for Commodities (CFC), Rome

Caprino G, Tagliaferri V (1995) Damage development in drilling glass fiber reinforced plastics. Int J Mach Tools Manuf 35:817–829

Chandramohan D, Marimuthu K (2011) Drilling of natural fiber particle reinforced polymer composite material. Int J Adv Eng Res Stud 1(1):134–145

Drzal L, Mohanty D, Burgueno R, Misra M (2004) Biobased structural composite materials for housing and infrastructure applications: opportunities and challenges. Proceedings of the NSF housing research agenda workshop, vol 2, pp 129–140

Friend CA, Clyne RW, Valentine GG (1973) Machining graphite composite materials. In: Noton BR (ed) Composite materials in engineering design. ASM, Washington, pp 217–224

Groover MP (1996) Fundamentals of modern manufacturing: materials, processes, and systems. Prentice Hall, New York, pp 543–611

Hamdoun Z, Guillaumat L, Lataillade JL (2004) Influence of the drilling on the fatigue behavior of carbon epoxy laminates. ECCM 11, Rhodes

Hashish M (1989) Machining of advanced composites with abrasive water jets. Manuf Rev 2(2):142–160

Hocheng H (1990) A failure analysis of water jet drilling in composite laminates. Int J Mach Tools Manuf 30(3):423–429

Hu N, Zhang L (2001) Grind ability of unidirectional carbon fiber-reinforced plastics. ICCM-13, Beijing

Jayabal S, Natarajan U (2010) Optimization of thrust force, torque, and tool wear in drilling of coir fiber-reinforced composites using Nelder–Mead and genetic algorithm methods. Int J Adv Manuf Technol 51:371–381

Jayabal S, Natarajan U (2011) Drilling analysis of coir–fibre-reinforced polyester composites, Indian Academy of Sciences. Bull Mater Sci 34(7):1563–1567

Klocke F, Koenig W, Rummenhoeller S, Wuertz C (1998) Milling of advanced composites, machining of ceramics and composites. Ed Marcel Dekker, New York, pp 249–266

Klocke F, Wurtz C (1998a) Comparison of techniques for the machining of thermoplastic fiber reinforced plastics. ECCM-8, Naples, vol 2, pp 729–735

Klocke F, Wurtz C (1998b) The use of PCD tools for machining fiber reinforced Materials. ECCM-8, Naples, vol 2, pp 509–515

Kremer D (1981) The state of the art of ultrasonic machining. Ann CIRP 30(1):107–110

Lau WS, Wang M, Lee WB (1990) Electrical discharge machining of carbon fiber composite materials. Int J Mach Tools Manuf 30(2):297–308

Pan CT, Hocheng H (1998) Prediction of extent of heat affected zone in laser grooving of unidirectionally fiber-reinforced plastics. J Eng Mater Technol 120:321–327

Pentland EW, Ektermanis JA (1965) Improving ultrasonic machining rates—some feasibility studies. J Eng Ind-T ASME 87:39–46

Piquet R, Ferret B, Lachaud F, Swider P (2000) Experimental analysis of drilling damage in thin carbon/epoxy plate using special drills. Compos Part A 31:1107–1115

Rakesh PK, Singh I, Kumar D (2012) Drilling of composite laminates with solid and hollow drill point geometries. J Compos Mater 46(25):3173–3180

Ramulu M (1998) Cutting-edge wear of polycrystalline diamond inserts in machining of fibrous composite material, machining of ceramics and composites. Ed. Marcel Dekker, New York, pp 357–410

Ramulu M, Faridnia M, Garbini JL, Jorgensen JE (1989) Machining of graphite/epoxy materials with polycrystalline diamond tools, machining characteristics of advanced materials. Winter annual meeting of ASME, pp 33–39

Rios-Soberanis CR, Cruz-Estrada RH, Rodriguez-Laviada J, Perez-Pacheco E (2012) Study of mechanical behavior of textile reinforced composite materials. Dyna 176:115–123

Rodden WSO, Kudesia SS, Hand DP, Jones DC (2002) A comprehensive study of the long pulse Nd:YAG laser drilling of multi-layer carbon fibre composites. Opt Commun 210(3–6):319–328

Sadat AB (1991) Machining of composites. Int Encycl Comp 3:95–102

Schwartz MM (1992) Composite materials handbook, 2nd edn. Mc Graw Hill Inc., USA, pp 34–35

Strong AB (1991) Damage control. Int Encycl Compos 2:1–6

Tagliaferri V, DiIlio A, Crivelli Visconti I (1985) Laser cutting of fiber reinforced polyesters. Compos 16(4):317–325
Tsao CC, Hocheng H (2004) Analysis of delamination in drilling composite materials using core-center drill. 11th ICCE, Hilton Head
Tuersley IP, Hoult TP, Pashby IR (1998) Nd-YAG laser machining of SiC fibre/borosilicate glass composites. Part II—the effect of process variables. Compos Part A 29(8):955–964
Venkateshwaran N, ElayaPerumal A (2013) Hole quality evaluation of natural fiber composite using image analysis technique. J Reinf Plast Compos 32(16):1188–1197
Wern CW, Ramulu M, Shukla A (1994) Investigation of stresses in the orthogonal cutting of fibre-reinforced plastics. Exp Mech 36:33–41

Chapter 8
Yarn Flax Fibres for Polymer-Coated Sutures and Hand Layup Polymer Composite Laminates

T.C. Fong, N. Saba, C.K. Liew, R. De Silva, M. Enamul Hoque, and K.L. Goh

Abstract Flax (*Linum usitatissimum*) is a blue-flowered herbaceous plant that widely grows in temperate zones as a source of both fibres and oil. While it is well known that flax fibres have been used for making linen, from a biomedical perspective, probably one of the more important applications of flax fibres is for making surgical sutures. In addition to employing flax fibres for polymer-coated sutures, more recently, flax fibres have been proposed for reinforcing polymer composites, such as hand layup laminated scaffolds in tissue engineering. This chapter explores the mechanics of flax fibres based on recent findings from studies conducted by the authors as well as from the literature. The focus is on the effects of moisture on the mechanical properties of yarn flax fibres as well as the possible dependence on knot geometry. The findings are applied to support arguments for the design considerations of polymer-treated flax fibre for, e.g., sutures as well as hand layup polymer composite laminates.

T.C. Fong
School of Mechanical & Systems Engineering, Newcastle University,
Newcastle Upon Tyne NE1 7RU, UK

N. Saba
Laboratory of Biocomposite Technology, Institute of Tropical Forestry and Forest Products (INTROP), Universiti Putra Malaysia, 43400 Serdang, Selangor, Malaysia

C.K. Liew
Singapore Institute of Technology, 10 Dover Road, Singapore 138683, Singapore

R. De Silva
School of Engineering, Monash University Malaysia,
Jalan Lagoon Selatan, 47500 Bandar Sunway, Selangor Darul Ehsan, Malaysia

M. Enamul Hoque
Department of Biomedical Engineering, University of King Faisal, Al-Hofuf, Al-Hassa, Saudi Arabia

K.L. Goh (✉)
School of Mechanical & Systems Engineering, Newcastle University,
Newcastle Upon Tyne NE1 7RU, UK

NUInternational Singapore Pte Ltd, Singapore 569830, Singapore
e-mail: kheng-lim.goh@ncl.ac.uk

M.S. Salit et al. (eds.), *Manufacturing of Natural Fibre Reinforced Polymer Composites*, DOI 10.1007/978-3-319-07944-8_8

8.1 Introduction

Flax (Latin name: *Linum usitatissimum*) is a blue-flowered herbaceous plant that widely grows in temperate zones as a source of both fibres and seed oil. For a long time, flax fibre is well known in the textile industry, i.e., for spinning into yarn to make linen (European Confederation of Flax and Hemp 2012). More recently, there is a great interest in using flax fibres to replace glass fibres for reinforcing polymer-based materials. This is not just because the mechanical properties of the flax fibre are comparable to that of the traditional glass fibres (Moothoo et al. 2014; Shalwan and Yousif 2013); the other advantages over glass fibres are low cost, low density, comparable specific tensile properties, nonabrasive to the equipment, non-irritation to skin, reduced energy consumption, less health risk, renewability, recyclability, and biodegradability (Ku et al. 2011). Thus, flax fibres have also been proposed for biomedical applications, such as for reinforcing biocomposites for tissue engineering (Alix et al. 2008; Cheung et al. 2009), because they are, to a large extent, biocompatible with the tissues of the human body (Cheung et al. 2009; Kunert-Keil et al. 2012). For instance, fibroblasts are known to proliferate well on scaffolds made from flax fibre blended in polylactide (PLA) or polycaprolactone (PCL) as a matrix (Kunert-Keil et al. 2012). There are many studies on flax/thermoplastic composites, with polypropylene (PP) being the most suitable thermoplastic matrix for flax-reinforced composites due to its various advantages, such as low density, low thermal expansion, good resistance to water, and recyclability (Zhu et al. 2013 and therein). However, one major drawback of flax fibre-reinforced polymer composites is the incompatibility between the hydrophilic fibres and the hydrophobic polymeric matrices. Thus, should water gets into the composite, this could drastically alter the mechanical properties of the composite for what it was intended for. Water absorption in flax-reinforced PP composites occurs by diffusing directly into the matrix, through the imperfections, like pores and cracks, which could negatively affect the mechanical properties. Not only will moisture alter the mechanical properties of the fibre, but the adhesion between hydrophilic flax fibres and hydrophobic PP is of critical concern because this is important for effective stress transfer from the PP matrix to the fibre (Goh et al. 2004a, b, 2010). A simple approach to overcome this limitation is to boil the flax fibre, followed by chemical treatment to improve adhesion between the fibre and the matrix (Zhu et al. 2013).

Alternatively, flax fibres are suitable for making surgical sutures (Abdessalem et al. 2009; Greenberg and Clark 2009), having satisfied many of the key criteria for sutures, namely, tensile strength, extensibility, fatigue properties, inertness, biocompatibility, ease of sterilization, and, lastly, good knot performance during the healing process (Abdessalem et al. 2009; Greenberg and Clark 2009; Lim et al. 2011). Generally, the surgical sutures take the form of monofilament, braided, and pseudo-monofilament (Abdessalem et al. 2009; Greenberg and Clark 2009). Depending on the materials used for the sutures, sutures can be classified as absorbable or nonabsorbable (Abdessalem et al. 2009; Greenberg and Clark 2009). The absorbable sutures are intended as temporary structures for holding the wound closed but will "biodegrade" into the body with time, as the wound heals (Abdessalem et al. 2009;

Greenberg and Clark 2009). On the other hand, nonabsorbable sutures, such as those made from yarn flax fibres, have to be removed physically or left alone for an extended period of time (Abdessalem et al. 2009; Greenberg and Clark 2009). Flax fibres have to be yarn before they are further processed into a medical grade suture. They are also treated with polymer-based materials such as silicone in polyvinyl solution (B Braun Sutures Linatrix Brochure 2014); the coating on the fibre is intended to protect the fibre from damages. More importantly, the (tensile) fracture strength of linen sutures was found to increase (by 10–20 %) when the sutures are moist, suggesting that a moist suture can take up high stress while holding the wound (Cherif et al. 2013). Detailed analysis of the kinetics of water penetration into the flax fibre is reported by researchers (Cherif et al. 2013; Alix et al. 2008).

This chapter is concerned with the manufacturing of flax composites and mechanical properties of yarn flax fibres (i.e., fibre bundles) in the hydrated and dehydrated states. Here, the discussion will be based on the findings from a recent study carried out by the authors. In this study, yarn flax fibres were subjected to tensile test to rupture using an in-house built horizontal micro-tensile test rig, mounted on an inverted microscope to provide real-time microscopic observation of the process of rupture. The advantage of a horizontal test rig is that the yarn flax fibres could be hydrated continuously in a Petri dish (which formed part of the rig) during testing. The aim of this study is to add further insights to our current understanding of the underlying effects of moisture on the mechanical properties by drawing the reader to the evaluation of the results for composite design consideration. From the manufacturer's perspective, the arguments presented in the design consideration are important for ensuring effective optimization of the manufacturing process of polymer-coated flax fibres or hand layup polymer-based composites reinforced by flax fibres. We would like to add that presently, there are many studies on flax fibres, covering an extensive range of issues, from the structural properties (from the molecular level to the whole fibre), biological mechanisms (from the genetic to biochemistry), to engineering applications (from sutures to biocomposites). More importantly, these studies have provided many insights concerning the mechanisms regulating the physical properties of flax fibre; our intent is not to cover all of them here but to highlight key insights that can lead to possible connections to the underlying causes of the changes in the fibre mechanical properties due to moisture effects. Where appropriate, we would suggest to the reader to examine the relevant literature for further details.

8.2 Rupture Morphology

8.2.1 Structure of Flax Fibre

Flax fibre may be considered as a multilayered, hierarchical structural composite which may be described by a model shown in Fig. 8.1. The stem is where the fibres are located, in between the bark and xylem region (Gibson 2012). Typically, the

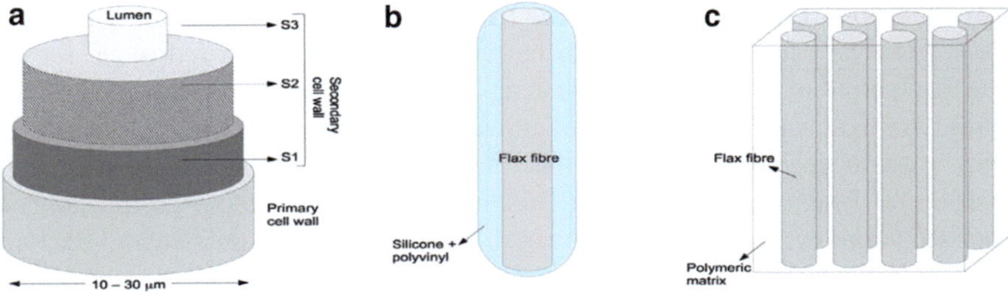

Fig. 8.1 Schematics of (**a**) flax fibres from the stem to the fibre, (**b**) a coated flax fibre arising from treatment with, e.g., silicone and polyvinyl solution (such as B Braun Sutures Linatrix Brochure 2014), (**c**) flax fibres reinforcing polymeric matrix. The diameter of flax fibres varies from 10 to 30 μm with a mean of 17 μm (Hassan et al. 2010)

hierarchical structure features 20 bundles in each stem section and 10–40 fibres in each bundle; these bundles are held together by a pectic middle lamella (Gibson 2012). Similar to the other plant fibres, there is a primary cell wall and a thicker secondary cell wall; the latter is made up of three sublayers known as S1, S2, and S3 (Gibson 2012). The cell walls comprise cellulose microfibrils (subunits of the flax fibre), spiraling about the fibre axis, which are embedded in pectic matrix (Gibson 2012). The microfibrils in the secondary cell walls are positioned at about 10° off the longitudinal axis (Gibson 2012).

Flax fibres are extracted from the stem by laying the stems in the soil so that microorganisms in the ground can break down the pectin to separate the flax bundles from the xylem and bark (Tahir et al. 2009). Thereafter, the raw fibres (which are not continuous) and shives are mechanically extracted from the stems; the fibres are "hackled" into long tows, aligned into a continuous tape (Tahir et al. 2009). From the tape, single flax fibres are extracted with the aid of enzymes or other mechanical means before they are drawn into yarns (Tahir et al. 2009).

8.2.2 *Effects of Moisture*

We have assessed the response of wet and dry flax fibres to an external tensile load acting on them. These fibres are in the form of yarn as surgical sutures are for increased stiffness compared to single long fibres. We have found that there are appreciable differences in the rupture morphology of the knot-free fibres for dry versus wet condition. The knot-free (dry) yarn fibre experiences brittle rupture. Here, the rupture site exhibits a somewhat "clean" cut, i.e., the fibres are broken into two across the rupture site (Fig. 8.2a). When wet, the deforming fibre bundle initially undergoes defibrillation; at large deformation, fibre pullout occurs (hereafter referred to as the defibrillation-pullout mode). We have observed that the rupture site reveals a somewhat tapered profile at the respective ends of the rupture (Fig. 8.2b). These conclusions surrounding the differences in the response of wet versus dry flax fibres are important from the perspective of reinforcing

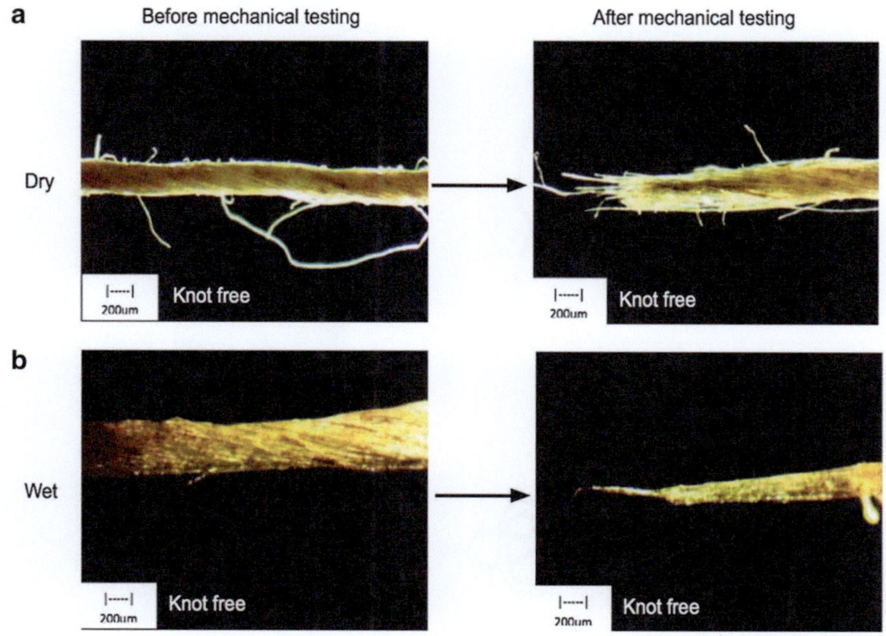

Fig. 8.2 Rupture morphology of knot-free yarn flax fibre: (**a**) dry fibre, (**b**) wet fibre

polymer-based composites. In particular, if water enters the polymer composites that are reinforced by yarn flax fibres, it could be expected that the composite fractures with fibre pull out from the bundle when the tensile load acting on the composite increases until it reaches a sufficiently large magnitude.

8.2.3 Effects of Knot Geometry

Knots are created to secure the flax fibres during the suturing process. Yarn flax sutures exhibit high flexibility but possess rougher surface, when compared to monofilaments (Abdessalem et al. 2009). The roughness underlies the microscopic crimps that tend to concentrate stress when the suture holding the wound is loaded (i.e., during the healing process) and, consequently, increases the possibility of breaking (Abdessalem et al. 2009). On the other hand, monofilament sutures have a smoother surface; knots created on monofilament sutures can be easily untied (Abdessalem et al. 2009). Knots will slip under a sufficiently high load or under a constant load over a period of time. Slippage can also be reduced by implementing additional throws during knotting. However, implementing additional throws could lead to an increased duration of the surgical operation and pose a greater risk of infection to the patient (Lim et al. 2011).

We have assessed the response of yarn flax fibres for two elementary types of knots known as overhand knot and overhand loops (Fig. 8.3a, b). In the case of the

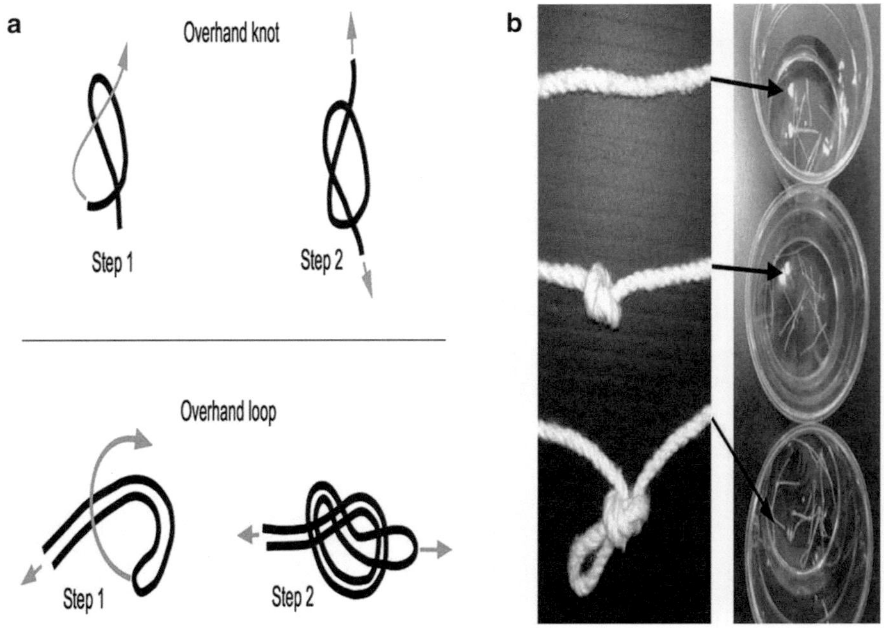

Fig. 8.3 Knots. (**a**) Schematics of overhand knot and loop. (**b**) Experimental setup for moisture treatment of knot-free flax fibres as well as flax fibres with overhand knot and overhand loop

overhand knot, we have found that there are appreciable differences in the rupture morphology for dry versus wet yarn flax fibres. In particular, the rupture morphology for the dry fibre is dominated by brittle failure, arising from fibres breaking into two at the rupture site. However, for the wet yarn fibre, the fibre bundle fails by defibrillation-pullout; the rupture site reveals a somewhat tapered profile for the "entrance" portion (Fig. 8.4a, b). Additionally, in both dry and wet cases, rupture occurs at the "entrance" to the knot (Fig. 8.5a, b).

On the other hand, in both dry and wet cases, the overhand loop yarn flax fibre displays little differences in the rupture morphology. The rupture morphology in both cases appears to be dominated by brittle failure and defibrillation-pullout; the former dominates the rupture mode. In both cases, rupture occurs at the entrance portion, similar to the overhand knot case.

8.3 Mechanical Properties

8.3.1 Stress Versus Strain Curves

We have assessed the stress–strain profile of yarn flax fibres subjected to a tensile load for both dry and wet cases. Typical stress versus strain curves for knot-free yarn flax fibres are shown in Fig. 8.6 for a dry specimen and a wet specimen.

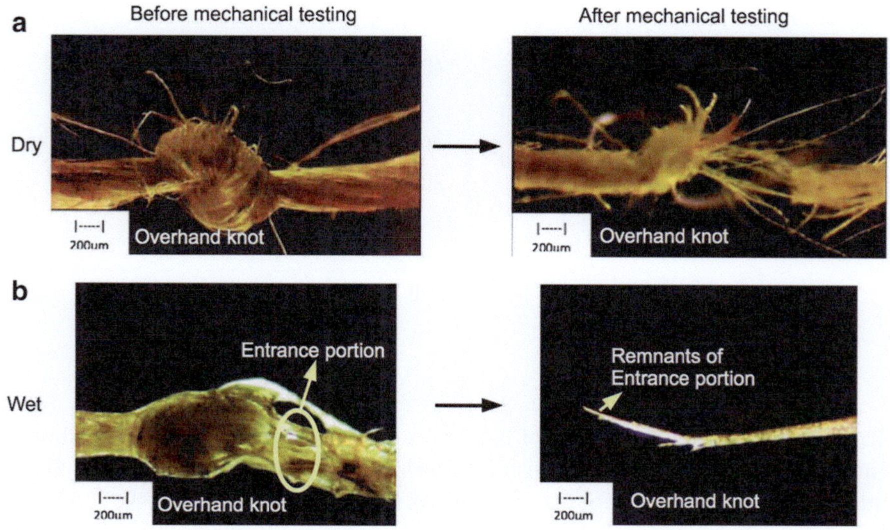

Fig. 8.4 Rupture morphology of overhand knot yarn flax fibre: (**a**) dry fibre, (**b**) wet fibre

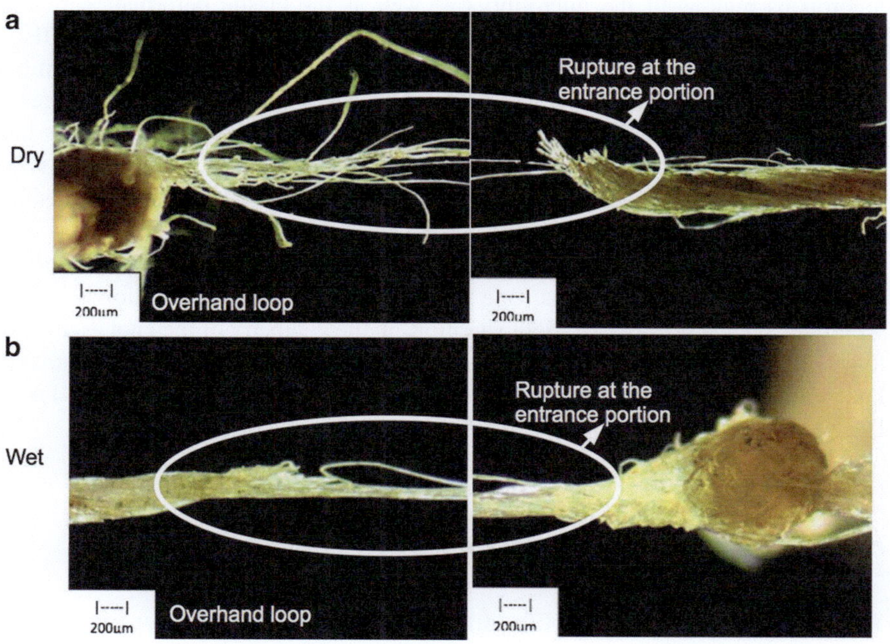

Fig. 8.5 Rupture morphology of overhand loop yarn flax fibre

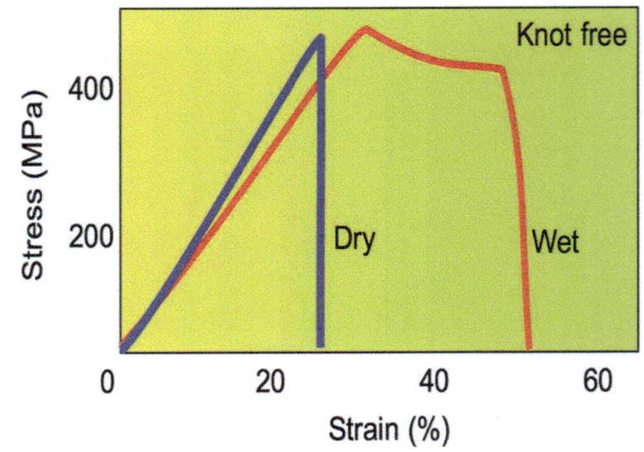

Fig. 8.6 Typical plots of stress versus strain for knot-free yarn flax fibres. *Blue line*, dry specimen; *red line*, wet specimen

In the dry case, the deformation of flax fibres involves an initial region describing a linear stress versus strain relationship (Cherif et al. 2013). It is thought that this is attributed to the stretching of the cell walls as well as the alignment of the cellulose-based microfibrils (Sect. 8.2.1) along the flax fibre axis (Gibson 2012). The next region is described by a small nonlinear stress versus strain relationship, which reaches a point of maximum stress (Cherif et al. 2013). This is attributed to the alignment of the cellulose-based microfibrils and the viscoelastic deformation of the amorphous regions of the microfibrils as well as the elastic deformation of the microfibrils as the interface between the microfibrils is sheared (Gibson 2012). Finally the yarn fibre ruptures into two. The wet yarn flax fibre shares a similar profile to that of the dry yarn flax fibre but only up to the point of maximum stress. Beyond the maximum stress point, the stress drops a little and plateaus off as the strain increases up to a point before dropping rapidly when the rupture process is completed. The plateau region is attributed to the process of defibrillation, i.e., sliding of the individual flax fibres in the twisted bundle.

We have found that dry specimens exhibit smaller variability in the stress versus strain curve than wet specimens. One reason is that the rate of relaxation of stress within the knot is sensitive to slight variations in the geometry of the knot as no two knots are the same. Interestingly, the stress versus strain curves for the overhand knot and overhand loop yarn flax fibres share a similar profile to those of the knot-free yarn flax fibres.

8.3.2 *Stiffness, Fracture Strength, and Fracture Strain*

We have quantified the extent of the differences in the mechanical properties of dry versus wet woven flax fibres. Figure 8.7 shows scatter plots of the respective mechanical properties of knot-free, overhand knot, and overhand loop woven flax fibres derived for the dry and wet cases. In all the cases considered here, one-way ANOVA revealed that a significant difference is observed among the three different knot types. In particular, the elastic modulus of knot-free woven flax fibres is higher

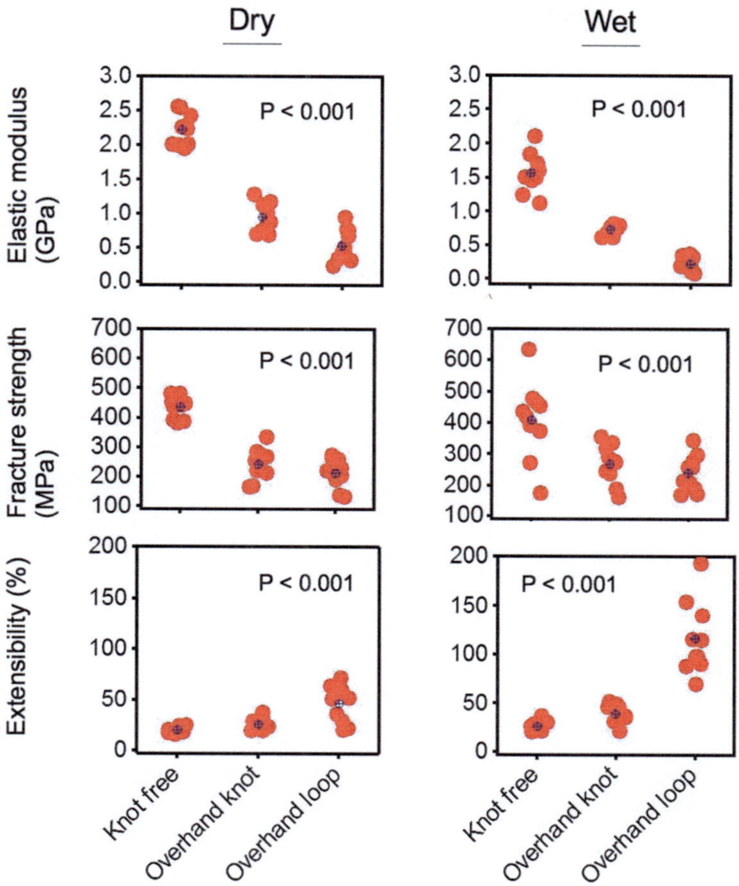

Fig. 8.7 The mechanical properties of dry versus wet yarn flax fibres. The elastic modulus was determined from the linear region of the stress versus strain curve (see Fig. 8.6); the fracture strength referred to the maximum stress on the stress versus strain curve; the extensibility referred to the maximum strain (which may be corresponding to the maximum stress point). One-way analysis of variance (ANOVA) was carried out to test for differences in the respective mechanical properties for knot-free, overhand knot, and overhand loop yarn fibres; here "P" refers to the *P* value generated from the ANOVA test

than those with knots; there is no difference in the elastic modulus of woven fibres with overhand knot versus loop (Fig. 8.8). Similar trend was observed for the strength of these fibres. However, the extensibility of overhand loop is larger than those of knot-free and overhand knot woven fibres. Two-factor ANOVA revealed that there was no significant difference between the strengths of dry versus wet woven fibres. Also, there were no significant interactions between the state of hydration and knot type. For the case of stiffness and extensibility, there was significant difference (1) between the stiffness of dry versus wet woven fibres and (2) among the different knot geometries. Significant effects on the stiffness were observed for the interaction

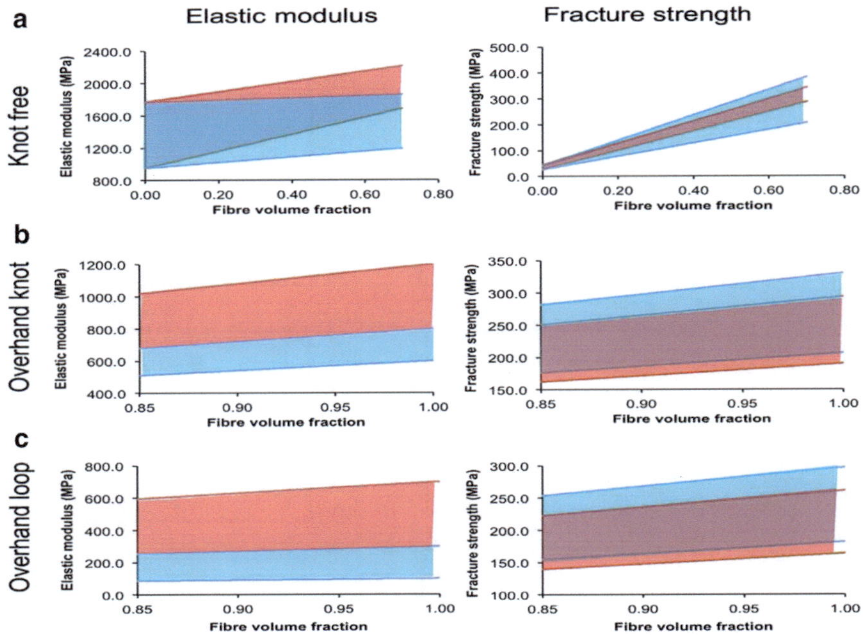

Fig. 8.8 Predictions of elastic modulus and fracture strength of yarn flax fibre blended (**a**) with polypropylene (PP) for knot-free fibres and with silicone for (**b**) overhand knot fibres and (**c**) overhand loop fibres. In (**a**), we consider the range of volume fraction of fibre from 0 (no fibre) to 0.70. In (**b**) and (**c**), we consider the range of volume fraction of fibre from 0.85 to 1.00. The pair of *blue* (*red*) lines represents the upper and lower extremes of the respective mechanical properties for wet (dry) fibres. The corresponding *blue* (*red*) regions enclosed by the pair of *blue* (*red*) lines indicate the possible values of the mechanical properties of the composite with wet (dry) fibres

Table 8.1 Summary of the mechanical properties of dry and wet yarn flax fibre in the dry

	Knot-free	Overhand knot	Overhand loop
Dry yarn flax fibres			
Elastic modulus (GPa)	2.2±0.2	1.0±0.2	0.5±0.2
Fracture strength (MPa)	436.2±36.4	242.3±52.1	212.6±48.7
Extensibility (%)	19.8±3.2	26.1±5.9	46.2±18.4
Wet yarn flax fibres			
Elastic modulus (GPa)	1.6±0.3	0.7±0.1	0.2±0.1
Fracture strength (MPa)	407.6±123.1	268.7±61.9	239.8±57.8
Extensibility (%)	26.3±5.5	39.3±9.9	116.5±36.9

between the state of hydration and knot geometry. A summary of the mechanical properties of woven flax fibre in the dry and wet states can be found in Table 8.1.

It is instructive to note that previous studies have shown that the fracture (tensile) strength, extensibility, and elastic modulus of single flax fibres in the dry state range from 300–1,500 MPa, 1.3–10.0 %, and 24–80 GPa, respectively (Cheung et al. 2009). Our results of the fracture strength of yarn flax fibres lie somewhat at the lower extreme of these values which is to be expected as the fibres are not aligned in the direction of the load. The trade-off is relatively higher stiffness and extensibility in the twisted fibres compared to single fibres. Our results indicate that a dry knotted yarn flax fibre is about half as strong as a dry knot-free specimen. However, there are no appreciable differences in the fracture strength of wet and dry yarn flax fibres when the knot is present. In the case of stiffness and extensibility, the characteristics of these properties are attributed to the mechanics and high degree of freedom in fibre–fibre sliding when the bundle is undergoing deformation.

When the knotted yarn is subjected to tensile forces, the knotted region experiences shear forces, due to the geometry of the knot (arising from bends) (Pieranski et al. 2001). Initially, constriction on the knot occurs followed by the generation of tension with the fibre when the knot has reached its limit of tightening (Pieranski et al. 2001). After this point, a rotation of the specimen about its longitudinal axis occurs; the internal forces of the friction within the knot are redistributed in the tightly intertwined region of the knot (Pieranski et al. 2001). We have observed that the overhand loop yields appreciable change in the rotational direction as compared to the overhand knot. This could be attributed to higher frictional forces within the knot of the overhand loop as compared to that of the overhand knot. With a greater angular change, the stress is transferred from the knot to the fibre region outside the knot. Thus, the overall effect of knotted structures on the mechanical response (i.e., tension in the fibre and the bending forces and frictional forces within the knot) of the knot is sensitivity to the knot geometry (Pieranski et al. 2001). Sharp bends in the knot could lead to high stress concentration at the bends; implementing a knot with more gradual angles could reduce the stress concentration (Pieranski et al. 2001).

Finally, to highlight the variability of the respective mechanical properties in the dry state (Table 8.1) while a proportion of the variability could be attributed to experimental errors arising from clamping effects and nonuniformity of the yarn flax fibres, the other sources could be attributed to the cultivation process and the plant's natural environment. It is important to enable sufficient control of the condition so that the flax will exhibit smaller structural and mechanical properties variability.

8.4 Hand Layup Technique

The prime principles govern the assortment of composite fabrication process contingent on the cost, production rate, strength, size, and shape requests of the applications needed (Dipa and Rout 2005). Moreover, thermosets are processed by simple techniques such as hand layup, pre-preg, casting, hot press, resin transfer, extrusion, injection molding, sheet molding (SMC)/bulk molding (BMC) for chopped and short fibres, pultrusion, and filament winding for continuous fibres (Ray and Bousmina 2005).

Hand layup technique is considered as the most conventional and simplest open molding, low volume production method used in composite industries for making different composite products among spray-up and filament winding. The hand layout

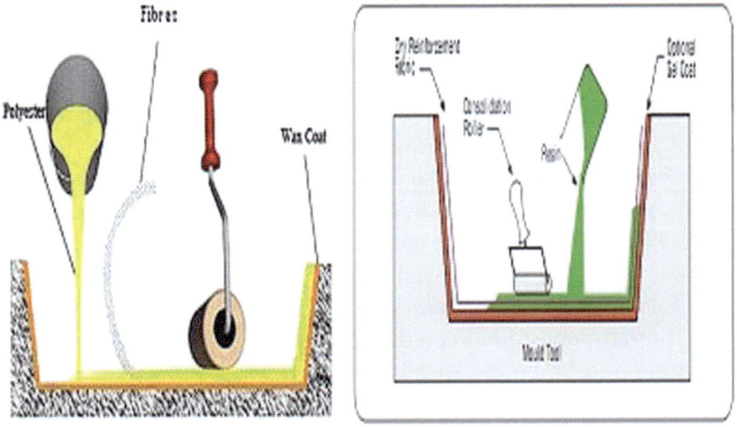

Fig. 8.9 Shows the typical hand layup technique

technique merely requires one-sided cost-effective tooling for infusion of resin through the reinforcement fabric over very short distances on the surface of the mold by technician in (hand layup) or drenched through atomizer in the modified hand layup (spray-up), that are quite similar to other composites manufacturing techniques. Figure 8.9 shows the typical hand layup technique. Moreover, for improved and highly eminent surface, pigmented gel coat is applied on the surface of the mold and resin is smeared and forced through distinct layers that are frequently lower than 1 mm thick. The crack propagation and crack initiation are greatly influenced by layup geometry of a composite, resulting in some highly notch-sensitive laminates, whereas others are completely insensitive to the presence of stress concentrators (Soutis 2009).

8.4.1 Equipment's of Hand Layup

Hand layup fabrications in most of the cases enjoy same molds and tooling that are highly simple apparatus/equipment as in the RTM (resin transfer molding specimens). The rollers are readily available from suppliers of composite materials; typically the rollers feature 25.4 mm in diameter and 89 mm in width. The thermoplastic squeegees and bristle brushes were economical stuffs easily collected from local hardware store. Figure 8.10 shows the tools of hand layup.

8.4.2 Raw Materials Used in Hand Layup Method

Hand layup technique can be carried out for a variety of matrix and reinforcement on both large and small scale effectively. Table 8.2 shows the materials widely used in this method.

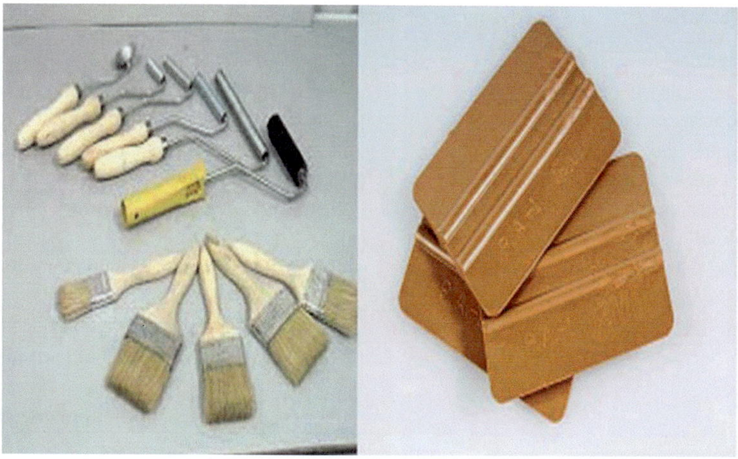

Fig. 8.10 Shows the tools of hand layup

Table 8.2 Different materials used in hand layup method

Polymeric matrix	Reinforcements in unidirectional/randomly mat, woven fabric form
Epoxy	Glass fibre
Polyester	Carbon fibre
Polyvinyl ester	Heavy aramid fibre
Phenolic resin	Kevlar fibre
Unsaturated polyester	Natural plant fibres (sisal, banana, kenaf, jute, hemp, flax, coir, etc.)
Polyurethane resin	

8.4.3 Methodology of Hand Layup

Fibres typically in the form of woven, knitted, stitched, or bonded fabrics are permeated inside the resin through hand. The imposing of resin into the fabrics is usually skilled by brushes, rollers, nip-roller-type impregnators, and rotating rollers. Laminates are then left to cure under normal or room temperature.

Overall, the two major steps involved in this manufacturing technique are: (1) fibre impregnation and stacking and (2) curing procedure (Avila and Morais 2005).

Steps of hand layup process: It can be carried out in the following steps:

Mold cleaning: Dust and dirt are removed from mold with soft cloth.

Smearing resin: Generally with the brush, resin is smeared on mold similar to painting a wall.

Fibre layup: Chopped fibres or fibre mats are laid out or laminated over the resin layer on the mold.

Table 8.3 Advantages and disadvantages of hand layup method

Advantages	Disadvantages
Extensively and effectively used manufacturing technique	Health and safety hazards during curing due to release of hazardous volatiles for operators
Equipment's, technique, processing steps, and principles simple and easy to learn	Resin contents, mixing and laminate quality dependent on the skills of laminators
The infrastructural requirement and tooling are minimal	Resin should have low viscosity
Comparatively longer and higher fibres used to spray layup	Expensive extraction systems for polyesters and vinyl esters
Great choice of material and suppliers brands	Labor intensive and production rate is less
	Results low volume output of parts and high cycle times
	Process gives inconsistent fibre orientations

Smearing resin: Reapply resin by brushing over the strand of fibre layer.

Squeeze action: Roller applied to move over the resin–fibre–resin layers to eliminate air sandwiched between layers, called squeeze action, which is usually used to compact the fabricated products.

Remold: The last step, remolding is done after curing at room temperature to remove the sheet from mold.

The key advantage to this manufacturing process is its skill or tendency of fabricating relatively large and complex parts in the first speedy preliminary step. Furthermore, Table 8.3 tabulates the advantages and disadvantages of these fabricating methods. Moreover, research study reveals that porosity content does not show significant courtesy in the hand layup. Alternatives to this method involve "pre-preg" material that is purchased with resin previously saturated inside the fibre mat. The researcher evaluates that problems existing in layup process can be avoided by increasing the production rate of sheets through automatic layup process by converting the existing layup process (Chaple et al. 2013).

8.4.4 Applications

The undoubted application which permits the fabrication of wide diversity of shapes is its extreme flexibility (Avila and Morais 2005). Hand layup is one of the presently proven and established low-cost and effective processes for widely constructing turbine blades composite and further extensive structural products such as airframe components, boats, truck and car body components, tanks, bath wares, housing, swimming pools, ducts decks and hulls for leisure boats, wind turbine blades, architectural moldings, etc.

Nowadays, fibre-reinforced plastic (FRP) products are generally manufactured by the hand layup or spray-up process. The quality and application of the obtained final fabricated reinforced composite in aircraft construction (Soutis 2005) involved optimum toughness which depends on the way in which considered fibres and resins are pooled in the layup. Some of the important works on flax fibre as reinforcements in polymeric matrix through traditional hand layup technique are given in Table 8.4.

Table 8.4 Reported work on flax composites fabricated by hand layup method

Fibre reinforcement/polymer	Manufacturing techniques	References
Flax/tannin resins (bio-resins)	Hand layup, compression molding	Zhu et al. (2013)
Flax and banana fibres/epoxy resin	Hand layup	Srinivasan et al. (2014a, b)
Flax fibre/(UP), (VE), (UP/VE), (EP)	Hand layup, RTM	Goutianos et al., (2006)
Flax woven, knitted/epoxy	Hand layup	Muralidhar et al. (2012)
Jute–flax-glass fibre/epoxy	Hand layup	Ramnath et al. (2014)
Flax fibre/epoxy	Hand layup	Bertomeu et al. (2012)
Flax plain laminate-weft rib knitted hybrid/epoxy	Hand layup	Muralidhar et al. (2013a, b)
Flax/epoxy laminates	Hand layup	Santulli (2009)
Bidirectional flax/epoxy	Initial hand layup phase followed by vacuum bagging technique	Bella et al. (2010); Yuhazri and Phongsakorn (2010)
Flax-plain weave/epoxy	Hand layup	Muralidhar (2013a, b)
Flax and jute fibre/epoxy	Hand layup	Dhakal et al. (2013)
Bidirectional untreated flax–sisal fabrics/cementitious matrices	Hand layup	Olivito et al. (2014)
Flax-E-glass fibres/epoxy	Hand layup	Santulli et al. (2005)
Flax fibre/melamine-formaldehyde	Hand layup	Hagstrand and Oksman (2001)
Flax–kenaf hybrid/epoxy	Hand layup	Srinivasan et al. (2014a, b)
Flax–banana hybrid/epoxy	Hand layup	Srinivasan et al. (2014a, b)
Flax laminates, glass fibre, flax-glass fibres hybrids/modified urethane, polyester	Hand layup followed by vacuum bagging	Rao et al. (2012)
Flax FRP tube-coir fibre/(EP), (PC)	Hand layup	Yan et al. (2012)
Flax FRP tube-coconut fibre/concrete beams	Hand layup	Yan and Chouw (2013)
Flax fibre/recycled high-density polyethylene	Hand layup, compression molding technique	Singleton et al. (2003)
Aligned flax fibres/(UP), (EP)	RTM and hand layup	Goutianos et al. (2007)
Empty flax/epoxy tubes composite, polyurethane foam-filled flax fabric/epoxy	Hand layup	Yan et al. (2014a, b)
Flax FRP/(CFRC)	Hand layup, vacuum bagging	Yan et al. (2014a, b)

8.5 Design Considerations

8.5.1 Models

This section presents simple mathematical models that lead to order-of-magnitude predictions for the stiffness and fracture strength of yarn flax fibre composites. As an example, we shall consider two different applications of yarn flax fibres: (1) hand layup yarn flax fibre-reinforced polymer composites and (2) polymer-based treatment of yarn flax fibre for sutures. For case 1, the polymer matrix in question is polypropylene (PP) (Biagiotti et al. 2004; Anderson et al. 2006); for case 2, the polymer to be used to treat the flax is silicone (a major constituent of the coating) (B Braun Suture Linatrix 2014). One of the motivations here is to develop mathematical descriptions for these composites in a consistent manner to allow comparison of the composition of yarn fibre in the composite. Another motivation is to establish the bounds in which the design of these composites can be considered.

Let us begin with the stiffness, E, of the composite material. Following an argument established in an earlier chapter ("Oil Palm Fibre for Polymer Composites: Influence of Moisture Content"), with regard to the yarn flax fibre-reinforced polymer composites (i.e., case 1), we consider unidirectional yarn flax fibres, with length comparable to the overall dimension of the composite, blended in the polymer composite. Accordingly, the composite stiffness, E, is directly dependent on the stiffness of the yarn flax fibre, E_f; the stiffness of the polymer, E_m; and the volume fraction of the fibre, V_f (where the volume fraction of the PP$=1-V_f$). In both cases (i.e., 1 and 2), the expression of E, given by the general model (Anderson et al. 2006)

$$E = \eta_{oE}\eta_{fE}E_f V_f + E_m\left(1-V_f\right), \tag{8.1}$$

can be estimated to order of magnitude by replacing the fibre orientation factor, η_{oE}, by unity and by replacing and the fibre efficiency factor, η_{fE}, by unity. One then finds that

$$E = E_f V_f + E_m\left(1-V_f\right). \tag{8.2}$$

The fracture strength, σ_U, of a yarn flax fibre-reinforced PP composite or silicone-treated woven flax fibre, given by the general model (Anderson et al. 2006)

$$\sigma_U = \eta_{fS}\sigma_{Uf}V_f + \sigma_{Um}\left(1-V_f\right), \tag{8.3}$$

where σ_{Uf} is the fibre strength, σ_{Um} is the stress in the polymer matrix at the fibre failure strain, and η_{fS} is the fibre efficiency (arising from the association with the fracture strength of the composite), can be estimated to order of magnitude by replacing the fibre orientation factor, η_{fS}, by unity. In both cases (i.e., 1 and 2), one then finds that

$$\sigma_U = \sigma_{Uf}V_f + \sigma_{Um}\left(1-V_f\right). \tag{8.4}$$

8.5.2 Polypropylene Composites

With regard to case 1, i.e., hand layup yarn flax fibre-reinforced PP composites, the value of E_m that establishes the lower bound of Eq. (8.2) is found by setting $E_m = 0.53$ MPa (Anderson et al. 2006); similarly the value of E_m that establishes the upper bound of this expression is found by setting $E_m = 0.64$ MPa (Anderson et al. 2006; Ku et al. 2011). The value of σ_{Um} that establishes the lower bound of Eq. (8.4) is found by setting $\sigma_{Um} = 26.0$ MPa (Anderson et al. 2006; Ku et al. 2011); similarly the value of σ_{Um} that establishes the upper bound of this expression is found by setting $\sigma_{Um} = 41.4$ MPa (Anderson et al. 2006; Ku et al. 2011). To further our arguments of the prediction of the extreme values of the composite stiffness and strength, we evaluate the extremes of the variability of the elastic modulus of the knot-free yarn flax fibres which we have expressed in terms of mean ± SD. We argue that the value of E_f that establishes the lower bound of Eq. (8.2) is numerically equal to 2,000 MPa (dry) and 1,300 MPa (wet); the value of E_f that establishes the upper bound of this expression is found by setting $E_f = 2{,}400$ MPa (dry) and 1,900 MPa (wet). Similar argument applies for the fracture strength of the composite, giving the value of $\sigma_{Uf} = 472.6$ MPa (dry) and 530.7 MPa (wet) for the upper bound and 400.0 MPa (dry) and 285.0 MPa (wet) for the lower bound.

Thus, when these values, i.e., E_f and E_m, are substituted back into Eq. (8.2) (NB: for σ_{Uf} and σ_{Um}, we substitute the values back into Eq. (8.4)), and considering V_f from 0.0 to 0.70 (Ku et al. 2011; Zhu et al. 2013), one arrives at the results displayed in Fig. 8.10a for case 1. At low V_f (i.e., up to 0.10), it can be seen that there is a considerable overlap in the elastic modulus of the PP composite for the dry versus wet fibres, suggesting that there is little effect of moisture on the composite stiffness. As V_f increases, the overlapped region decreases; the general trend reveals that the elastic modulus of the PP composite with dry fibres is higher than that of the composite with wet fibres. Thus, as the fibre content increases in the composite, the composite becomes more sensitive to the effects of moisture in that it leads to a lower stiffness when the fibres are wet. As for the fracture strength of the composite, there are little effects of moisture on the composite at low V_f; both fracture strengths of wet and dry fibre-reinforced PP composite reveal considerable overlap in the variability. However, as V_f increases, (1) the variability of strengths in the wet fibre-reinforced composite increases and (2) the region of variability for the wet fibre-reinforced composite encompasses the region of the dry fibre-reinforced composite. This suggests that the wet fibre-reinforced composite leads to highly unpredictable strengths, although there are possibilities that the strengths could be comparable to those of the dry fibre-reinforced composite.

8.5.3 Sutures

With regard to case 2, i.e., the silicone-treated yarn flax fibres, the value of the silicone stiffness, E_m, that establishes the lower bound of Eq. (8.2) is found by setting $E_m = 0.95$ GPa (Hatamleh and Watts 2010); similarly the value of E_m that establishes the upper bound of this expression is found by setting $E_m = 1.77$ GPa (Hatamleh and

Watts 2010). In the previous paragraph, we find that the value of E_f that establishes the lower bound of Eq. (8.2) is found by setting $E_f = 2{,}000$ MPa (dry) and 1,300 (wet); similarly the value of E_f that establishes the upper bound of this expression is found by setting $E_f = 2{,}400$ MPa (dry) and 1,900 MPa (wet).

Finally, when these values are substituted back into Eq. (8.2), and considering V_f from 0.85 to 1.00 (in other words, the silicone content ranges from 0.15, i.e. "lightly treated", to 0, i.e. "no treatment"), one arrives at the results displayed in Fig. 8.10b, c. It turns out that the overall trends (dry and wet) of the performance of overhand knot versus overhand loop yarn flax fibres are the same. In all cases, the magnitude of the respective mechanical properties (i.e., elastic modulus and fracture strength) increases with decreasing coating content. In particular, the elastic modulus of the dry knotted fibres is higher than those of the wet fibres; there was no appreciable overlapping region of moduli from the dry and wet cases. This suggests that the elastic moduli of the knotted fibres are sensitive to moisture. However, the same may not be said of the fracture strength. There is a considerable overlap of fracture strengths from both wet and dry fibres. Interestingly, the general trend of fracture strengths is such that the upper extreme of fracture strength is dominated by wet fibres while the lower extreme of fracture strength is dominated by dry fibres.

8.5.4 *Limitations*

We note that Fig. 8.10 provides an indication of the range of possible values of the stiffness and strength of cases 1 and 2, with respect to the fibre volume fraction. Such a plot can also help to illustrate the sensitivity of the stiffness and strength of the composite to moisture, as depicted by the extent of the regions of overlap for the wet versus dry fibres in the composite. In reality the relationships between the respective mechanical properties and the fibre volume fraction (or otherwise measured in weight percent) depart somewhat from the linear relationship between the respective mechanical property and fibre volume fraction. Based on an extensive survey of existing natural fibre-reinforced polymer composites, e.g., coir/PP, palm/PP, Kenaf/PP, and palm/rubber (Shalwan and Yousif 2013), it is observed that while it is possible to increase the volume fraction to achieve high composite strength, there exists a critical volume fraction (ranges from 5 to 50 % depending on the composite type) which will result in an optimal composite strength, but beyond the critical volume fraction, the strength decreases. Thus, the simple order-of-magnitude approach overestimates the magnitude of the mechanical property at large fibre volume fraction.

8.6 Conclusion and Future Perspective

Moisture in the yarn flax fibre has a significant influence on the mechanical performance of the fibres. Dry yarn flax fibres exhibit somewhat brittle fracture. Wet yarn flax fibres exhibit fibre debonding leading to fibre pullout. While both wet and dry knotted fibres fail by rupture at the entrance to the knot, the fracture morphology at the ruptured ends differ, i.e., dry ones exhibit brittle fracture and wet ones debonding followed by pullout. These variations in the fracture morphology underscore the differences in the mechanical properties of the wet and dry yarn flax fibres. Of note the stiffness and strength of dry fibres are larger than the wet ones (although the differences are not appreciably large). However, the wet fibres exhibit higher extensibility than the dry ones. Simple mathematical models that lead to order-of-magnitude predictions for the stiffness and fracture strength of hand layup yarn flax fibre-reinforced PP composites and silicone-treated woven flax fibres show that the stiffness and strength of the composites are sensitive to moisture, as depicted by the extent of the regions of overlap from the respective composite containing wet versus dry fibres. These discussions offer a simple methodology for engineering hand layup yarn flax fibre-reinforced polymer composites and polymer-treated yarn flax fibre sutures with mechanical properties tailored for the intended purpose.

References

Abdessalem SB, Debbab F, Jedda H (2009) Tensile and knot performance of polyester braided sutures. Text Res J 79:247

Alix S, Marais S, Morvan C, Lebrun L (2008) Biocomposite materials from flax plants: preparation and properties. Compos A Appl Sci Manuf 39:1793–1801

Andersons J, Sparnins E, Joffe R (2006) Stiffness and Strength of Flax Fiber/Polymer Matrix Composites, Polymer Composites, 27, pp 221–229

Avila AF, Morais DTS (2005) A multiscale investigation based on variance analysis for hand lay-up composite manufacturing. Compos Sci Technol 65:827–838

Bella GD, Fiore V, Valenza A (2010) Effect of a real weight and chemical treatment on the mechanical properties of bidirectional flax fabrics reinforced composites. Mater Des 31(9): 4098–4103

Bertomeu D, García-Sanoguera D, Fenollar O, Boronat T, Balart R (2012) Use of eco-friendly epoxy resins from renewable resources as potential substitutes of petrochemical epoxy resins for ambient cured composites with flax reinforcements. Polym Compos 33(5):683–692

Biagiotti J, Puglia D, Torre L, Kenny JM (2004) A systematic investigation on the influence of the chemical treatment of natural fibers on the properties of their polymer matrix composites. Polym Compos 25:470–479

B Braun Sutures Linatrix Brochure (2014). http://www.bbraun.no/documents/Products/Linatrix_11.04.pdf. Retrieved 22 Aug 2014

Chaple AR, Khedakar SS, Dharmadhikari SR, Chaple NR (2013) Newly developed automatic lay-up process for manufacturing of FRP sheets. Int J Comput Eng Res (ijceronline.com) 3

Cherif ZE, Poilane C, Falher T, Vivet A, Ouail N, Doudou BB, Chen J (2013) Influence of textile treatment on mechanical and sorption properties of flax/epoxy composites. Polym Compos 34:1761–1773

Cheung HY, Ho MP, Lau KT, Cardona F, Hui D (2009) Natural fibre-reinforced composites for bioengineering and environmental engineering applications. Compos B Eng 40:655–663

Dhakal HN, Zhang ZY, Bennett N, Lopez-Arraiza A, Vallejo FJ (2013) Effects of water immersion ageing on the mechanical properties of flax and jute fibre biocomposites evaluated by nanoindentation and flexural testing. J Compos Mater 48:1399–1406

Dipa R, Rout J (2005) Thermoset biocomposites. In: Natural fibers biopolymers and biocomposites

European Confederation of Flax and Hemp (2012) In: Reux F (ed) Flax and hemp fibres: a natural solution for the composite industry. JEC Composites
Gibson LJ (2012) The hierarchical structure and mechanics of plant materials. J R Soc Interface. doi:10.1098/rsif.2012.0341
Goh KL, Aspden RM, Mathias KJ, Hukins DWL (2004a) Finite-element analysis of the effect of material properties and fibre shape on stresses in an elastic fibre embedded in an elastic matrix in a fibre-composite material. Proc R Soc Lond Math Phys Eng Sci 460:2339
Goh KL, Aspden RM, Hukins DWL (2004b) Review: finite element analysis of stress transfer in short-fibre composite materials. Compos Sci Technol 64:1091
Goh KL, Meakin JR, Hukins DWL (2010) Influence of fibre taper on the interfacial shear stress in fibre-reinforced composite materials during elastic stress transfer. Compos Interface 17:75
Goutianos S, Peijs T, Nystrom B, Skrifvars M (2006) Development of flax fibre based textile reinforcements for composite applications. Appl Compos Mater 13(4):199–215
Goutianos S, Peijs T, Nystrom B, Skrifvars M (2007) Textile reinforcements based on aligned flax fibres for structural composites. http://www.risoe.dk
Greenberg JA, Clark RM (2009) 'Advances in Suture Material for Obstetric and Gynecologic Surgery', Reviews in Obstetrics & Gynecology, 2, pp146–158
Hagstrand PO, Oksman K (2001) Mechanical properties and morphology of flax fiber reinforced melamine–formaldehyde composites. Polym Compos 22(4):568–578
Hassan A, Salema AA, Ani FN, Abu Bakar A (2010) A Review on Oil Palm Empty Fruit Bunch Fiber-Reinforced Polymer Composite Materials, Polymer Composites, 31,2079–2101
Hatamleh MM, Watts DC (2010) Mechanical properties and bonding of maxillofacial silicone elastomers. Dent Mater 26:185–191. http://www.fibre-reinforced-plastic.com. http://www.net-composites.com
Ku H, Wang H, Pattarachaiyakoop N, Trada M (2011) A review on the tensile properties of natural fiber reinforced polymer composites. Compos B Eng 42:856–873
Kunert-Keil C, Gredes T, Meyer A, Wróbel-Kwiatkowska M, Dominiak M, Gedrange T (2012) The survival and proliferation of fibroblasts on biocomposites containing genetically modified flax fibers: an in vitro study. Ann Anat 194:513–517
Lim BH, Ooi LH, Chou SM, Goh KL (2011) Biomechanical properties of extensor tendon repair using the six-strand single-loop suture technique: a comparative analysis with three other techniques in cadaveric models. J Mech Med Biol 11:845–855
Moothoo J, Allaoui S, Ouagne P, Soulat D (2014) A study of the tensile behaviour of flax tows and their potential for composite processing. Mater Des 55:764–772
Muralidhar BA (2013a) Study of flax hybrid preforms reinforced epoxy composites. Mater Des 52:835–840
Muralidhar BA (2013b) Tensile and compressive properties of flax-plain weave preform reinforced epoxy composites. J Reinf Plast Compos 32:207–213
Muralidhar BA, Giridev VR, Raghunathan K (2012) Flexural and impact properties of flax woven, knitted and sequentially stacked knitted/woven preform reinforced epoxy composites. J Reinf Plast Compos 31:379–388
Olivito RS, Cevallos OA, Carrozzini A (2014) Development of durable cementitious composites using sisal and flax fabrics for reinforcement of masonry structures. Mater Des 57:258–268
Pieranski P, Kasas S, Dietler G, Dubochet J, Stasiak A (2001) Localization of breakage points in knotted strings. New J Phys 3:10.1–10.13
Ramnath BV, Elanchezhian C, Nirmal PV, Kumar GP, Kumar VS, Karthick S, Rajesh S, Suresh K (2014) Experimental investigation of mechanical behavior of Jute–Flax based glass fiber reinforced composite. Fibers Polym 15(6):1251–1262
Rao S, Bhardwaj A, Beehag A, Bhattacharyya D (2012) Performance of flax laminates and their hybrids. Adv Mater Res 410:114–117. Trans Tech Publications, Switzerland
Ray SS, Bousmina M (2005) Biodegradable polymers and their layered silicate nanocomposites: in greening the 21st century materials world, Progress in Materials Science, 50, 962–1079
Santulli C (2009) Falling weight impact damage characterisation on flax/epoxy laminates. Int J Mater Prod Technol 36. Inderscience Enterprises Ltd

Santulli C, Janssen M, Jeronimidis G (2005) Partial replacement of E-glass fibers with flax fibers in composites and effect on falling weight impact performance. J Mater Sci 40(13):3581–3585

Shalwan A, Yousif BF (2013) In state of art: mechanical and tribological behaviour of polymeric composites based on natural fibres. Mater Des 48:14–24

Singleton ACN, Baillie CA, Beaumont PWR, Peijs T (2003) On the mechanical properties, deformation and fracture of a natural fibre/recycled polymer composite. Compos B Eng 34:519–526

Soutis C (2005) Fibre reinforced composites in aircraft construction. Prog Aerosp Sci 41(2): 143–151

Soutis C (2009) Recent advances in building with composites. Plast Rubber Compos 38:359–366. http://www.materials.uoi

Srinivasan VS, Boopathy SR, Sangeetha D, Ramnath BV (2014a) Evaluation of mechanical and thermal properties of banana–flax based natural fibre composite. Mater Des 60:620–627

Srinivasan VS, Boopathy SR, Ramnath BV (2014) Investigation on shear behaviour of flax-kenaf hybrid composite. Adv Mater Res 1051:139–142. Trans Tech Publications, Switzerland. http://www.scientific.net

Tahir PM, Ahmed AB, SaifulAzry SOA, Ahmed Z (2009) Retting process of some bast plant fibres and its effect on fibre quality: a review. BioResources 6:5260–5281

Yan L, Chouw N (2013) A comparative study of steel reinforced concrete and flax fibre reinforced polymer tube confined coconut fibre reinforced concrete beams. J Reinf Plast Compos 32(16):1155–1164

Yan L, Yuan X, Nguyen C, Chouw N (2012) Compressive behaviour of flax FRP tube confined coir-fibre reinforced concrete. In: Barros J et al (eds) BEFIB2012 – fibre reinforced concrete. www.researchgate.net

Yan L, Chouw N, Jayaraman K (2014a) Effect of column parameters on flax FRP confined coir fibre reinforced concrete. Construct Build Mater 55:299–312

Yan L, Chouw N, Jayaraman K (2014b) Effect of triggering and polyurethane foam-filler on axial crushing of natural flax/epoxy composite tubes. Mater Des 56:528–541

Yuhazri M, Phongsakorn PT, Sihombing H (2010) A comparison process between vacuum infusion and hand lay-up method toward kenaf/polyester composites. Int J Basic Appl Sci (IJBAS-IJENS) 10(3):63–66

Zhu J, Zhu H, Njuguna J, Abhyankar H (2013) Recent development of flax fibre and their reinforced composites based on different polymeric. Materials 6:6171–6198

Chapter 9
Effect of Processing Conditions on the Mechanical and Morphological Properties of Composites Reinforced by Natural Fibres

Abouelkacem Qaiss, Rachid Bouhfid, and Hamid Essabir

Abstract In the last years, the exploitation of different natural fibres in the plastic industry has become increasingly essential for the introduction of new composites and products. The properties of polymers reinforced with natural fibre composites are generally governed by the fibre and matrix properties, the compatibility between the compounds, the process of treatment of fibres, and the manufacturing process of composites. Enhancements and innovations in manufacturing technology and assembly of fibre-reinforced polymer composite materials and structures are required to achieve the objectives of cost and performance to allow wider adoption in many sectors. In this chapter, natural fibres from doum, coir, and bagasse were added to polypropylene by twin-screw extrusion and molded by injection machine, to study the influence of manufacturing process especially during the extrusion and injection processes. The influence of extruder screw configuration (corotating and counterrotating), screw speed (60, 80, and 120 rpm), and temperature (190, 200, and 210 °C) was studied on the pressure of matter and torque, and then an optimized configuration was selected to evaluate the mechanical properties of injected composite. On the other hand, the preparation of the injected composite samples was performed by adopting different placing directions according to flow and tensile directions: parallel and transversal placements and the effect of fibre orientation in the mechanical properties of manufactured composites were evaluated. Advanced knowledge of the relationship between structure, composition, and characteristics of material composite based on natural fibres made possible the development of high-performance materials with excellent mechanical properties. We have demonstrated some aspects concerning the influence of the manufacturing process of the fibres in the thermoplastic matrix.

Keywords Doum fibre • Coir fibre • Bagasse fibre • Composites • Twin-screw extrusion • Processing conditions

A. Qaiss (✉) • R. Bouhfid • H. Essabir
Moroccan Foundation for Advanced Science, Innovation and Research, Rabat, Morocco
e-mail: a.qaiss@mascir.com

M.S. Salit et al. (eds.), *Manufacturing of Natural Fibre Reinforced Polymer Composites*, DOI 10.1007/978-3-319-07944-8_9

9.1 Introduction

For several years, the appearance of some concepts such as sustainable development, industrial ecology, and green chemistry has pushed manufacturers to develop new materials from renewable resource (Unterweger et al. 2014; Zakikhani et al. 2014; Pan and Zhong 2014). Wherever advancing technology has created a need for combinations of properties no single material can provide, composites are becoming the material of choice (Pan and Zhong 2014; Rojo et al. 2015). By dispersing fibres or particles of one substance in a matrix, or binder, of another, the designer of a composite can reach properties neither material shows on its own. The future polymeric products must be processed and developed taking into account the environmental impact from the raw materials used in the synthesis, until the end of product life (Shouha et al. 2014; Merkel et al. 2014; Węcławski et al. 2014). Developments in the production of composite concern the improvement of the evaluation of structural defects to ensure durability, reliability, cost reduction, and increased production rates. For the strengthening of polymers, natural filler has significant specific mechanical properties and many advantages. Several worldwide laboratories have begun work on the development of new composite materials based on thermoplastic matrices reinforced with natural fillers (bio-fillers) (Merkel et al. 2014; Węcławski et al. 2014). These bio-fillers are renewable resources, biodegradable, less abrasive, neutral as CO_2 emissions into the atmosphere, and requiring little energy to be produced. These fibres can replace glass fibres in many areas or allow reaching new markets (Arrakhiz et al. 2012a, b, 2013a).

Composite materials are composed of two materials of different properties but complementary. The reinforcement usually allows support of the load of the structure, reducing thermal stress, and ensures rigidity and macroscopic resistance. The matrix serves to link the reinforcing providing protection against external agents and transfer loads to the reinforcement via the connection between the filler and the matrix (Arrakhiz et al. 2012a, b, 2013a). The fibres used for the reinforcement of polymer matrix are generally classified into two groups: continuous fibres and discontinuous fibres. Continuous reinforced composites are reserved for applications with high performance. The costs of raw materials, methods of implementation, and low production capacity make these composites expensive and are limited to industries such as aeronautics and aerospace. Carbon fibres (high modulus and high strength) (Shi et al. 2014) and Kevlar (very light) (Shi et al. 2014; Ou et al. 2010) are more used in such composites. Composites reinforced with discontinuous fibre are used to bridge the gap between the properties of composites with long fibres (continuous) (Malha et al. 2013) and non-reinforced polymers used in applications with low loads. If the fibres are sufficiently long, the stiffness of the composite may approach that of a continuous fibre system. On the other hand, the implementation of the similar non-reinforced polymers is always possible, by injection, extrusion, or compression, allowing large production capacity (Fernandes et al. 2014; El-Sabbagh et al. 2014; Valente et al. 2011). The final properties of an advanced composite are shaped not only by the kind of matrix, reinforcing materials it contains (fillers), and synergy between the fillers and the matrix but also by the filler size, filler morphology, filler loading, dispersion/distribution of fillers into the matrix, and the manufacturing process (implementation condition, used process) (Salleh et al. 2014).

It is known that characteristics of composite materials depend essentially on technology of manufacturing processes of materials. The processing of composite materials is also one of the largest research areas where the academic community and industry should make a concerted effort (Feldmann and Bledzki 2014; Al-Maadeed et al. 2014). The development of composite materials is a complicated process and requires the simultaneous consideration of different parameters such as the geometry of the components, the volume of production, the type of reinforcement and the matrix, the requirements of tooling, and process market economy (Al-Maadeed et al. 2014). Improvements and innovation in manufacturing and assembling techniques for fibre-reinforced polymer composite materials and structures are necessary to achieve materials with low cost and high performance to allow wider adoption in many industrial sectors (Gamon et al. 2013; Ho et al. 2012). This chapter highlights the various aspects of secondary processing of polymer matrix composites.

9.2 Processing of Polymer Composites Based on Short Fibre

Polymer composite materials based on natural fibre have been fabricated by using several manufacturing techniques which are designed for conventional fibre-reinforced polymer composites. These techniques comprise compounding, extrusion, and injection molding, compression molding, resin transfer molding (RTM), and vacuum infusion. It is known that characteristics of composite materials depend strongly on technology of manufacturing processes. The extrusion, injection, and compression moldings are the key process to produce polymer/filler composites with a high homogeneity and a good dispersion/distribution of fillers into thermoplastic matrix (Gamon et al. 2013; Ho et al. 2012). It is important to note that the process parameters are the rotating speed of the mixer, the temperature, and time. Therefore, the investigation of the techniques is very important for the research and application of thermoplastic composites.

9.2.1 Compounding

In the compounding step, the composite is manufactured by blending the filler with the thermoplastic matrix. Usually, the compounded material can be molded or pressed into an end product or formed into pellets for further processing. The composite products can be manufactured using thermoforming, extrusion, calendaring, injection molding, or compression molding. Most polymer composites are made by extrusion, in which the fillers are incorporated into the molten thermoplastic using a screw system (Nekhlaoui et al. 2014a; Arrakhiz et al. 2013b). When more complex forms are desired, other methods such as injection or compression molding can be used. Injection molding is the process used in the automotive industry to produce plastic products. A major advantage of this manufacturing technique may be found in the ability to form complex shapes, at a high rate of production, due to the cyclical nature of this process. During the process of injection molding of thermoplastic materials reinforced with fibres, the fibre orientation is produced by the flow conditions. The molded product is anisotropic due to the fibre orientation. The material properties and characteristics of the final part are determined by the fibre orientation (Kim and Lee 2014; Notta et al. 2014; Hashimoto et al. 2012; Oumer and Mamat 2012).

9.2.2 Extrusion Condition: Screw Configuration and Temperature

Generally, the mechanical properties of composite materials depend strongly on the fibre properties, fibre content, fibre orientation, fibre distribution/dispersion into polymer matrix, interfacial adhesion, and the manufacturing process (Essabir et al. 2013a; Kakou et al. 2014). During the extrusion process, the torque was measured to compare the extrusion force required to transmit and blend the filler and matrix. Also, the matter pressure was controlled, as a response of compound state into the die. This parameter can be influenced by many variables such as the geometry of the die, screw configuration, screw speed, the fibre content, the temperature, or the resulted viscosity of the compound.

In this study, the machine used in the optimization of extrusion process parameter is the micro-extruder Thermo Scientific HAAKE Minilab II. This device was designed for mixing small-volume samples (7 cm^3). It is a conical twin-screw extruder with a recirculation channel. Two pressure sensors are positioned in the recirculation channel for evaluating the pressure in a given velocity gradient. It can be used easily with corotating and counterrotating screws. In general, for a given geometry of the screws, the micro-extruder is used to control various parameters such as temperature, mixing time, and screw speed. In this section, the effect of temperature, screw speed, and screw configuration (corotating and counterrotating screws) on the torque and pressure of the matter was studied.

For various concentrations (5, 10, and 15 wt%) of alkali-treated fibre (bagasse, coir, and doum), three different temperatures were tested (190, 200, and 210 °C), and for each temperature three screw speeds (n) were tested: 60, 80, and 120 rpm. Tables 9.1 and 9.2 exhibit the evolution of torque (N cm) and pressure of matter (bars) for all composites based on natural fibres (bagasse, coir, and doum) under different temperatures and screw speeds. Results show for all types of fibres and for two screw configurations (corotating and counterrotating screws) that a linear torque increases with fibre concentration and screw speed in the compound. The more fibre added and the increase in screw speed, the more torque needed and electricity consumed in the

Table 9.1 Counterrotating extrusion screw parameters for various fibre types and loadings versus a screw speed and temperature

Bagasse fibre					Coir fibre					Doum fibre				
T (°C)	Speed (rpm)	Content (wt%)	Torque (N cm)	Pressure (bars)	*T* (°C)	Speed (rpm)	Content (wt%)	Torque (N cm)	Pressure (bars)	*T* (°C)	Speed (rpm)	Content (wt%)	Torque (N cm)	Pressure (bars)
190	60	0	33	6	190	60	0	33	6	190	60	0	33	6
		5	32	6			5	33.5	8			5	39	7
		10	34.5	7.5			10	34.5	8			10	39	8
		15	39	9			15	39	9			15	42.5	9.5
	80	0	42	8		80	0	42	8		80	0	42	8
		5	43	8			5	45	10			5	50	10
		10	44	9.5			10	47	10			10	54	10.5
		15	49	10			15	53	11			15	58	11.5
	120	0	60	11		120	0	60	11		120	0	60	11
		5	62	11			5	72	13.5			5	71	12
		10	70	13			10	79	14			10	80	13
		15	77.5	14			15	84	14			15	83	15
200	60	0	30	5	200	60	0	30	5	200	60	0	30	5
		5	29	5			5	37	7			5	32	6
		10	31	6			10	41	7			10	35	7
		15	33	7			15	45	7.5			15	35	8
	80	0	44	8		80	0	44	8		80	0	44	8
		5	43	8			5	47	8.5			5	44	8
		10	45	8			10	47	8.5			10	46	8.5
		15	45	9			15	48	9			15	48	9
	120	0	64	10		120	0	64			120	0	64	10
		5	62	10			5	77	11			5	64	10
		10	72	10.5			10	77.5	11.5			10	67	11
		15	90	12			15	77.5	12			15	69	12

(continued)

Table 9.1 (continued)

Bagasse fibre					Coir fibre					Doum fibre				
T (°C)	Speed (rpm)	Content (wt%)	Torque (N cm)	Pressure (bars)	T (°C)	Speed (rpm)	Content (wt%)	Torque (N cm)	Pressure (bars)	T (°C)	Speed (rpm)	Content (wt%)	Torque (N cm)	Pressure (bars)
210	60	0	40	5	210	60	0	40		210	60	0	40	5
		5	38	5.5			5	24	5			5	28	5
		10	40	6			10	28	5			10	32	6
		15	45	6.5			15	31	5.5			15	32	7
	80	0	46	6		80	0	46			80	0	46	6
		5	45	6.5			5	37	7			5	39	7
		10	45	6.5			10	37	7			10	41	8
		15	47	7			15	38	7.5			15	46	9.5
	120	0	74	8		120	0	74			120	0	74	8
		5	72	8			5	57	8.5			5	58	9
		10	74	9			10	57	8.5			10	60	10
		15	74	10			15	64	9.5			15	62	11

Table 9.2 Corotating extrusion screw parameters for various fibre types and loadings versus a screw speed and temperature

Bagasse fibre					Coir fibre					Doum fibre				
T (°C)	Speed (rpm)	Content (wt%)	Torque (N cm)	Pressure (bars)	*T* (°C)	Speed (rpm)	Content (wt%)	Torque (N cm)	Pressure (bars)	*T* (°C)	Speed (rpm)	Content (wt%)	Torque (N cm)	Pressure (bars)
190	60	0	44	9	190	60	0	44	9	190	60	0	44	9
		5	40	9			5	41	9.5			5	38	10
		10	42	10			10	43	9.5			10	45	11
		15	49	11			15	45	11			15	48	11.5
	80	0	62	11		80	0	62	12		80	0		12
		5	59	11.5			5	59	12			5	58	12
		10	61	12			10	60	12			10	63	13
		15	70	13			15	63	12			15	67	13.5
	120	0	43	8		120	0	43	8		120	0	43	8
		5	84	14			5	87	14			5	88	15
		10	89	15			10	94	14.5			10	95	16
		15	91	15			15	94	15.5			15	97	16
200	60	0	40	9	200	60	0	40	9	200	60	0	40	9
		5	34	7.5			5	41	8			5	35	7.5
		10	39	8			10	41	8.5			10	39	8.5
		15	41	9			15	44	9			15	41	10
	80	0	60	11		80	0	60	11		80	0	60	11
		5	51	9			5	58	10			5	51	10
		10	51	10.5			10	59	10			10	57	10
		15	55	11			15	61	10			15	60	10.5
	120	0	77	12.5		120	0	77	12.5		120	0	77	12.5
		5	81	13			5	74	13			5	65	13
		10	85	13.5			10	79	13			10	68	13.5
		15	87	14			15	79	13			15	69	13.5

(continued)

Table 9.2 (continued)

Bagasse fibre					Coir fibre					Doum fibre				
T (°C)	Speed (rpm)	Content (wt%)	Torque (N cm)	Pressure (bars)	*T* (°C)	Speed (rpm)	Content (wt%)	Torque (N cm)	Pressure (bars)	*T* (°C)	Speed (rpm)	Content (wt%)	Torque (N cm)	Pressure (bars)
210	60	0	39	7	210	60	0	39	7	210	60	0	39	7
		5	30	7			5	37	7			5	34	7
		10	35	7.5			10	39	7.5			10	35	7
		15	40	8			15	39	7.5			15	37	8
	80	0	47	7.5		80	0	47	7.5		80	0	47	7.5
		5	50	8.5			5	47	7.5			5	48	9
		10	51	9.5			10	49	8			10	48	9
		15	55	10			15	50	8.5			15	52	10
	120	0	67	10		120	0	67	10		120	0	67	10
		5	63	11			5	69	11			5	69	11
		10	67	11			10	71	11			10	72	11
		15	67	11			15	71	11			15	78	12.5

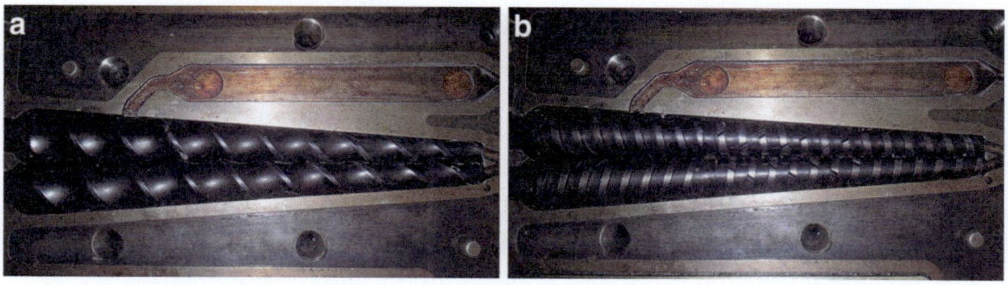

Fig. 9.1 Twin-screw extruder configuration: (**a**) corotating screw, (**b**) counterrotating screw

compound preparation. The temperature profile has an impact not only on the thermal properties of composites (in terms of degradation of fibre and matrix) but also on the manufacturing parameter. Results (Tables 9.1 and 9.2) show that increasing temperature, the torque values are decreased and at higher temperature the compound becomes more fluid (less viscous) and the compound requires low torque. Blending PP with natural fibres at different conditions (screw speed and fibre concentration) also led to a pressure rise. This result comes to join the torque study in confirming the influence of implementation parameter on manufacturing process parameter.

Moreover, it was observed that the corotating screw configuration shows a higher torque and pressure for all composite systems under different parameters (temperature and screw speed) compared to counterrotating screws. Figure 9.1 shows the screw geometry in corotating and counterrotating screw configuration. It was observed that in the two configurations, the screws are conical with the same dimension, but the difference is in the screw pitch length. The corotating screw configuration has a larger pitch than the counterrotating configuration which leads to a higher amount of mixed matter which leads to a higher pressure and torque. As screw speed and temperature were kept constant, torque and pressure of the compound must depend to two different parameters, i.e., screw geometry and the compound viscosity. First, if the feed rate was maintained constant, it was related to the matter weight. It is known that PP and fibres have different densities that could cause volume variations depending on the fibre/polymer ratio; however, the screw geometry has an important impact in terms of the amount of the transported, compressed, and mixed matter. In the case of counterrotating screw, the matter was more homogenized which allows a high shear stress, compared to corotating configuration when the matter undergoes less shear stress due to the large pitch of screw, leading to higher pressure and torque of corotating configuration.

9.2.3 Melt Rheology

Another parameter was the viscosity of matter. Therefore, the rheological behavior of produced composites was analyzed to complete the observation made during compounding. Rheology is the science that studies the response of materials (solid, liquid, gas) under applied stresses and deformations. It has been widely used to determine the rheological properties of the material such as viscosity and the elastic and viscous modulus, which lead to assess the percolated network structure, dispersion state of filler into the matrix, and the interaction between components (Elkhaoulani et al. 2013; Fernandes et al. 2013; Kabir et al. 2012). Many authors have studied the rule of rheological analysis as a potentially valuable technique to model and evaluate the microstructure, dispersion/distribution status, and the interfacial properties between fillers and the polymer matrix.

Using the MCR 500 (Physica) rheometer equipped with CTD600 device, previous strain sweep test at 1 Hz was applied to materials in order to determine strain at linear viscoelastic behavior; 5 % was chosen as strain. Frequency sweeps ranging from 500 to 0.05 Hz were applied. The measurements were carried out at 200 °C under small-amplitude oscillatory shear mode using parallel plate–plate geometry (25 mm diameter). The 2 mm thick sample disks were used for all tests. The dynamic rheological properties of composites were measured to characterize the dispersion state of the fibre within the polymer matrix, the degree of interaction between fibres and the polymer matrix, and the viscosity of the melt. In addition, the processing behavior of the polymer composites could also be obtained from their rheological characteristics. The dynamic rheological properties of three composites (PP/doum, PP/coir, PP/bagasse) were measured, and the PP/doum was taken as an example. Figure 9.2 shows the changes of the dynamic moduli (G′ and G″) and viscosity of

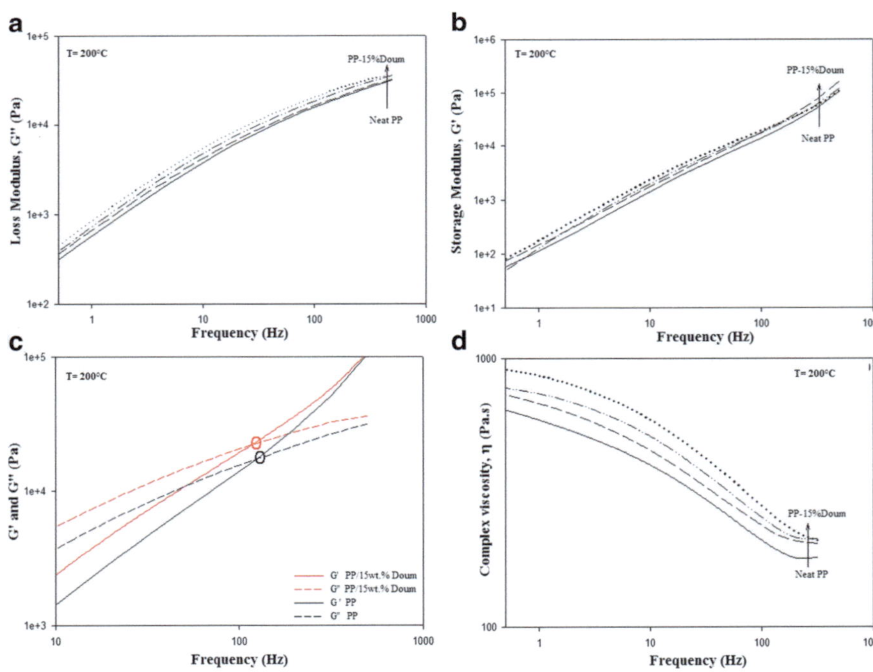

Fig. 9.2 Melt rheological properties as a function of fibre content and frequency at temperature of 200 °C: (**a**) loss modulus, (**b**) storage modulus, (**c**) crossover point, and (**d**) complex viscosity

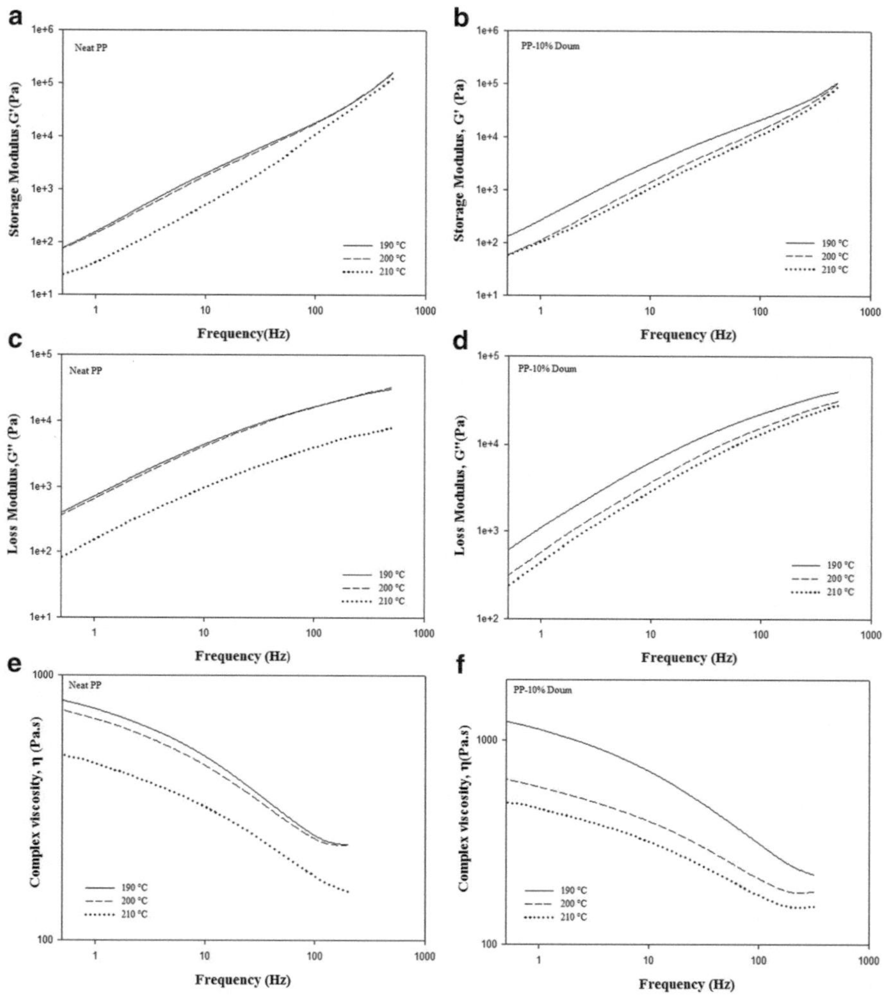

Fig. 9.3 Melt rheological properties as a function of frequency and temperature (**a**, **a**′): storage modulus, (**b**, **b**′) loss modulus, and (**c**, **c**′) complex viscosity

the melt as a function of doum fibre content and frequency. It can be obviously seen that the incorporation of fibres increases the dynamic properties of composites. This increase in complex modulus is attributed to the addition of rigid fibres leading to a change in the polymer chain displacement and the change in the molecular dynamics (Nekhlaoui et al. 2014a); this result reflects also the good interfacial adhesion between alkali-treated fibres and polymer matrix; removal of noncellulose compounds during alkali treatment exposes hydroxyl groups on the fibre's surface and favors interfacial bonding between the fibre's surface and polymeric matrix (Boujmal et al. 2014); this improvement in the interfacial adhesion affects the final properties of manufactured composites. In addition, as can be seen from Fig. 9.3, the storage modulus of the composites (5, 10, and 15 wt% fibre content) is higher than that of neat PP, indicating the stress transfer from the matrix to the fibres. It was also observed from Fig. 9.3 that, increasing frequency, both G′ and G″ increase; the noted result is related to the relaxation time response of polymeric chains (Essabir et al. 2013b, c). At high frequencies, the material behavior will be more like a solid because the polymeric chains will not get enough time to attend permanent deformation (irreversible flow) (Essabir et al. 2013b, c). But at low frequency, the polymer chains have enough time to relax. On the other hand, at higher frequencies, the elastic behavior is dominant compared to the viscous ones, whereas at low frequencies the viscous behavior is the predominant. It is important to notice that with addition of fibre, the storage and loss moduli increase, while the crossover frequency where G′ = G″ decreased. Figure 9.2c illustrates the crossover frequency, and it is noted that the crossover frequency values go from 125.99 to 116.98 Hz when doum fibre content increased from 0 to 15 wt%, respectively. This reduction in crossover frequency indicates that G″ becomes smaller than G′ and shows a plateau-like nonterminal behavior which implies elastic characteristic of the melt. This elastic behavior at lower frequency can be explained by a good interaction between the fibre and matrix which is enhanced by the chemical modification of fibres.

The complex viscosity curves as a function of fibre content and frequency are shown in Fig. 9.2d. These measurements showed a marked increase in complex viscosity with the introduction of fibres from 0 to 15 wt%. As can be seen in the figure, at 15 wt% fibre content, the complex viscosity of the melt is 100 times higher than the neat PP viscosity which makes the blending and processing of composite more difficult. Indeed, the distribution and dispersion of solid and rigid fibres into molten polymer prevent its flow (increase in the flow resistance) which is reflected by an increase in viscosity. It was found in previous works (Gamon et al. 2013) that the viscosity is closely related to the fibre–fibre interaction, fibre–matrix interaction, and fibre orientation. Moreover, it was observed that the melt viscosity decreases with increasing frequency, indicating the existence of a yield stress. At low frequency, shear rate was insufficient to ensure the mobility of the system. The presence of fibres perturbs normal polymer flow and hinders the mobility of chain segments, the fibres are randomly oriented, and polymer chains are entangled, and the fibre–fibre interaction and the frictions were more important. Gamon et al. (2013) observed that a good affinity between the compounds leads to a higher viscosity. In our case, the used alkali treatment enhances the compatibility between the fibre and matrix and ensures a good interfacial adhesion.

However, at higher shear rate, the fibres got oriented in the flow direction, and the fibre–fibre interaction was diminished; the lower perturbations in melt flow resulted in viscosity decrease. In conclusion, the increase of viscosity with fibre concentration is most likely due to an increase of the matter pressure and torque observed during the compounding process.

The problem in the compounding process is the difficulty to compound the fibres with thermoplastic polymer at high processing temperatures without thermally decomposing the natural fibre. Thus, the viscosity of the matter is also an important parameter in the manufacturing process; high viscosity makes the processing and blending of material composites more difficult. It is known that the temperature affects significantly the rheological dynamic properties (G′, G′, and complex viscosity) of the polymer composites. Figure 9.3 shows the changes in the dynamic properties of manufactured composites compared to neat PP as a function of temperature (Tazi et al. 2014; Parparita et al. 2014). As can be seen in Fig. 9.3, the dynamic modulus decreases with increase in temperature; this may be due to the softening of materials which led to the increase in the ability of the matrix chains to move freely. Moreover, it was observed that the viscosity decreases significantly in the composites compared to neat PP when temperature increases (Fig. 9.3c, c′). The presence of fibres makes the viscosity more sensitive to temperature; this is may be due to the internal humidity of fibres (Tazi et al. 2014; Parparita et al. 2014).

9.3 Composite Manufacturing

9.3.1 Extrusion

The thermoplastic polymer composites reinforced by short fibre are generally manufactured by extrusion processes and compression or injection molding (Arrakhiz et al. 2013b; Qaiss et al. 2014). These material composites provide improvements in mechanical properties such as strength, stiffness, and impact strength and have the advantage that can be shaped by the same processes as thermoplastic unfilled and without large modification tools. In every composite, there is an optimal fibre content where mechanical and rheological properties of composites are desirable. This optimum is affected by diverse parameters, among which are the fibre and matrix nature, the fibre aspect ratio, the distribution/dispersion of the fibre into a matrix (fibre agglomeration), the fibre–matrix interfacial adhesion, the fibre orientation, and the processing technique. A good dispersion and distribution of the fibres into the matrix can be achieved by effective compounding of the various components and by a suitable compounding process (extrusion) (Essabir et al. 2013a, b; Arrakhiz et al. 2013c; Nekhlaoui et al. 2014b; Sebaibi et al. 2014). The interfacial adhesion can be enhanced by using physical, chemical modification or by coupling agents or compatibilizers. However, a fibre modification and correct processing will produce homogeneous structure and fibre matrix interface that improve the final properties of manufactured composite. The mechanical properties for composite materials are carried out from tensile, flexural, compression, torsion, and dynamic analysis. The uniaxial tensile test is the most frequent mechanical test used for the characterization of materials. The tensile test is used to determine the following characteristics: The Young's modulus or stiffness modulus (E in MPa), the tensile strength or maximum stress (reached during the stress) (σM in MPa), the strain at yield (εM in mm/mm), and the plastic energy. Young modulus is determinated from stress versus strain curve, between 0.0025 and 0.005 % strain, according to ISO 527-1 (2012), which is seen to be linear for these tapes.

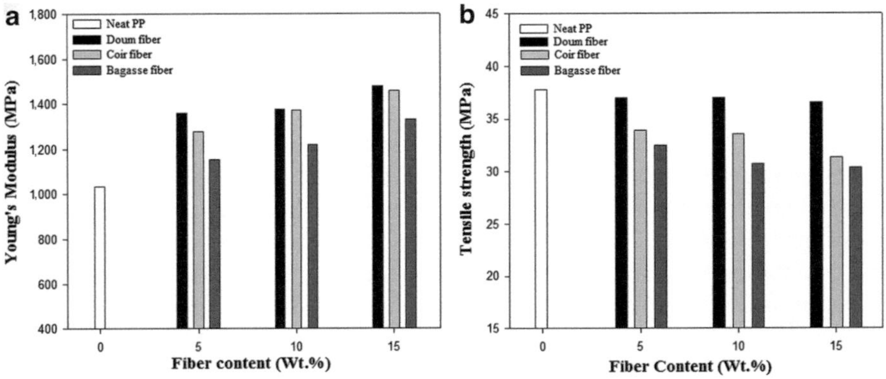

Fig. 9.4 Tensile properties of various fibre content for each fibre type as function of fibre content: (**a**) Young's modulus and (**b**) tensile strength

Figure 9.4 shows the influence of fibre loading and fibre species (doum, coir, and bagasse,) on the mechanical properties of polymer matrix. All fibre was alkali treated to remove noncellulosic components from the fibre surface and to enhance the compatibility between the hydrophilic fibres and the hydrophobic matrix. When compared to neat PP, the trend clearly shows an increase in the Young's modulus with results depending on the fibre type and loading. For all used fibres (Doum, Coir and Bagasse), the maximum Young's modulus gain is found in composites having 15 wt% fibre content. The Young's modulus shows a remarkable improvement from 1,034 MPa for neat PP to a higher value of 1,479 MPa for a 15 wt% doum fibre composite (Fig. 9.4a). This level of enhancement in the Young's modulus is assigned to a high quality of stress transfer from the matrix to fillers as natural fibres (Qaiss et al. 2014). A good dispersion and distribution of fibres into matrix ensured by the optimized manufacturing process parameter used, including screw configuration (corotating), screw speed (80 rpm), and temperature profile (200 °C), also explain this increase in the Young's modulus. Moreover, from Fig. 9.4b, tensile strength shows a decreased trend with the incorporation of fibres, and the maximum tensile strength loss was found at 15 wt%. The losses measured in tensile strength compared to neat PP are 4 % for the doum composite, 17 % for a coir composite, and 19 % for the bagasse composite. This decrease in tensile strength is quite a typical characteristic of natural fibre-reinforced composites (Qaiss et al. 2014). However, it was observed that the tensile strength of doum composite remains almost constant for all content. This is due to the good wetting of the treated fibre by the PP matrix, which leads to a good interfacial adhesion between compounds (Qaiss et al. 2014).

As a conclusion, the process used for composite production has a high importance too. Good fibre dispersion is needed to aim good material performance. Fibre orientation play a role in the mechanical properties of the materials, the longitudinal direction of fibres (compared to the stress direction) provides à higher properties (Wang et al. 2011). Twin-screw extrusion is a high shear process that can help to match good fibre dispersion (Arrakhiz et al. 2012c).

9.3.2 Injection

Injection molding consists in injecting using a rotating screw, a molten polymer in a cavity whose walls are kept at a temperature below the solidifying temperature or glass transition temperature of the polymer used. It is composed generally of two functional groups (Fig. 9.5): the injection group and the mold group. The injection process comprises a first step of mixing and plasticizing composite granules when the screw plasticized is rotated, when the screw plasticized is rotated and the molten polymer or polymer composite is forced through an orifice (gate) into a mold cavity, which is, namely, a filling stage, where it solidifies under pressure in the shape of the mold cavity, and then injecting the melt into the cavity by translation of the polymer screws, after cooling the polymer material in the mold and the final product is ejected (Arrakhiz et al. 2012c; Essabir et al. 2014).

During the injection processes, each fibre is transported by the flow, and its direction evolves according to the stress imposed by the matrix and the die walls. However, the fibres orient themselves and show a privileged orientation, which results in an anisotropic property of final composite materials. The anisotropy induced by the presence of fibres is an important property that must be taken into account in the design of materials and mold. In effect, the final characteristics and properties of the material composites are determined by the fibre orientation. At the same time, induced by the heterogeneous distribution and nonhomogeneous concentration of fibre length in the flow is often a source of defects that can induce early fatigue of the composite material. It is therefore essential to study the effect of the fibre orientation on mechanical properties of the manufactured composites.

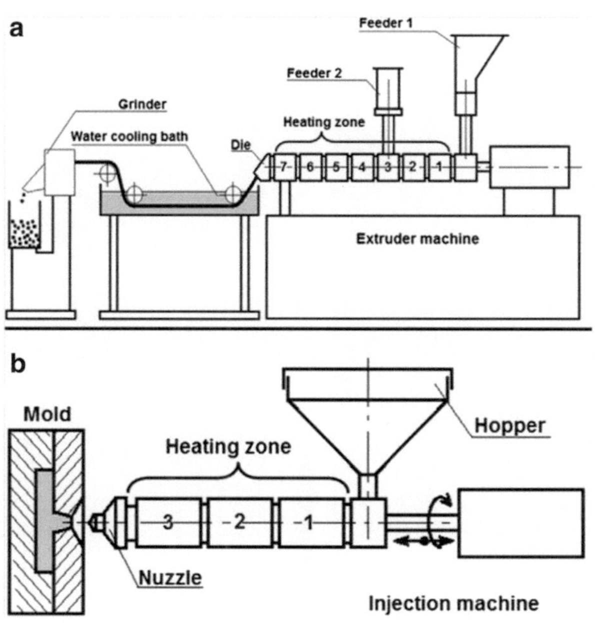

Fig. 9.5 (**a**) Extrusion process and (**b**) injection process

In this chapter, the composite is prepared by extrusion and injection molding process. Composites of polypropylene (PP) filled with *5*, *10*, and *15* wt% of alkali doum fibre were blended using a LEISTRITZ ZSE-18 twin-screw extruder (LEISTRITZ EXTRUSIONSTECHNIK GMBH, Germany). The melt-extruded composites were cooled in a water bath and then pelletized into granules of about *2* mm length, from which test samples were molded using an injection molding machine (ENGEL e-Victory). The principle of operation of the extruder and injection is shown in Fig. 9.5.

9.3.3 Mechanical Properties: Fibre Orientations

The compounded materials (polypropylene (PP) filled with *5*, *10*, and *15* wt% of alkali doum fibre) are also molded by injection process to the tensile specimens as well as the rectangular plate's samples of $100 \times 100 \times 3$ mm^3 (Fig. 9.6). However, it is known that during the injection process, the fibres are transported by the flow and its direction evolves according to the stress imposed by the matrix and the die walls.

In order to investigate the effect of the fibre orientation distribution on the tensile properties, each rectangular sample was cut into five specimens according to ISO 527-3 ($60 \times 8 \times 2$ mm^3) in different placing (Fig. 9.6). Two types of fibre orientation distribution were induced by adopting different placing directions. Two cases were considered for the placing directions according to the flow and tensile direction: parallel placements where the flow direction and the tensile direction are parallel (P1, P2, P3, P4, and P5) and transverse placements where the flow direction and tensile directions are transversal (T1, T2, T3, T4, and T5). Figure 9.6 illustrates each of the placing directions adopted for the manufacture of the specimens. The tensile

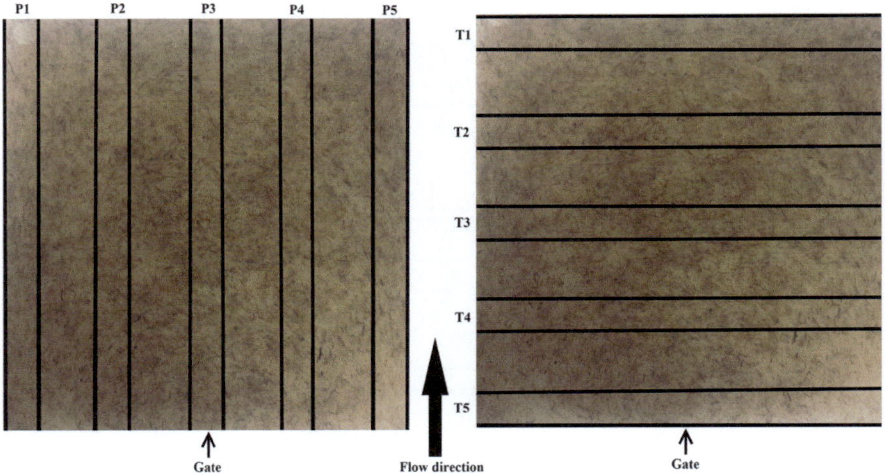

Fig. 9.6 Cutoff mechanical samples (along the parallel and transversal direction of the flow and tensile direction) of the composite

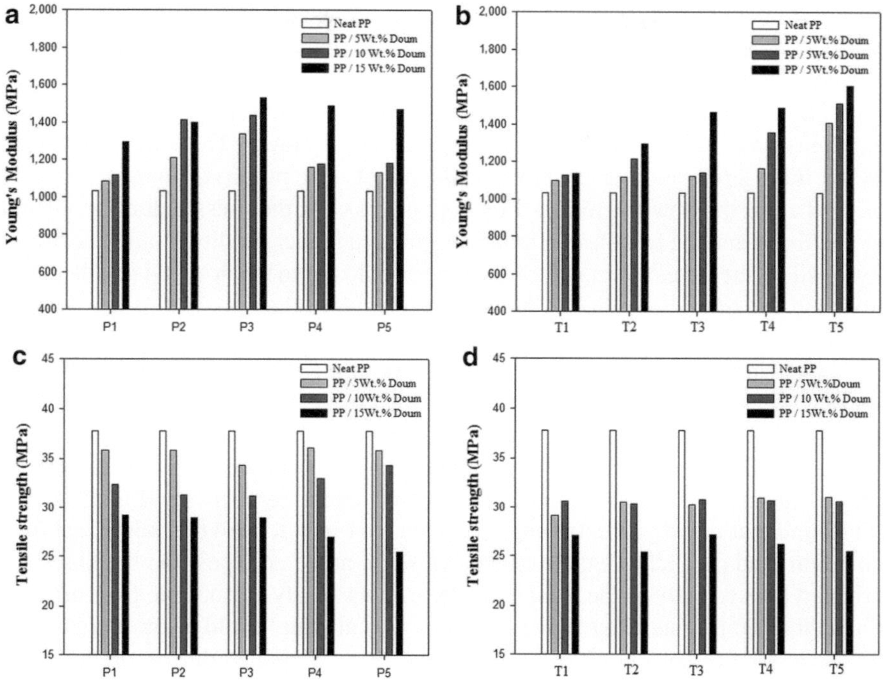

Fig. 9.7 Young's modulus and tensile strength of different doum fibre content, adopting different placing directions: (**a**, **b**) parallel placements where the flow direction and the tensile direction are parallel and (**a**′, **b**′) transverse placements where the flow direction and tensile directions are transversal

properties of samples were carried out under room temperature conditions (25 °C, 40 % relative humidity). Tensile tests were carried out according to ISO 527-3 using a universal testing machine (Instron 8821S, USA) at a crosshead speed of 5 mm/min using 5 kN load cell without extensometer.

Many factors influence fibre orientation in injection molding. Numerous studies have examined the orientation of fibres in short fibre-reinforced thermoplastic moldings (Nekhlaoui et al. 2014b; Sebaibi et al. 2014; Meyer et al. 2014). The modulus of short fibre-reinforced plastics is significantly affected by the fibre orientation against the loading direction, and the fibre orientation was generated by the polymer flow during the molding process. For this reason, it is important for strength prediction to analyze the fibre orientation in the reinforced plastics.

Figure 9.7 shows the evolution of tensile properties as the Young's modulus and tensile strength as a function of fibre loading and fibre direction against the flow and tensile direction. In the first observation, it is clear that the fibre content influences the tensile properties; this is due to the reinforcing effect of natural fibre. In addition, the increased fibre content did not bring a homogeneous distribution and could not contribute to the tensile property in longitudinal direction to achieve a maximum value. However, from Fig. 9.7, the Young's modulus of the manufactured composite is influenced by flow direction against the tensile direction. In the case of parallel placements where the flow direction and the tensile direction are parallel (P1, P2, P3, P4, and P5) and the P3 placement shows the higher Young's modulus for all fibre content, this position corresponds to the core of samples; however, at the extremity position (P1 and P5), the Young's modulus is lower. This can be explained by the fibre orientation at different positions; at core position, fibres are mostly aligned along the flow and tensile direction; in this case, the stress is directly applied to the fibres, and the response of materials shows a higher rigidity. While at extremity position, the fibres are mostly aligned perpendicular to the flow and tensile direction, and the stress is applied to the interface between the fibre and matrix leading to the lowest Young's modulus.

In the transverse configuration where the flow direction and tensile directions are transversal, it was found from Fig. 9.7a′ that the Young's modulus decreases by moving away from the injection point (gate) from maximum value at T5 position to lower one at T1 position, for all fibre content. The losses measured in tensile strength are 22, 29, and 20 % at 5, 10, and 15 wt% fibre content, respectively (Fig. 9.7b, b′). It is found that in the gate extremity, fibres tend to orient toward the transversal flow direction (and parallel to tensile direction), while at the core position the fibres are oriented transversally to the tensile direction. This highly anisotropic fibre orientation in the composites away from the injection point (gate) could be the main reason for the lower mechanical property, where the stress is applied directly to the interface between the fibre and matrix.

Figure 9.7b, b′ shows the tensile strength of manufactured composites as a function of flow and tensile direction. It is noted that the tensile strength curve shows a stabilized trend at various positions and for two configurations (parallel and transverse to flow and tensile direction). However, it was concluded that the tensile strength has not been affected by fibre orientation, because the tensile strength gives an idea of the degree of interfacial adhesion; however, in our case, the interfacial adhesion between fibres and matrix was assured by chemical treatment of fibres which ensure a good affinity and compatibility between them.

9.4 Conclusion and Perspective

The aim of this chapter was to study the processing of natural fibre-reinforced thermoplastic matrix with twin-screw extrusion and injection process using polypropylene as the matrix polymer and alkali-treated fibres (doum, coir, and bagasse). In the first part of the work, natural fibre-reinforced polypropylene composites were manufactured using twin-screw extrusion. The effect of twin-screw extrusion conditions as temperature, screw configuration and screw speed on the manufacturing process parameter on the extruder device as matter pressure and torque during the compounding process of composites was investigated. The second part of the work focused on the fibre orientation during the injection process; during the injection processes, the fibres are transported by the flow, and it orient themselves leading which to an anisotropic property of final composite materials. The study showed that a better reinforcement is achieved by using the alkali-treated fibres (doum, bagasse, and coir) with a higher aspect ratio, when the manufacturing process was optimized. The extrusion conditions have a remarkable influence of the processing, and the corotating screw configuration shows an optimized value in terms of pressure and torque. However, the study of fibre orientation shows that the increased fibre content did not bring a homogeneous distribution. In addition, it was found that for parallel placements at core position, fibres are mostly aligned along the flow and tensile direction and at extremity position the fibres are mostly aligned perpendicular to the flow and to the tensile direction, while, for the transversal placements, fibres tend to orient parallel to the tensile direction in the gate extremity and transversally in the core position.

As a second step of this study and due to anisotropic properties of the molded composites, it is very important to clarify the relationship between the fibre orientation and mechanical properties by using a distribution of fibre orientation by a numerical analysis method performed by the numerical simulation in comparison with the results obtained from the image processing.

References

Al-Maadeed MA, Shabana YM, Khanam PN (2014) Processing, characterization and modeling of recycled polypropylene/glass fibre/wood flour composites. Mater Des 58:374–380

Arrakhiz FZ, El Achaby M, Kakou AC, Vaudreuil S, Benmoussa K, Bouhfid R, Fassi-Fehri O, Qaiss A (2012a) Mechanical properties of high density polyethylene reinforced with chemically modified coir fibers: impact of chemical treatments. Mater Des 37:379–383

Arrakhiz FZ, Elachaby M, Bouhfid R, Vaudreuil S, Essassi M, Qaiss A (2012b) Mechanical and thermal properties of polypropylene reinforced with alfa fiber under different chemical treatment. Mater Des 35:318–322

Arrakhiz FZ, El Achaby M, Benmoussa K, Bouhfid R, Essassi EM, Qaiss A (2012c) Evaluation of mechanical and thermal properties of Pine cone fibers reinforced compatibilized polypropylene. Mater Des 40:528–535

Arrakhiz FZ, Malha M, Bouhfid R, Benmoussa K, Qaiss A (2013a) Tensile, flexural and torsional properties of chemically treated alfa, coir and bagasse reinforced polypropylene. Compos Part B Eng 47:35–41

Arrakhiz FZ, Benmoussa K, Bouhfid R, Qaiss A (2013b) Pine cone fiber/clay hybrid composite: mechanical and thermal properties. Mater Des 50:376–381

Arrakhiz FZ, El Achaby M, Malha M, Bensalah MO, Fassi-Fehri O, Bouhfid R, Benmoussa K, Qaiss A (2013c) Mechanical and thermal properties of natural fibers reinforced polymer composites: doum/low density polyethylene. Mater Des 43:200–205

Boujmal R, Essabir H, Nekhlaoui S, Bensalah MO, Bouhfid R, Qaiss A (2014) Composite from polypropylene and henna fiber: structural, mechanical and thermal properties. J Biobased Mater Bioenergy 8:246–252

Elkhaoulani A, Arrakhiz FZ, Benmoussa K, Bouhfid R, Qaiss A (2013) Mechanical and thermal properties of polymer composite based on natural fibers: Moroccan hemp fibers/polypropylene. Mater Des 49:203–208

El-Sabbagh A, Steuernagel L, Ziegmann G, Meiners D, Toepfer O (2014) Processing parameters and characterisation of flax fibre reinforced engineering plastic composites with flame retardant fillers. Compos Part B Eng 62:12–18

Essabir H, Hilali E, Elgharad A, El Minor H, Imad A, Elamraoui A, Al Gaoudi O (2013a) Mechanical and thermal properties of bio-composites based on polypropylene reinforced with nut-shells of Argan particles. Mater Des 49:442–448

Essabir H, Nekhlaoui S, Malha M, Bensalah MO, Arrakhiz FZ, Qaiss A, Bouhfid R (2013b) Bio-composites based on polypropylene reinforced with almond shells particles: mechanical and thermal properties. Mater Des 51:225–230

Essabir H, Elkhaoulani A, Benmoussa K, Bouhfid R, Arrakhiz FZ, Qaiss A (2013c) Dynamic mechanical thermal behavior analysis of doum fibers reinforced polypropylene composites. Mater Des 51:780–788

Essabir H, Bensalah MO, Bouhfid R, Qaiss A (2014) Fabrication and characterization of apricot shells particles reinforced high density polyethylene based bio-composites: mechanical and thermal properties. J Biobased Mater Bioenergy 8:344–351

Feldmann M, Bledzki AK (2014) Bio-based polyamides reinforced with cellulosic fibres – processing and properties. Compos Sci Technol 100:113–120

Fernandes EM, Mano JF, Reis RL (2013) Hybrid cork–polymer composites containing sisal fibre: morphology, effect of the fibre treatment on the mechanical properties and tensile failure prediction. Compos Struct 105:153–162

Fernandes EM, Correlo VM, Mano JF, Reis RL (2014) Polypropylene-based cork–polymer composites: processing parameters and properties. Compos Part B Eng 66:210–223

Gamon G, Evon P, Rigal L (2013) Twin-screw extrusion impact on natural fibre morphology and material properties in poly(lactic acid) based biocomposites. Ind Crop Prod 46:173–185

Hashimoto M, Okabe T, Sasayama T, Matsutani H, Nishikawa M (2012) Prediction of tensile strength of discontinuous carbon fiber/polypropylene composite with fiber orientation distribution. Compos Part A Appl Sci 43:1791–1799

Ho M, Wang H, Lee JH, Ho C, Lau K, Leng J, Hui D (2012) Critical factors on manufacturing processes of natural fibre composites. Compos Part B Eng 43:3549–3562

ISO 527-1 (2012) Plastics—determination of tensile properties—part 1: general principles

Kabir MM, Wang H, Lau KT, Cardona F, Aravinthan T (2012) Mechanical properties of chemically-treated hemp fibre reinforced sandwich composites. Compos Part B Eng 43: 159–169

Kakou CA, Arrakhiz FZ, Trokourey A, Bouhfid R, Qaiss A, Rodrigue D (2014) Influence of coupling agent content on the properties of high density polyethylene composites reinforced with oil palm fibers. Mater Des 63:641–649

Kim JW, Lee DG (2014) Study on the fiber orientation during compression molding of reinforced thermoplastic composites. Int J Precis Eng Manuf Green Technol 1:335–339

Malha M, Nekhlaoui S, Essabir H, Benmoussa K, Bensalah M-O, Arrakhiz F-Z, Bouhfid R, Qaiss A (2013) Mechanical and thermal properties of compatibilized polypropylene reinforced by woven doum. J Appl Polym Sci 130(6):4347–4356

Merkel K, Rydarowski H, Kazimierczak J, Bloda A (2014) Processing and characterization of reinforced polyethylene composites made with lignocellulosic fibres isolated from waste plant biomass such as hemp. Compos Part B Eng 67:138–144

Meyer KJ, Hofmann JT, Baird DG (2014) Prediction of short glass fiber orientation in the filling of an end-gated plaque. Compos Part A Appl Sci 62:77–86

Nekhlaoui S, Essabir H, Bensalah MO, Fassi-Fehri O, Qaiss A, Bouhfid R (2014a) Fracture study of the composite using essential work of fracture method: PP–SEBS–g–MA/E1 clay. Mater Des 53:741–748

Nekhlaoui S, Essabir H, Kunal D, Sonakshi M, Bensalah MO, Bouhfid R, Qaiss A (2014b) Comparative study for the talc and two kinds of Moroccan clay as reinforcements in polypropylene-SEBS-g-MA matrix. Polym Compos. doi:10.1002/pc.22986

Notta-Cuvier D, Lauro F, Bennani B (2014) An original approach for mechanical modelling of short-fibre reinforced composites with complex distributions of fibre orientation. Compos Part A Appl Sci 62:60–66
Ou R, Zhao H, Sui S, Song Y, Wang Q (2010) Reinforcing effects of Kevlar fiber on the mechanical properties of wood-flour/high-density-polyethylene composites. Compos Part A Appl Sci 41:1272–1278
Oumer AN, Mamat O (2012) A study of fiber orientation in short fiber-reinforced composites with simultaneous mold filling and phase change effects. Compos Part B Eng 43:1087–1094
Pan Y, Zhong Z (2014) Modeling of the mechanical degradation induced by moisture absorption in short natural fiber reinforced composites. Compos Sci Technol 103:22–27
Parparita E, Darie RN, Popescu C-M, Uddin Md A, Vasile C (2014) Structure–morphology–mechanical properties relationship of some polypropylene/lignocellulosic composites. Mater Des 56:763–772
Qaiss AEK, Bouhfid R, Essabir H (2014) Natural fibers reinforced polymeric matrix: thermal, mechanical and interfacial properties. Springer, Switzerland
Rojo E, Alonso MV, Oliet M, Del Saz-Orozco B, Rodriguez F (2015) Effect of fiber loading on the properties of treated cellulose fiber-reinforced phenolic composites. Compos Part B Eng 68:185–192
Salleh FM, Hassan A, Yahya R, Azzahari AD (2014) Effects of extrusion temperature on the rheological, dynamic mechanical and tensile properties of kenaf fiber/HDPE composites. Compos Part B Eng 58:259–266
Sebaibi N, Benzerzour M, Abriak NE (2014) Influence of the distribution and orientation of fibres in a reinforced concrete with waste fibres and powders. Constr Build Mater 65:254–263
Shi S, Sun Z, Hu X, Chen H (2014) Carbon-fiber and aluminum-honeycomb sandwich composites with and without Kevlar-fiber interfacial toughening. Compos Part A Appl Sci 67:102–110
Shouha P, Swain M, Ellakwa A (2014) The effect of fiber aspect ratio and volume loading on the flexural properties of flowable dental composite. Dent Mater 30:1234–1244
Tazi M, Erchiqui F, Godard F, Kaddami H, Ajji A (2014) Characterization of rheological and thermophysical properties of HDPE–wood composite. J Appl Polym Sci. doi:10.1002/app.40495
Unterweger C, Brüggemann O, Fürst C (2014) Effects of different fibers on the properties of short-fiber-reinforced polypropylene composites. Compos Sci Technol 103:49–55
Valente M, Sarasini F, Marra F, Tirillò J, Pulci G (2011) Hybrid recycled glass fiber/wood flour thermoplastic composites: manufacturing and mechanical characterization. Compos Part A Appl Sci 42:649–657
Wang J, Geng C, Luo F, Liu Y, Wang K, Fu Q, He B (2011) Shear induced fiber orientation, fiber breakage and matrix molecular orientation in long glass fiber reinforced polypropylene composites. Mater Sci Eng A 528:3169–3176
Węcławski BT, Fan M, Hui D (2014) Compressive behaviour of natural fibre composite. Compos Part B Eng 67:183–191
Zakikhani P, Zahari R, Sultan MTH, Majid DL (2014) Extraction and preparation of bamboo fibre-reinforced composites. Mater Des 63:820–828

Chapter 10
Manufacturing of Long Puchika Grass Fibre Reinforced Polyester Composites: Assessment Under Mechanical and Dielectric Loading

Nadendla Srinivasababu

Abstract The introduction of the 'sustainable development' concept changed the mindset of environmental concerned people to introduce the new materials reinforced with natural renewable substances. An attempt is made in this work to introduce a new natural fibre extracted from the puchika grass, and its physical and mechanical performance is evaluated. Variation of chemicals, concentration and soaking time on this natural fibre resulted in the diversified characteristics of the fibre surface and are observed using scanning electron microscope. Fibre in its raw form and chemically treated is reinforced into the polyester matrix at various contents to assess their mechanical performance under tensile, flexural and impact loadings. Puchika grass CT-7 FRP composites have exhibited the highest tensile strength (58.09) and modulus (1.313 GPa) at maximum fibre content. The composites reinforced with puchika grass CT-3 and CT-7 at maximum fibre volume fraction resulted in maximum flexural strength of 74.93 MPa and 7.51 GPa, respectively. The Charpy impact strength of the material is 65.01 kJ/m^2 at 45.95 % fibre volume fraction. The insulation ability of the composites is also assessed after dielectric strength test, and the highest value achieved is 9.67 kV/mm at very low fibre content.

Keywords Puchika grass fibre • Polyester resin • Mechanical properties • Dielectric strength • Scanning electron micrograph

10.1 Introduction

The substances originated from the nature are called natural materials which are available in renewable and non-renewable forms. Some potential natural renewable resource/biomaterials that are grown extensively in the tropical regions of India like

N. Srinivasababu (✉)
Department of Mechanical Engineering,
Vignan's Lara Institute of Technology and Science,
Vadlamudi, Guntur 522213, Andhra Pradesh, India
e-mail: cnjlms22@yahoo.co.in

M.S. Salit et al. (eds.), *Manufacturing of Natural Fibre Reinforced Polymer Composites*, DOI 10.1007/978-3-319-07944-8_10

Kerala are fruits, plants and trees that exist in diversified forms. In the ancient times, Indians had very intact relationship with the natural world, be it for gathering of food, clothing and shelter to protect them from the surrounding weathering conditions. So, for their house (huts) construction, various stems of the plants and grasses were utilized along with the black sand which are evidenced from the history and are seen in some of the villages in and around the state of Andhra Pradesh and the other parts of India. For instance, coconut one-seeded drupe (dry) fire was utilized from a long time for the making of ropes used for the various household and agricultural applications; leaves of the Palmyra palm were used for house cover; the young grown plants of various varieties were used for nutrition of the sand. Another important area of utilization of natural fibres is in the textile industry to make shirts, bags and other degradable items which are of the present interest to the public. Several researchers have struggled a lot to exploit various natural fibres in chopped, short and long forms to reinforce them into a variety of matrices like polyester, vinyl ester, epoxy and thermoplastic materials for the making of the composite materials. The use of natural fibres dates as far back as 3,000 years ago to ancient Egypt, where the straw was reinforced into clay (composite). The composites were reinforced with natural fibres either in as is or in modified form (physical, chemical, etc.) to attain the target properties. The green reinforcements are of low price, abundantly available, renewable in nature and of high specific properties.

The role of fibres and their applications as reinforcement in composites were neatly exploited (Thomas and Pothan 2009). The German auto manufacturers Mercedes, BMW, Audi and Volkswagen have taken the initiative to make interior and exterior parts using natural fibres as reinforcement in the composites made by injection moulding. Using 35 % Baypreg F semi-rigid elastomer from Bayer and a 65 % blend of flax, hemp and sisal, an inner door panel for the 1999 S-Class Mercedes-Benz was made in Germany.

These situations have given the author an opportunity to do review on nearly 1,500 natural fibre reinforced polymer composites articles. Based on the article survey, the author understood that almost all the natural fibres were extracted by 'retting and manual extraction method', so he introduced various new methods (Srinivasababu et al. 2010, 2012, 2014) of extraction of various new natural fibres, and the gist is given in Table 10.1, and the procedures were published elsewhere.

A specific review is conducted on various grass fibre reinforced polymer composites along with the regular study, and the key points are given in the order of grass composites that possess polyester matrix, thermoplastic material, epoxy and

Table 10.1 Novel methods of extraction of natural fibres (Srinivasababu et al. 2010, 2012, 2014)

Rolling and splitting (R and S)	Pure splitting (PS)	Hot water immersion (HWI)
Turmeric stem (TS)	Palm tree sprout leaf (PTSL)	Palmyra palm empty fruit bunch (EFB)
Turmeric petiole (TP)	Indian date leaf (IDL)	
	Broom grass (BG)	
	Sacred grass	
	Palm petiole	

rubber. Indian grass fibre reinforced soy protein composites were analysed for their mechanical properties when the fibre was chemically treated (Liu et al. 2004a, b). The influence of chemically extracted elephant grass fibre on tensile strength and modulus was determined and is 58 and 41 % more than the untreated fibre (Murali Mohan Rao et al. 2007). Flexural strength and modulus of the polyester composites reinforced with chemically extracted wild cane grass fibre had shown 7 and 17 % more than the composites reinforced with untreated fibre (Ratna Prasad et al. 2011). The influence of hybridization effect was studied and reported that snake grass/banana and snake grass/coir fibre reinforced polyester composites have better performance than snake grass FRP composites (Satish Kumar et al. 2013). Waste broom grass fibre has a tensile strength of 297.58 MPa, 18.28 GPa (Ramanaiah et al. 2012). The increase in snake grass fibre content resulted in an increase in tensile and flexural properties (Satish Kumar and Navaneethakrishnan 2012).

The degradation of Bermuda grass and orchard grass was evaluated gravimetrically and by SEM and TEM (Danny Akin and Luanne Rigsby 1985). An analysis was made to use switch grass as a potential bioenergy feedstock (McLaughlin et al. 1996). The effect of morphology, structure and properties of Indian grass/native grass of the USA was reported (Liu et al. 2004a, b). Alkali treatment resulted in the improvement of thermal stability, tensile properties and crystallinity of the Napier grass fibres (Obi Reddy et al. 2009). Napier grass fibre (native African) strands treated with 10 % NaOH showed optimum tensile strength, modulus and percentage elongation with an improvement of 51.9, 47.3 and 12.1 %, respectively (Kommula et al. 2013a, b).

Natural rubber composites reinforced with 400 mesh grass (*Cyperus tegetum* Roxb) fibre had shown superior mechanical properties (De et al. 2004). Further rubber and grass fibre interface was modified, and improved bonding was achieved by the addition of resorcinol formaldehyde latex as bonding agent for the formulation (De et al. 2006). Big bluestem grass fibre reinforced high-density polyethylene composites were analysed for their mechanical properties (Liu et al. 2007). The composites reinforced with 1 % alkali-treated grass fibre and 55 % phenol formaldehyde resin had shown high tensile strength, whereas the composites prepared from 5 % alkali-treated grass fibre and 55 % resin had exhibited maximum flexural properties (De et al. 2007). Untreated and alkali-treated long Napier grass fibre reinforced epoxy composites (Kommula et al. 2013a, b).

Hence an attempt is made in this work to introduce a new natural fibre extracted from the puchika grass which is used in as is condition and treated in the matrix for making composites. The composites are analysed for mechanical and dielectric properties.

10.2 Materials, Methods of Manufacturing and Testing

In this section the focus is made on describing the materials used, method of manufacturing employed to extract the fibre and to make the composites and testing procedures followed.

Fig. 10.1 Dried puchika grass

10.2.1 Materials

The materials considered in the present work are puchika grass fibre in as is condition and modified chemically and polyester resin as reinforcement and matrix, respectively.

10.2.1.1 Puchika Grass Fibre

Puchika grass available in Eluru, Andhra Pradesh, is collected from the field, and the crop is of 5–6.2 ft. in height. Initially the grass is exposed to the sunlight for the first-stage removal of moisture, and the dried grass is shown in Fig. 10.1. Then the fibre is extracted by PS method and the extracted fibre is visible as shown in Fig. 10.2, and the procedure of extraction is described in the subsequent section.

10.2.1.2 Ecmalon 4413 Matrix

The matrix used in the present work is Ecmalon 4413, which is purchased from Ecmas Resins Pvt. Ltd., Hyderabad, Telangana, India. The resin is a general purpose unsaturated polyester resin of medium reactivity and has medium viscosity, and the mouldings are rigid. This kind of matrix is ideally suited for mouldings using wet lay-up technique. The appearance of the liquid is clear to light yellow. As per the specifications given by the manufacturer, the properties of the liquid resin are given in Table 10.2.

Fig. 10.2 Extracted puchika grass by pure splitting method

Table 10.2 Properties of Ecmalon 4413 liquid resin

Property	Value
Viscosity at 25 °C (cps)	500 (Brookfield viscometer)
Specific gravity (25/25 °C)	1.13
Acid value (mgKOH/g)	25
Volatiles at 150 °C (%)	35
Gel time at 25 °C (minutes)	20

10.2.2 Methods of Manufacturing

The methods of extraction of fibre and fabrication of the composites are discussed in the forthcoming sections.

10.2.2.1 Pure Splitting Method of Puchika Grass Fibre

Initially one end of the grass is positioned under the needle of 0.25 mm diameter and is moved along its length to split into two. Care must be taken during the passage and straight path is to be maintained during its travel, so that the grass is exactly separated into fibres. The procedure is repeated for entire grass is converted to fibres. Palm tree sprout leaf fibres were extracted by the same method and were firstly introduced.

10.2.2.2 Chemical Processing of Puchika Grass Fibre

The role of alkali, permanganate treatments is given as follows. In 1919 Beckmann proposed the first method of NaOH treatment which consists of straw that was treated with 1.5 % NaOH solution and washed with huge quantity of water.

Table 10.3 Puchika grass chemical treatment procedures

Fibre	Chemical used for treatment	Soaking time (h)	Concentration (M)	Fibre name after chemical treatment
Puchika grass	NaOH	7	0.275	Puchika grass CT-1
Puchika grass	NaOH	14	0.275	Puchika grass CT-2
Puchika grass	NaOH	28	0.275	Puchika grass CT-3
Puchika grass	$KMnO_4$	7	0.03164	Puchika grass CT-4
Puchika grass	$KMnO_4$	7	0.03164	Puchika grass CT-5[a]
Puchika grass	$KMnO_4$	14	0.03164	Puchika grass CT-6[a]
Puchika grass	$KMnO_4$	28	0.03164	Puchika grass CT-7[a]

[a]Puchika grass fibre is pretreated with NaOH

Subsequently in 1964 Wilson and Pigden developed new methods of treatment with aqueous solutions of NaOH (Galletti 1991). The consumption and conversion of crystalline cellulose through mercerization leads to the formation of amorphous cellulose. In the network structure, the hydrogen bond removal is the key step. Alkali cellulose is formed as a result of NaOH penetration into the crystalline region of cellulose I. In the next phase non-reacted cellulose is percolated which results in the formation of cellulose II. The elementary cells are modified during the transformation of cellulose I to II which is an irreversible exothermic reaction. The effect of alkali treatment depends on concentration, time, temperature, etc. Permanganate treatment results in the formation of cellulose radicals, through the formation of MnO_3^- ion. This radical will enhance the interlocking at the fibre matrix interface (Frederick Wallenberger 2004).

Hence the fibre is subjected to various chemical treatments under different conditions and is described in Table 10.3. Chemical treatment procedures are adopted from the knowledge of literature review and the experience of the author in previous works.

Then the fibre in as is condition and treated condition is dried in NSW-143 Oven Universal, supplied by Narang Scientific Works Pvt. Ltd., New Delhi, to make the fibre free from moisture.

10.2.2.3 Rolled Hand Lay-Up Method of Manufacturing Composites

The entire composites are fabricated by hand lay-up technique for various kinds of mechanical and electric tests, viz., tensile, flexural, impact and dielectric, at different volume fractions of the fibre as per the procedure of regular hand lay-up with slight modifications (Srinivasababu et al. 2010). The extracted puchika grass fibres are very stiff and look like a stick. The chemically processed fibres are trying to mingle with each other which is quite undesirable, since the present work concentrates on understanding complete ability of the reinforcement in its longitudinal direction. Hence with the help of a 25 mm diameter mild steel roller, the entire mould surface is rolled slowly and manually from one end to the other end.

This resulted in proper settlement of the fibre in the mould with correct alignment. Then the rest of the procedure is followed according to the regular procedure of hand lay-up technique, and the composites are fabricated.

10.2.3 Testing Procedures

Initially the fibres were examined for its properties, and then the composites were evaluated under various loading conditions.

10.2.3.1 Fibre Density and Tensile Test

Density of the fibre is determined by ASTM D3800-99 Procedure A-Buoyancy (Archimedes) method. In order to assess the capabilities of the reinforcement, the tensile properties of the fibre are determined by single fibre tensile test (SFTT) according to ASTM C 1557-03$^{\epsilon 1}$. Ecmalon 4413 unsaturated polyester resin is procured from Ecmas Resins Pvt. Ltd., Hyderabad.

10.2.3.2 Composite Test

Then the manufactured composites are taken out from the mould, ground using a belt-type grinding machine and conditioned as per ASTM procedure in NSW-178 Environmental Chamber, procured from Narang Scientific Works Pvt. Ltd., New Delhi.

Using PC 2000 Electronic Tensometer, supplied by Kudale Instruments Pvt. Ltd., Pune, tensile and flexural tests are conducted, whereas the impact test is conducted on Computerized Izod/Charpy Impact Tester, which can be bought from International Equipments, Bombay. The notch for impact test is cut using a notch cutter precisely. The dielectric test is performed on a dielectric strength tester.

10.3 Results and Discussion

Initially using the JEOL JSM-5350A, the morphological study of the puchika grass fibres is examined with the kind permission of the Department of Physics, University of Pune, Pune. The study concentrates only to understand the surface of the fibre before and after treatments. From Fig. 10.3 it is understandable that the structure resembles a honeycomb, and the treatment resulted in the suppression of the structure which is visible as shown in Figs. 10.4, 10.5, 10.6, 10.7, 10.8, 10.9 and 10.10.

Puchika grass fibre in untreated and treated condition is used to determine their density, and the values range from 438.44 to 908.53 kg/m^3 and is graphically

Fig. 10.3 Scanning electron micrograph of puchika grass fibre

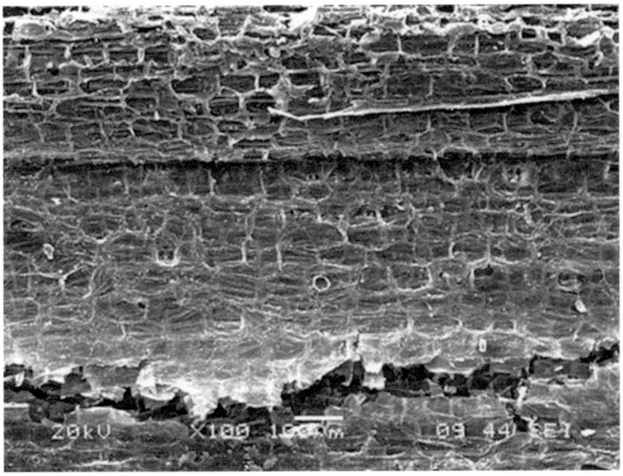

Fig. 10.4 Scanning electron micrograph of puchika grass CT-1 fibre

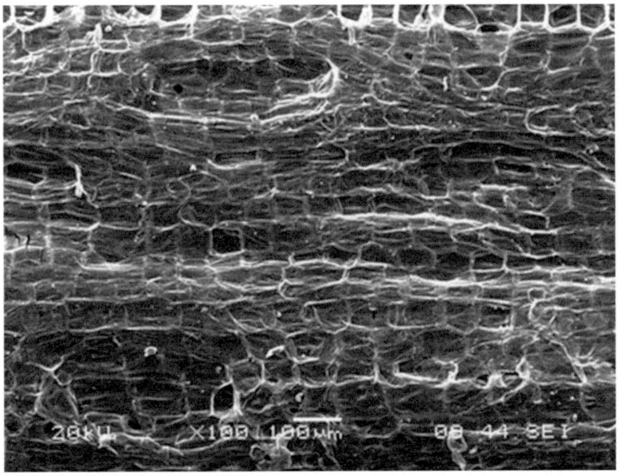

Fig. 10.5 Scanning electron micrograph of puchika grass CT-2 fibre

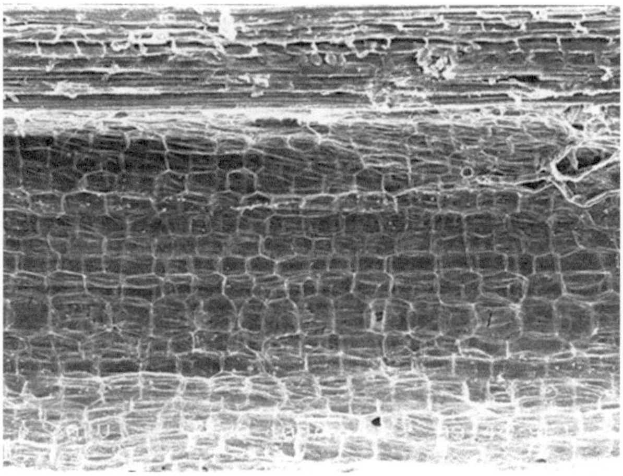

Fig. 10.6 Scanning electron micrograph of puchika grass CT-3 fibre

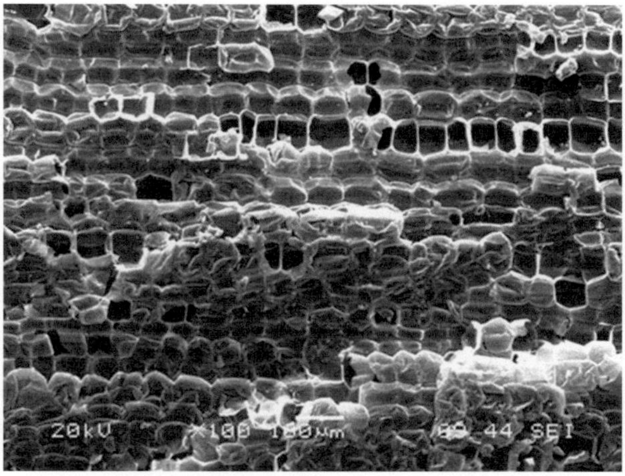

Fig. 10.7 Scanning electron micrograph of puchika grass CT-4 fibre

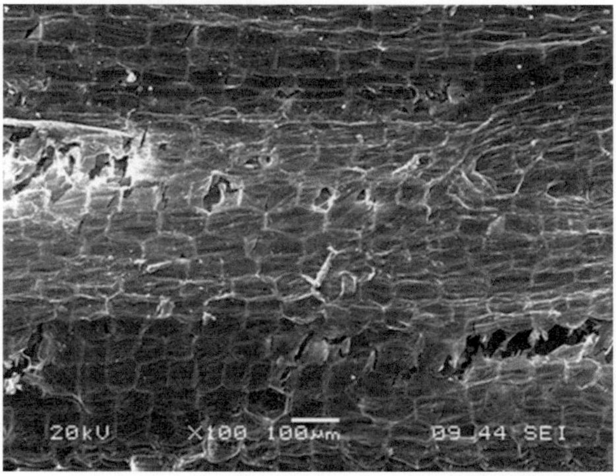

Fig. 10.8 Scanning electron micrograph of puchika grass CT-5 fibre

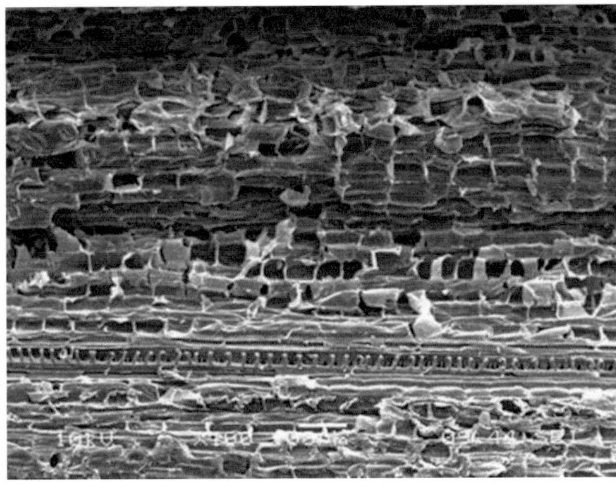

Fig. 10.9 Scanning electron micrograph of puchika grass CT-6 fibre

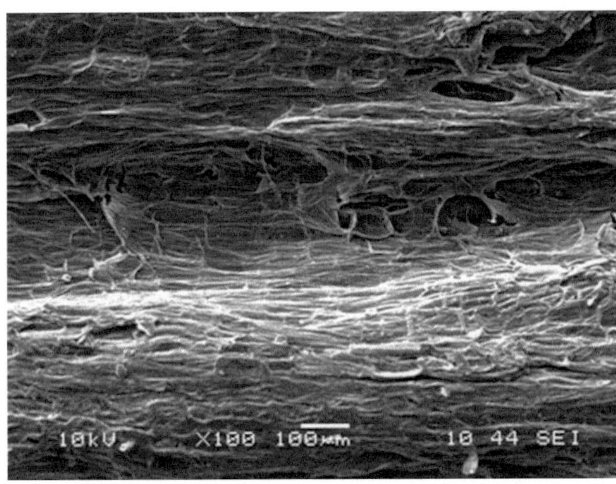

Fig. 10.10 Scanning electron micrograph of puchika grass CT-7 fibre

represented in Fig. 10.11. The reinforcement ability is examined from SFTT, and the highest value of tensile strength and modulus achieved is 126.62 MPa and 2.11 GPa in the case of Puchika grass CT-7 and CT-1 fibre, respectively. Chemical treatment obviously enhanced the tensile properties and is varied with respect to the treatment which is shown in Figs. 10.12 and 10.13.

The properties of the fibre had given enough confidence to reinforce into the polyester matrix, and thereby the prepared composites are subjected to various types of tests. Out of which the tensile test of the puchika grass CT-7 FRP composites according to ASTM D5083-02 has exhibited the highest tensile strength of 58.04 MPa and modulus of 1.31 GPa at maximum fibre volume fraction, i.e. 22.84 %. Influence of chemical treatment is visible as shown in Figs. 10.14 and

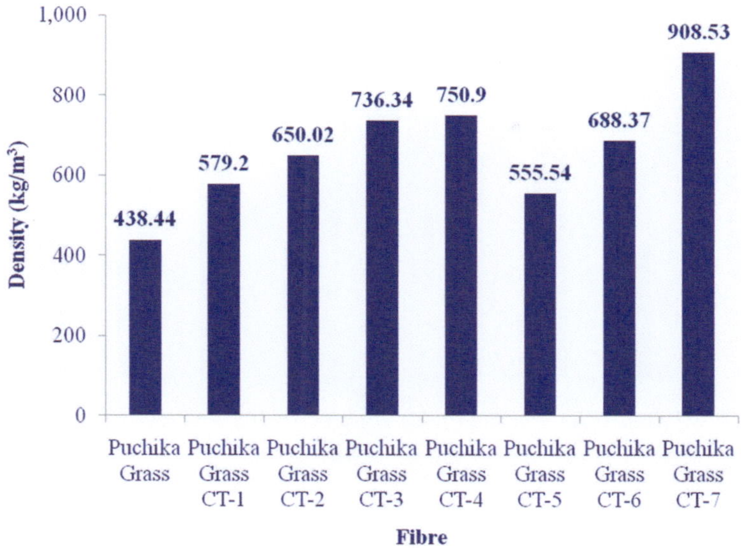

Fig. 10.11 Density of puchika grass fibre

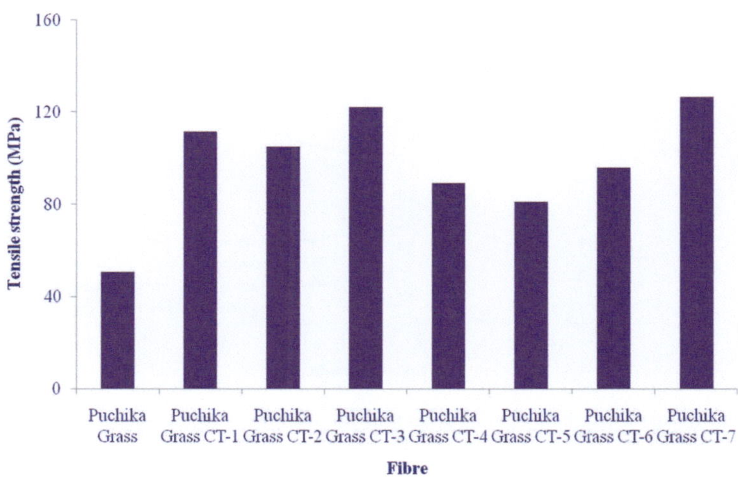

Fig. 10.12 Tensile strength of SFTT puchika grass fibre

10.15, where the tensile strength of all the treated fibre composites is increased with increase in fibre content which has not happened in the case of the untreated fibre composites due to lack of bonding. Tensile modulus of puchika grass CT-2 and CT-7 FRP composites is increased with increase in fibre volume fraction, whereas the composites reinforced with other treated fibres like CT-1, CT-3, CT-4, CT-5 and CT-6 had shown variation due to more elongation exhibited by the composites than the load.

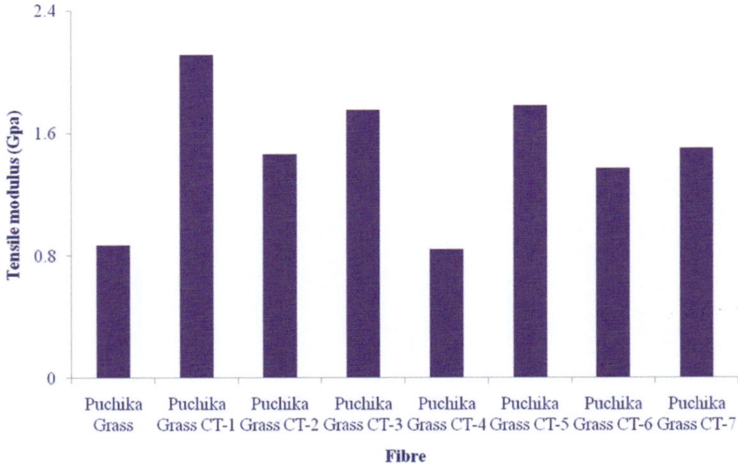

Fig. 10.13 Tensile modulus of SFTT puchika grass fibre

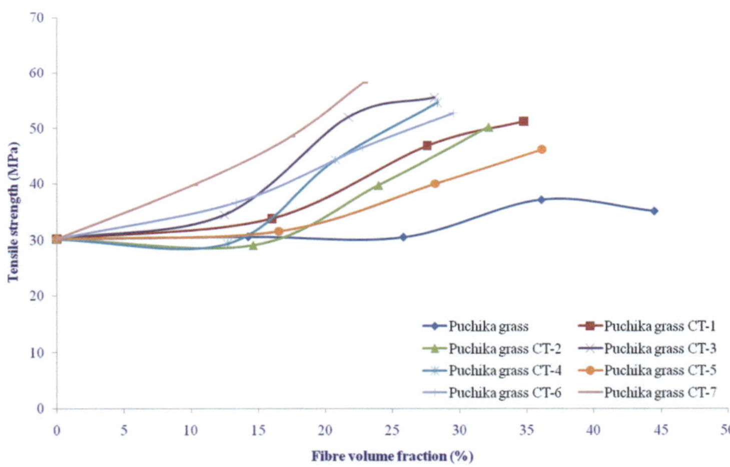

Fig. 10.14 Tensile strength of puchika grass fibre reinforced polyester composites

The alkali-treated puchika grass at 0.275 M concentration and 28 h soaking time, i.e. CT-3 FRP composites at 21.69 % fibre volume fraction, resulted in the tensile modulus of 1.314 GPa. All the composites failed due to tensile load only and there is no pull-out of the fibre visualized.

The composites under the bending load are examined using a three-point bend test, and their flexural properties are determined as per the procedures of ASTM D790-07ε^1. Because of the non-stick nature between the fibre and matrix, it is not allowed to test the puchika grass fibre beyond 39.87 % fibre volume fraction.

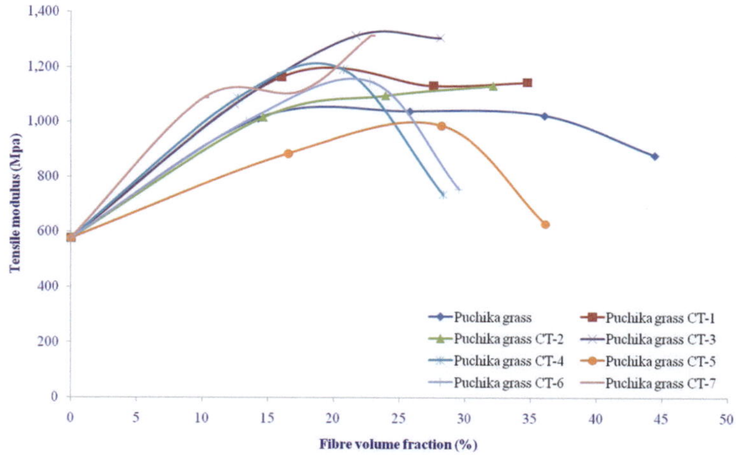

Fig. 10.15 Tensile modulus of puchika grass fibre reinforced polyester composites

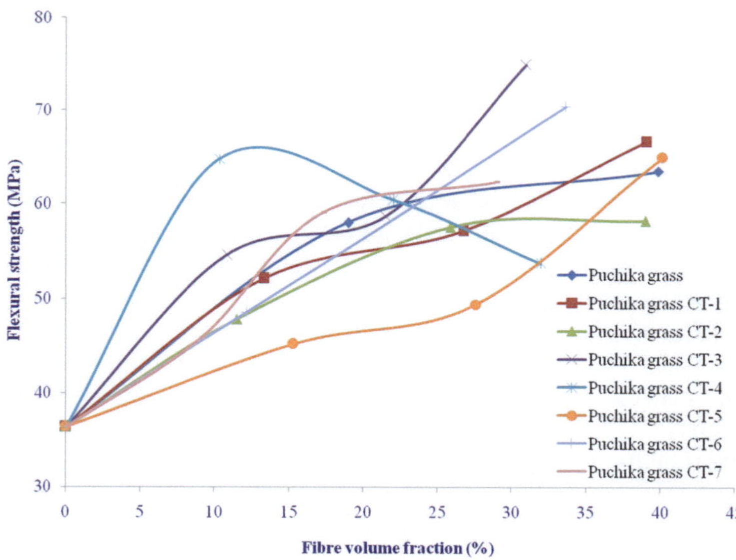

Fig. 10.16 Flexural strength of puchika grass fibre reinforced polyester composites

Except CT-4 all the puchika grass FRP composites have showed increase in flexural strength with increase in fibre volume fraction. During the bend test it is interesting to note that all the chemically treated puchika grass FRP composites failed in its outer layer which is quite desirable. The flexural modulus of all the puchika grass FRP composites is increased with increase in fibre volume fraction. The graphs drawn between the flexural strength and modulus against fibre volume fraction are represented in Figs. 10.16 and 10.17, respectively.

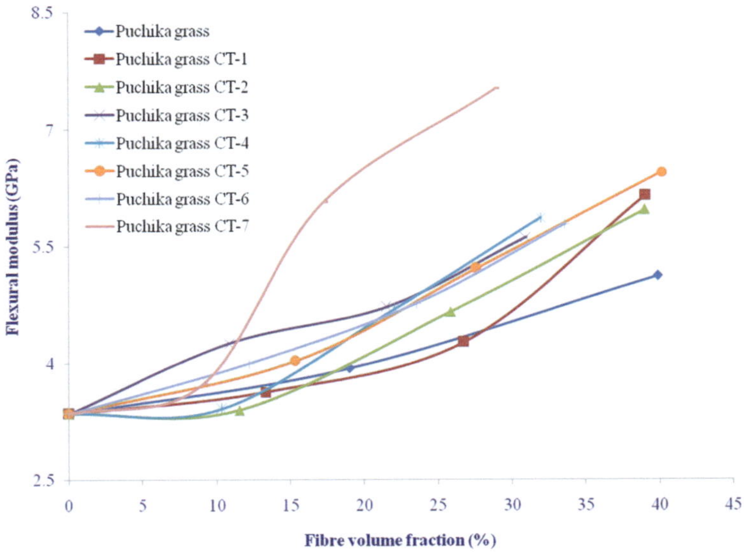

Fig. 10.17 Flexural modulus of puchika grass fibre reinforced polyester composites

The relative susceptibility of the component to failure due to the stress applied at higher resistance is called impact resistance, whereas the ability of the material to withstand the application of a sudden load which results in fracture is defined as an impact strength of the material. Hence Charpy impact test is conducted as per the standards of ASTM D6110-08 on puchika grass FRP composites, and its impact resistance and strength are determined which are graphically shown in Figs. 10.18 and 10.19, respectively. The impact resistance and strength of the composite material range from 18.44 to 744.37 J/m and 1.66 to 65.01 kJ/m^2, respectively.

The maximum field that can be applied to an insulating material without causing dielectric breakdown is called dielectric strength. So the insulation property of the puchika grass FRP composites is determined, and its dielectric strength decreased with increase in fibre content. The maximum dielectric strength achieved at 12.67 % fibre volume fraction is 9.67 kV/mm and is shown in Fig. 10.20.

10.4 Conclusions and Applications

The puchika grass fibre is extracted successfully by pure splitting method and the wet lay-up technique adopted for neat manufacturing of the composites. The initial phase of success is visualized from the experimental results of the composites which are reinforced with alkali-treated puchika grass fibres. Thereafter alkali-pretreated

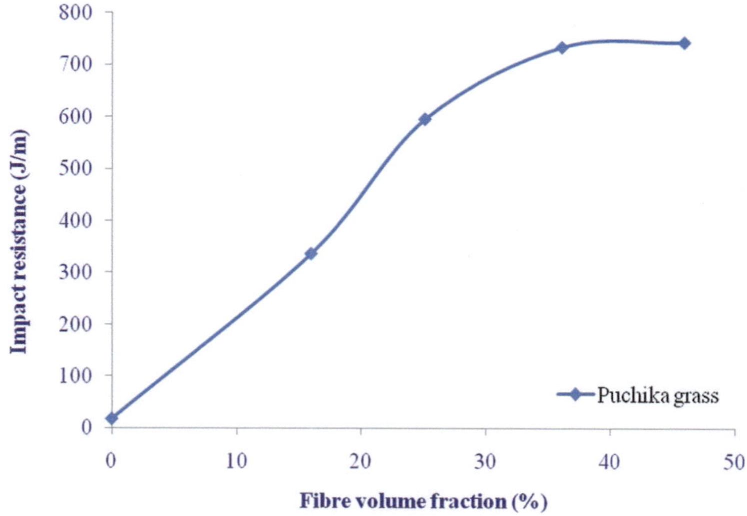

Fig. 10.18 Impact resistance of puchika grass fibre reinforced polyester composites

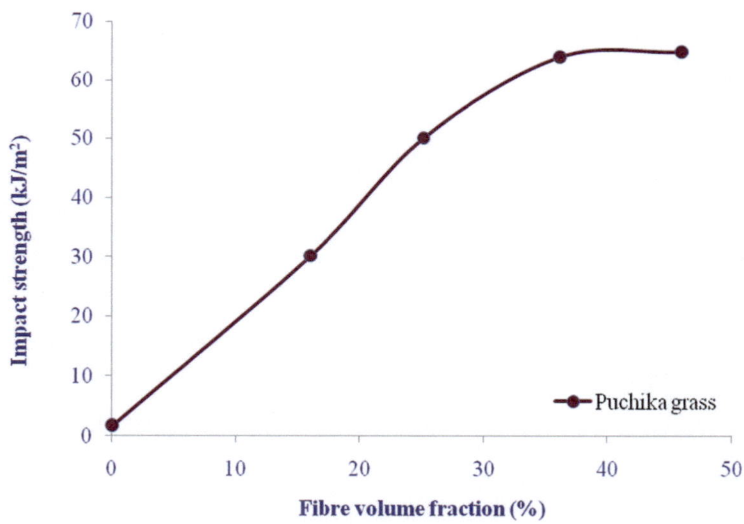

Fig. 10.19 Impact strength of puchika grass fibre reinforced polyester composites

and permanganate-treated puchika grass, i.e. CT-7 FRP composites, resulted in the achievement of good tensile properties through good interface formation. This kind of grass is more suitable in making door panels and various automobile parts. Puchika grass fibre polyester composites may also find an application where it needs lightweight and reasonable dielectric strength.

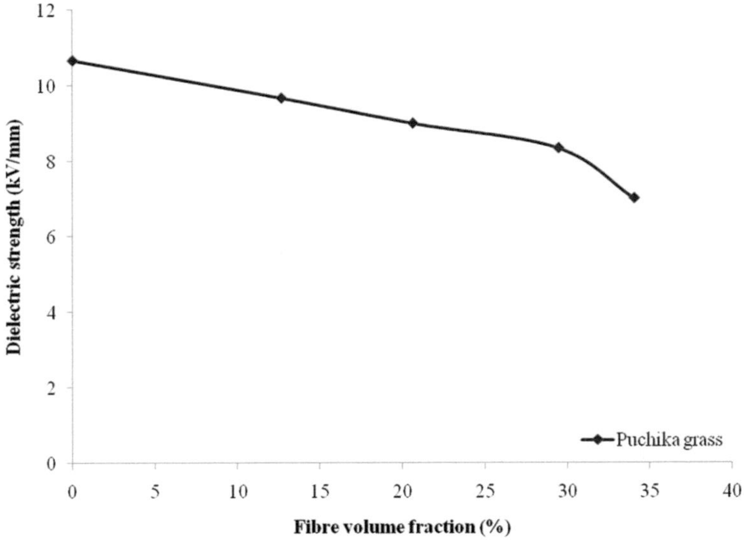

Fig. 10.20 Dielectric strength of puchika grass fibre reinforced polyester composites

References

Danny Akin E, Luanne Rigsby L (1985) Degradation of Bermuda and orchard grass by species of ruminal bacteria. Appl Environ Microbiol 50:825–830

De D, De D, Adhikari B (2004) The effect of grass fiber filler on curing characteristics and mechanical properties of natural rubber. Polym Adv Technol 15:708–715

De D, De D, Adhikari B (2006) Curing characteristics and mechanical properties of alkali-treated grass-fiber-filled natural rubber composites and effects of bonding agent. J Appl Polym Sci 101:3151–3160

De D, De D, Adhikari B (2007) Grass fiber reinforced phenol formaldehyde resin composite: preparation, characterization and evaluation of properties of composite. Polym Adv Technol 18:72–81

Frederick Wallenberger T (2004) Natural fibers, plastics and composites. Kluwer Academic Publishers, Norwell, pp 260, 263

Galletti GC (1991) Production and utilization of lignocellulosics. Elsevier, London, p 344

Kommula VP, Obi Reddy K, Shukla M, Marwala T (2013a) Tensile properties of long untreated and alkali treated Napier grass fiber strands/epoxy composites. International conference on chemical, mining, and metallurgical engineering, pp 269–274

Kommula VP, Obi Reddy K, Shukla M, Marwala T, Varada Rajulu A (2013b) Physico-chemical, tensile and thermal characterization of Napier grass (native African) fiber strands. Int J Polym Anal Charact 18:303–314

Liu W, Mohanty AK, Drzal LT, Askel P, Misra M (2004a) Effect of alkali treatment on the structure, morphology and thermal properties of native grass fibers as reinforcements for polymer matrix composites. J Mater Sci 39:1051–1054

Liu W, Amar Mohanty K, Askeland P, Lawrence Drzal T, Misra M (2004b) Influence of fiber surface treatment on properties of Indian grass fiber reinforced soy protein based biocomposites. Polymer 45:7589–7596

Liu W, Thayer K, Misra M, Drzal LT, Amar Mohanty K (2007) Processing and physical properties of native grass-reinforced biocomposites. Polym Eng Sci 47:969–976

McLaughlin SB, Samson R, Bransby D, Wiselogel A (1996) Evaluating physical, chemical and energetic properties of perennial grasses as biofuels. Bioenergy 1996-the seventh national bioenergy conference, pp 1–8

Murali Mohan Rao K, Ratna Prasad AV, Ranga Babu MNV, Mohana Rao K, Gupta AVSSKS (2007) Tensile properties of elephant grass fiber reinforced polyester composites. J Mater Sci 42:3266–3272

Obi Reddy K, Uma Maheswari C, Jeevan Prasad Reddy D, Varada Rajulu A (2009) Thermal properties of Napier grass fibers. Mater Lett 63:2390–2392

Ramanaiah K, Ratna Prasad AV, Hema Chandra Reddy K (2012) Thermal and mechanical properties of waste grass broom fiber-reinforced polyester composites. Mater Des 40:103–108

Ratna Prasad AV, Mohana Rao K, Gupta AVSSKS, Reddy BV (2011) A study on flexural properties of wildcane grass fiber-reinforced polyester composites. J Mater Sci 46:2627–2634

Satish Kumar TP, Navaneethakrishnan SS (2012) Tensile and flexural properties of snake grass natural fiber reinforced isophthalic polyester composites. Compos Sci Technol 72:1183–1190

Satish Kumar TP, Navaneethakrishnan SS, Kumar J (2013) Mechanical properties of randomly oriented snake grass fiber with banana and coir fiber-reinforced hybrid composites. J Compos Mater 47:2181–2191

Srinivasababu N, Murali Mohan Rao K, Suresh Kumar J (2010) Tensile properties of turmeric fibre reinforced polyester composites. Indian J Fibre Text Res 35:324–329

Srinivasababu N, Suresh Kumar J, Vijaya Kumar Reddy K (2012) Mechanical and dielectric properties of PTSL FRP composites. Adv Mater Res 585:311–316

Srinivasababu N, Suresh Kumar J, Vijaya Kumar Reddy K (2014) An experimental evaluation of mechanical and dielectric properties of one year dried palm fruit empty fruit bunch fibre reinforced polyester composites. Indian Sci Cruiser 28:33–38

Thomas S, Pothan LA (2009) Natural fibre reinforced polymer composites from macro to nano scale. Old City Pub Inc, pp 527–528

Chapter 11
Pultrusion Process of Natural Fibre-Reinforced Polymer Composites

A.M. Fairuz, S.M. Sapuan, E.S. Zainudin, and C.N.A. Jaafar

Abstract In this chapter a review of pultrusion process is presented. The process can be used to produce composites with reinforcements that are either synthetic fibres such as glass and carbon fibres or natural fibres like kenaf, jute, and hemp fibres. Selected previous work such as mechanical properties, processing parameters, and product development of pultruded natural fibre composites is reviewed. Special attention is paid on the development and properties of kenaf fibre composites fabricated from the pultrusion process.

Keywords Pultrusion • Natural fibre composites • Kenaf • Biopolymers • Product development

11.1 Introduction

Advances in composite material have made composite a very important class of material in many industries such as aerospace, automotive, oil and gas, and construction industries (Sapuan and Abdalla 1998; Davoodi et al. 2008; Hambali et al. 2009; Abdalla et al. 2007). In the recent years, there is a growing interest to use natural fibres as reinforcements for polymer composites due to many advantages that they offer such as abundance, environmentally friendly, low cost, low density,

A.M. Fairuz • E.S. Zainudin • C.N.A. Jaafar
Department of Mechanical and Manufacturing Engineering, Universiti Putra Malaysia (UPM), 43400 Serdang, Selangor, Malaysia

S.M. Sapuan (✉)
Department of Mechanical and Manufacturing Engineering, Universiti Putra Malaysia (UPM), 43400 Serdang, Selangor, Malaysia

Laboratory of Biocomposite Technology, Institute of Tropical Forestry and Forest Products, Universiti Putra Malaysia (UPM), 43400 Serdang, Selangor, Malaysia

Aerospace Manufacturing Research Centre, Faculty of Engineering, Universiti Putra Malaysia (UPM), 43400 Serdang, Selangor, Malaysia
e-mail: sapuan@upm.edu.my

M.S. Salit et al. (eds.), *Manufacturing of Natural Fibre Reinforced Polymer Composites*, DOI 10.1007/978-3-319-07944-8_11

renewability, recyclability, low energy consumption, biodegradability, and no harmful effect to health, and they have comparable specific properties (El-Shekeil et al. 2012; Anwar et al. 2009).

11.2 Pultrusion Process

Pultrusion is a composite fabrication technique where continuous fibres are impregnated with the thermoset polymer matrix through heated die to form composites (Fairuz et al. 2014a, b, c; Memon and Nakai 2013). It has the capability to run continuously to produce components with constant cross-sectional profile for large scale which is difficult using other composite fabrication techniques. The pultruded composites normally comprise as high as 70 % of fibre content, and the products are very stiff. In addition, the ease of fabrication and the cost-effectiveness of this process make it one of the important techniques over the others for the fabrication fibre composite products (Sharma et al. 1998a, b).

Figure 11.1 shows a schematic drawing of the pultrusion process. The process started with pulling the continuous fibres from fibre creel and passing them through a resin bath. The composites are shaped and cured in heated die and finally cut by a cutter.

Fibre creel (Fig. 11.2) is an equipment to sort the fibre bobbins so that the fibres are easily pulled without a knot, and the fibres can be glass fibres as well as natural fibres like, kenaf, jute, etc. The size of fibre creels is dependent on the quantity of the fibre strand used in production of the composite profiles. Each creel can support 100 fibre bobbins. In Fig. 11.2, glass and kenaf fibre bobbins are placed on the top and bottom parts of the creel, respectively.

The fibres are pulled from the creel into a resin bath (Fig. 11.3). The fibres are immersed into the polymer resin in the resin bath. Complete immersion of fibres in the resin is ensured to avoid incomplete fibre wetting. The polymer resins used in the pultrusion process are generally thermosetting polymers such as epoxy, polyester, vinyl ester, and phenolic, and their formulations are modified to suit the

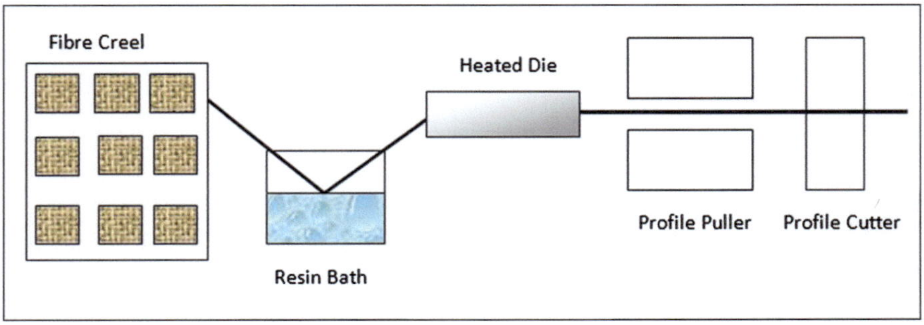

Fig. 11.1 Basic schematic diagram of pultrusion process

Fig. 11.2 Fibre creel (Courtesy of Rasidin Senawi of Innovative Pultrusion Sdn. Bhd., Seremban, Negeri Sembilan, Malaysia)

Fig. 11.3 Thermosetting polymer matrix resin bath (Courtesy of Rasidin Senawi of Innovative Pultrusion Sdn. Bhd., Seremban, Negeri Sembilan, Malaysia)

pultrusion process. Each resin has its own specific characteristic advantages in terms of processing, performance, and application. A low-viscosity resin is needed to facilitate ease of resin penetration into fibre yarn. Voids inside pultruded natural fibre composites are generally higher than that of pultruded synthetic fibre composites due to higher moisture content and poorer wetting in the former. Resin is formulated to contain the filler, catalyst, pigment, and release agent. The function of the filler in the resin is to reduce the use of resin and to fill in the porosity inside the composites.

The function of the fibre guide plate is to guide fibre strands so that they are aligned almost horizontally before entering the heated die. It also functions to eliminate excessive resin and to give a pressure to increase in wetting the composites (see Fig. 11.4). The excessive resins are dropped onto a tray and channeled back to the resin bath. The tray is placed at a slant, so the excessive resin can flow back to the resin bath easily.

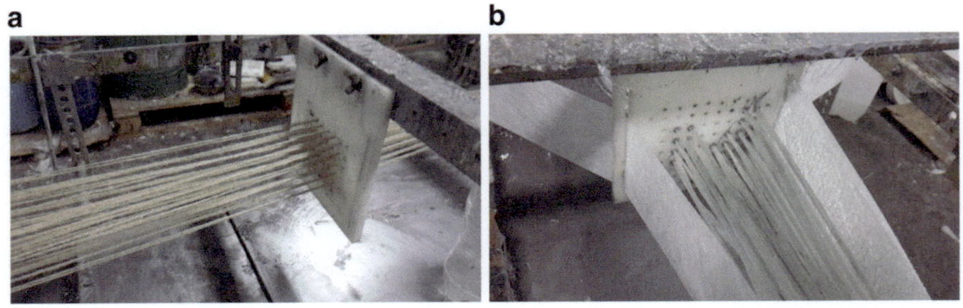

Fig. 11.4 (**a**) Guide plate for kenaf fibre yarn (**b**) Guide plate for glass fibre strand with glass stitch mat (Courtesy of Rasidin Senawi of Innovative Pultrusion Sdn. Bhd., Seremban, Negeri Sembilan, Malaysia)

Fig. 11.5 Composites before entering heated die (Courtesy of Rasidin Senawi of Innovative Pultrusion Sdn. Bhd., Seremban, Negeri Sembilan, Malaysia)

The die (Fig. 11.5) is heated by an aluminum heater block, and the temperature is controlled by the thermocouple attached to the die. The heat exposed inside the die is divided into two zones, i.e., gelation zone and curing zone. Thermocouple sensors interact with the die to ensure that the temperature is sufficient and to prevent the die from becoming overheated which may cause defects to the pultruded profiles (see Fig. 11.6). The closer the thermocouple sensors to the profile, the more accurate the die temperatures. The heater capacity or size is dependent upon the die size, and a die smaller in size needs a smaller amount of heat to achieve sufficient required temperature. Some of the die need more heater, or the heater should be designed to give more heat load into the die. A cured pultruded composite leaving the heated die is shown in Fig. 11.7.

The exothermic reaction of the resin starts when the composite reaches the initiation temperature at which the gelation of the resin takes place (Baran et al. 2013). The range of the temperature for the pultruded composite profile is generally between 100 and 200 °C. The temperature can be reduced when the amount of catalyst is increased, and it can be increased when the pulling speed is increased. The size of the pultruded composite profile also affects the temperature and pulling speed setting during the process. The temperature setting for the pultruded fibre composites has to be carefully chosen to prevent the loss in their properties.

Fig. 11.6 Heated die with heater block and thermocouple sensors (Courtesy of Rasidin Senawi of Innovative Pultrusion Sdn. Bhd., Seremban, Negeri Sembilan, Malaysia)

Fig. 11.7 Fully cured pultruded composite after leaving the heated die.(Courtesy of Rasidin Senawi of Innovative Pultrusion Sdn. Bhd., Seremban, Negeri Sembilan, Malaysia)

Increasing the catalyst or promoter in the matrix can reduce the exothermic reaction temperature.

The pultrusion puller is shown in Fig. 11.8, and the puller is pulling the pultruded profiles on a continuous basis. Initially, before the pullers can grip the pultruded profiles, the fibre yarn or rovings are pulled, manually assisted by one of the pullers. Manual pulling is continued until the pultruded composites reach the puller and a fully automated pulling can be set. The pulling process can be set by pulling the composites one at a time. The first puller pulls the composites until it reaches the maximum limit, and it is continued by the second puller, and at the same time the first puller is retracted to the original position. The second type of pulling is that both pullers are pulling the composites together at the same time and are retracted at the same time to the original position.

The cutter can be set to cut the composite profiles following a desired length required by the operator (see Fig. 11.9). The desired quantity of composite profiles

Fig. 11.8 Pultruded composite profile puller (Courtesy of Rasidin Senawi of Innovative Pultrusion Sdn. Bhd., Seremban, Negeri Sembilan, Malaysia)

Fig. 11.9 A cutter in a pultrusion machine (Courtesy of Rashidin Senawi of Innovative Pultrusion Sdn. Bhd., Seremban, Negeri Sembilan, Malaysia)

to be cut can also be set in a control monitor. The composite profiles are clamped on both sides, and the cutter cuts the composites without the need to stop the pulling process. Different types of pultruded composite profiles have a different puller rubber clamp, cutter cover, and cutter clamp. The cutter is connected to a vacuum system to suck the dust during cutting time and the dust trapped inside the vacuum bag.

Figure 11.10 shows the I-beam profile made from the pultrusion process for the construction industry. The pultruded composite step ladder shown in Fig. 11.11 is used to replace the aluminum ladder which has an advantage in terms of safety which can prevent electrical leakage that can be against electric shock which can prove to be fatal. Figure 11.12 shows the pultruded composite grating holder that can be used to replace the metal holder or clips to support or hold the composite.

Fig. 11.10 I beam profile for building construction (Courtesy of Rasidin Senawi of Innovative Pultrusion Sdn. Bhd., Seremban, Negeri Sembilan, Malaysia)

Fig. 11.11 Pultruded composite step ladder (Courtesy of Rasidin Senawi of Innovative Pultrusion Sdn. Bhd., Seremban, Negeri Sembilan, Malaysia)

Fig. 11.12 Pultruded composite grating holder (Courtesy of Rasidin Senawi of Innovative Pultrusion Sdn. Bhd., Seremban, Negeri Sembilan, Malaysia)

11.3 Materials Used in the Pultrusion Process

Fibres normally used in pultrusion include synthetic fibres such as glass, carbon, and aramid fibres as well as natural fibres like kenaf and jute fibres. The most commonly used synthetic fibre is a glass fibre due to mainly low cost of material. The fibre glass stand is available in different sizes and qualities such as E glass and S glass, with size ranging from 1,200 to 4,800 tex. A large pultruded composite profile used a large size of tex to minimize the number of bobbins used during the setup of a pultrusion process. Pultruded glass fibre composites are used in construction industries to produce beams and reeling, in oil and gas industries to produce deck platforms, and in house appliances like the step ladder.

Polymer matrices that are generally used as matrices in the pultrusion process are mainly thermosetting polymer materials. Polyester, vinyl ester, phenolic, and epoxy are examples of thermosetting polymers used as matrices in the pultrusion process (Vuppalapati et al. 2014). Thermosetting is normally modified to suit in the pultrusion process. Polyester is one of the thermosetting polymers mostly used in the pultrusion process because of its price and performance. It dominated the pultrusion production for general purpose application such as in the pedestrian walking deck.

Vinyl ester and epoxy are also polymers used to produce pultruded composite products such as grating deck in chemical and oil and gas industries. The advantages of these resins include better chemical and oil resistances and better performance compared to polyester in terms of mechanical properties. Phenolic is the cheapest resin available in the market to be used as matrix in the pultrusion process. Phenolic can withstand high temperature and produces less smoke when it is being burnt. However, the mechanical properties are the most inferior compared to other resins. Martin (2013) reported that urethane resin provides the highest glass loading compared to polyester and vinyl ester which leads to pultruded urethane composites to have flexural modulus properties which approach the properties of aluminum.

Bio-resins had also been used as matrices in pultrusion process. A significant number of research was devoted to the use of bio-based polymers as matrices in pultruded composites mainly utilizing soybean oils (Liang et al. 2005; Chandrashekhara 2004; Sundararaman et al. 2005; Vuppalapati et al. 2011, 2014; Zhu et al. 2001, 2004).

11.4 Pultrusion Process of Selected Natural Fibre Composites

In the past, many researchers have carried out research on natural fibre-reinforced polymer composites using the pultrusion process. Majority of the research work on pultruded natural fibre composites in the world was carried out by a research group at Universiti Sains Malaysia led by Professor Hazizan Md. Akil (Akil et al. 2009a, b, c, 2010a, b, 2014; Zamri et al. 2011, 2012, 2014; Mazuki et al. 2009a, b, 2011; Nosbi et al. 2009, 2010, 2011; Omar et al. 2010; Safiee et al. 2009a, b, 2011; Affzan et al. 2011). However, only selected papers from this research group are reviewed in detail here.

Akil et al. (2009a) investigated mechanical properties of pultruded jute composites after immersion of the composites in distilled water, seawater, and acidic solution. The flexural moduli of pultruded jute composites decreased with the increase of the immersion time because of the high absorption rate of the composites. The immersion in the seawater showed less effect on flexural modulus compared to distilled water and acidic solution.

A study on the degradation of compressive properties of pultruded kenaf fibre composites was carried out by Nosbi et al. (2010). The compressive properties of pultruded kenaf were performed by the immersion of the pultruded composites in different solutions, i.e., seawater, acid and distilled water. Immersion of pultruded composites for extended time reduced the compressive properties of the composites especially in seawater and acid. Mazuki et al. (2011) carried out a research investigation of the effect of water absorption on dynamic mechanical analysis of pultruded kenaf composites after immersion of various solutions such as seawater, rainwater, and distilled water. In DMA analysis, storage modulus, loss modulus, and tan δ were determined, and these properties were reduced after the pultruded composites were immersed in those solutions. The study was supported by scanning electron microscope (SEM) analysis.

Akil et al. (2010a, b, c) performed an experimental studies on determination on flexural properties of pultruded jute/glass and kenaf/glass polyester hybrid composites. The major tool used was acoustic emission system. Another work of Akil et al. (2014) reported the work on the effect of the environments on mechanical properties of pultruded jute/glass fibre unsaturated polyester hybrid composites. Water absorption of the pultruded composites was mainly investigated. Mechanical properties studied include tensile strength, tensile modulus, flexural strength, and modulus. From the studies, it is found that prolonged immersion of the pultruded composites resulted in the reduction of mechanical properties. The work was supported by SEM and acoustic emission analysis.

Safiee et al. (2011) investigated the compression and flexural and thermal properties of pultruded jute-reinforced unsaturated polyester composites. The thermal properties studied include thermomechanical analysis (TMA) and dynamic mechanical and thermal analysis (DMTA). The work was supported with optical and SEMs. A significant degree of matrix cracking and overall buckling took place due to poor adhesion between the polymer matrix and jute fibre.

Omar et al. (2010) studied dynamic mechanical properties of pultruded jute and kenaf composites using compression Split Hopkinson pressure bar technique. The cylindrical shape-pultruded jute fibre composites showed better dynamic compressive performance such as compressive modulus and strength compared to pultruded kenaf composites.

In a study conducted by Zamri et al. (2012), the effect of water absorption on compressive and flexural performance of pultruded jute–glass fibre-reinforced unsaturated polyester hybrid composites was investigated. The pultruded composites were immersions in three different solutions, i.e., distilled water, acid solution, and seawater. For compressive properties, the longer the immersion time, the poorer are the properties. Similar trend is observed for flexural properties.

In a recent paper by Zamri et al. (2014), they investigated pultruded kenaf-reinforced unsaturated polyester composites with different types of fibres, i.e., twisted yarn and single fibre yarn. The void content in the twisted yarn is higher than the single yarn because of poor impregnation for the twisted yarn. The flexural and compressive properties of the pultruded single fibre yarn and twisted yarn composites increased when the fibre loading was increased. The pultruded single fibre yarn kenaf composites with 70 % fibre loading show higher values for both compressive and flexural properties.

Malek et al. (2014) of Universiti Putra Malaysia studied pultruded kenaf glass-reinforced phenolic hybrid composites. They investigated the effect of additive in the composites, and the improvement in the wetting resin and matrix was observed. The increase in the flexural properties of pultruded kenaf glass composites was achieved when the additive was added in the resin formulation.

Nasir and Ghazali (2014) studied tribological properties of pultruded glass kenaf-reinforced polyester hybrid composites with a glass and kenaf ratio of 1:1 with 0.8 m/min pulling speed and 120 and 140 gelation and curing temperature, respectively. The work was compared with paddy fibre-reinforced polypropylene composites. The results of the tribological properties of both materials were compared. Pultruded glass kenaf polyester hybrid composites show lower wear rate compared with other composites.

Van de Velde and Kiekens (2001) carried out research work to determine tensile and flexural properties of the pultruded flax fibre-reinforced polypropylene composites. This work was quite unique in the sense that majority of the research in the past was devoted to thermosetting polymer matrices, but it used thermoplastic matrix. A pultrusion machine specially designed to cater for the thermoplastic matrices was successfully developed.

Similarly, Nguyen-Chung et al. (2007) investigated the feasibility of developed thermoplastic composites, i.e., pultruded flax-reinforced polypropylene composites. The flax was used in the form of commingled yarn. Simulation and experimental works were carried out. The pultruded composite profiles developed were rectangular solid profiles. Another work on pultruded thermoplastic composites was performed by Angelov et al. (2007). The pultrusion process was carried out using two different types of polypropylene–flax yarn. Polypropylene and flax fibres were spun together. The studies used the same parameter settings but different fibre loadings, i.e., 30 and 50 % of flax fibres. They found that slow pulling speed at higher temperature affected the flexural properties of pultruded composites. When the pulling speed was increased, the flexural modulus of the pultruded composites was decreased.

Lackey et al. (2008) had performed research on pultruded hemp-reinforced polyester composites. Single play hemp yarn and woven hemp fabric were used as reinforcement in the study. Polyester resin specially formulated for the pultrusion process was used in this study. The density of pultruded hemp yarn and woven hemp fabric–glass fibre-reinforced polyester hybrid composites was determined, and these densities were used to estimate the fibre volume contents of the hybrid hemp and E-glass composites. The pultruded hemp polyester composites were successfully fabricated with no damage resulted from mechanical working or processing temperatures.

Singh and Gupta (2005) reported their research work on pultruded jute fibre-reinforced phenolic composites for door frame application. Testings carried out in this study include tensile and flexural testings, effect of hydrothermal aging, and effect of weathering supported by SEM analysis. The performance-pultruded jute composites were compared with a wooden door frame, and the performance was noticed to be comparable.

Peng et al. (2012) studied mechanical and morphological properties of pultruded hemp/wool fibre-reinforced polyurethane composites, and their performance was compared with polyester and vinyl ester composites. The composites were prepared in the form of rods. Water absorption properties were also determined. From the study, it was concluded that pultruded hemp/wool composites demonstrated comparable properties to pultruded glass fibre composites.

Fairuz et al. (2014a, b) had carried out investigation on determining the most suitable processing parameters in pultruded kenaf yarn-reinforced vinyl ester composites with the circular profile. These optimum processing parameters were determined using the Taguchi design of experiments (DOEs). These processing

Table 11.1 An L9 orthogonal array for pultruded kenaf fibre-reinforced vinyl ester composites (Fairuz et al. 2014b)

Sample	Filler loading (%)	Gelation temperature (°C)	Curing temperature (°C)	Pulling speed (m/min)
1	10	80	140	0.2
2	10	100	160	0.4
3	10	120	180	0.6
4	30	80	160	0.6
5	30	100	180	0.2
6	30	120	140	0.4
7	50	80	180	0.4
8	50	100	140	0.6
9	50	120	160	0.2

parameters include pulling speed, gelation temperature, curing temperature, and filler loading. In their studies, DOE was performed as L_9 (3^4) orthogonal array and was calculated using Microsoft Excel which consists of the required number of experimental samples. Nine types of pultruded kenaf composite samples with four different pultrusion processing parameters (filler loading, gelation temperature, curing temperature, and pulling speed) and three levels of parameter values were selected (Table 11.1).

11.5 Conclusions

From this study, the following conclusions can be drawn:

- Pultrusion machine can be used to fabricate pultruded composite products either using conventional fibre or natural fibre composites.
- The matrix system for pultruded composites can be derived from synthetic or biopolymers.
- There is still a huge scope of work to optimize the processing parameters of pultruded natural fibre composites.
- Pultruded composites can also use thermosetting as well as thermoplastic polymers as matrices.
- The work on pultruded kenaf composites has been progressing ranging from characterization to product development.

Acknowledgment The researchers wish to thank Universiti Putra Malaysia for the funding support through the Research University Grant Scheme (RUGS 2–2012), project number: 05-02-12-1916RU.

References

Abdalla FH, Saad AM, Khalid YA, Sapuan SM, Hamouda AMS, Sahari BB, Hamdan MM (2007) Design and fabrication of low cost filament winding machine. Mater Des 28:234–239

Affzan MH, Akil HM, Ishak ZAM, Bakar AA, Abd Rahman MA (2011) The effect of natural weathering exposure on the properties of pultruded jute fibre reinforced unsaturated polyester composites. In: Proceedings of the 5th Asia-Europe symposium on processing and properties of reinforced polymers, 29 May–1 June, Dresden

Akil HM, Leong WC, Ishak ZAM, Bakar AA, Abd Rahman MA (2009a) Water absorption study on pultruded jute fibre reinforced unsaturated polyester composites. Compos Sci Technol 69:1942–1948

Akil HM, Bakar AA, Ishak ZAM, Mazuki AAM, Safiee S (2009b) Properties of pultruded jute and kenaf fiber reinforced unsaturated polyester composites. In: NATO workshop "Textile Composites", 18–21 May, Kiev

Akil HM, Bakar AA, Ishak ZAM, Mazuki AAM, Safiee S (2009c) Study on properties of pultruded jute fibre reinforced unsaturated polyester composites for outdoor application. In: Proceedings of international conference on kenaf & allied fibres 2009 (ICKAF 2009), 1–3 December, Kuala Lumpur

Akil HM, De Rosa IM, Santulli C, Sarasini F (2010a) Flexural behaviour of pultruded jute/glass and kenaf/glass hybrid composites monitored using acoustic emission. Mater Sci Eng A527:2942–2950

Akil HM, Cheng LW, Zamri MH, Bakar AA, Ishak ZAM (2010b) Water absorption study on pultruded jute/glass fibre reinforced unsaturated hybrid composites. In: Proceedings of the 4th World Engineering Congress, 2–5 August, Kuching

Akil HM, Sahnizam S, Ariffin ZAM, Bakar AA, Mazuki MAA (2010c) Study on dynamic mechanical properties of pultruded kenaf fiber reinforced composites. In: Proceedings of the 7th Asian-Australian conference on composite materials, 15–18 November, Taipei

Akil HM, Santulli C, Sarasini F, Tirillò J, Valente T (2014) Environmental effects on the mechanical behaviour of pultruded jute/glass fibre-reinforced polyester hybrid composites. Compos Sci Technol 94:62–70

Angelov I, Wiedmer S, Evstatiev M, Friedrich K, Mennig G (2007) Pultrusion of a flax/polypropylene yarn. Compos A: Appl Sci Manuf 38:1431–1438

Anwar UMK, Paridah MT, Hamdan H, Sapuan SM, Bakar ES (2009) Effect of pressing time on physical and mechanical properties of bamboo strip fibre reinforced phenolic composites. Ind Crop Prod 29:214–219

Baran I, Tutum CC, Hattel JH (2013) The effect of thermal contact resistance on the thermosetting pultrusion process. Compos B Eng 45:995–1000

Chandrashekhara K (2004) Energy efficient and sustainable building enclosures using bio-based materials. In: Proceedings of the NSF-PATH housing research workshop, Orlando, pp 187–191

Davoodi MM, Sapuan SM, Yunus R (2008) Conceptual design of a polymer composite automotive bumper absorber. Mater Des 29:1447–1452

El-Shekeil YA, Sapuan SM, Abdan K, Zainudin ES (2012) Influence of fiber content on the mechanical and thermal properties of kenaf fiber reinforced thermoplastic polyurethane composites. Mater Des 40:299–303

Fairuz AM, Sapuan SM, Zainudin ES, Jaafar CNA (2014a) Polymer composite manufacturing using a pultrusion process: a review. Am J Appl Sci 11:1798–1810

Fairuz AM, Sapuan SM, Zainudin ES, Aiza CN (2014b) Study of pultrusion process parameters. In: Proceedings of the postgraduate symposium on composites science and technology 2014 & 4th postgraduate seminar on natural fibre composites 2014, 28th January, Putrajaya, pp 116–120

Fairuz AM, Sapuan SM, Zainudin ES, Jaafar CAN (2014c) Optimization of pultrusion process for kenaf fibre reinforced vinyl ester composites. In: the 3rd international conference on design and concurrent engineering (iDECON 2014), Melaka, 22–23 September

Hambali A, Sapuan SM, Ismail N, Nukman Y (2009) Application of analytical hierarchy process in the design concept selection of automotive composite bumper beam during the conceptual design stage. Sci Res Essays 4:198–211

Lackey E, Vaughan JG, Inamdar K, Hancock B (2008) Statistical characterization of pultruded composites with natural fiber reinforcements – part A: fabrication. J Nat Fibers 4:73–87

Liang G, Garg A, Chandrashekhara K, Flanigan V, Kapila S (2005) Cure characterization of pultruded soy-based composites. J Reinf Plast Compos 24:1509–1520

Malek FHA, Zainudin ES, Paridah MT, Jawaid M (2014) The effect of additives on bending strength of pultruded hybrid reinforced resol type phenolic composite. Appl Mech Mater 564:418–421

Martin P (2013) Polyurethane pultrusion resin. Reinf Plast 57:13

Mazuki AAM, Safiee S, Akil HM, Bakar AA, Ishak ZAM (2009a) Study of water absorption behavior on pultruded jute fiber reinforced polyester composites using dynamic mechanical analysis. In: Proceedings of the 4th international conference of recent advances in materials, minerals & environment and the 2nd Asia symposium on material and processing, 1–3 June, Batu Ferringhi

Mazuki AAM, Akil HM, Bakar AA, Ishak ZAM, Safiee S (2009b) Effect of water absorption on the dynamic mechanical properties of pultruded jute fibre reinforced polyester composites. In: Proceedings of the international conference on composites or nano engineering. Composites and reinforced plastics (ICCE 17), 26 July–1 August, Honolulu

Mazuki AAM, Akil HM, Safiee S, Ishak ZAM, Bakar AA (2011) Degradation of dynamic mechanical properties of pultruded kenaf fibre reinforced composites after immersion in various solution. Compos B Eng 42:71–76

Memon A, Nakai A (2013) The processing design of jute spun yarn/PLA braided composite by pultrusion molding. Adv Mech Eng 2013:1–8. doi:10.1155/2013/816513

Nasir RM, Ghazali NM (2014) Tribological performance of paddy straw reinforced polypropylene (PSRP) and unidirectional glass-pultruded-kenaf (UGPK) composites. J Tribol 1:1–17

Nguyen-Chung T, Friedrich K, Mennig G (2007) Processability of pultrusion using natural fiber and thermoplastic matrix. Res Lett Mater Sci 2007:1–6. doi:10.1155/2007/37123

Nosbi N, Akil HM, Bakar AA, Ishak ZAM (2009) Influence of water absorption on mechanical properties of pultruded kenaf fiber reinforced polyester composites. In: Proceedings of the 4th international conference of recent advances in materials, minerals & environment and the 2nd Asia symposium on material and processing, 1–3 June, Batu Ferringhi

Nosbi N, Akil HM, Ishak ZAM, Bakar AA (2010) Degradation of compressive properties of pultruded kenaf fiber reinforced composites after immersion in various solutions. Mater Des 31:4960–4964

Nosbi N, Akil HM, Ishak ZAM, Bakar AA (2011) Effect of water absorption on the mechanical properties of pultruded kenaf fibre reinforced polyester composites. Adv Compos Lett 20:21–26

Omar MF, Akil HM, Ahmad ZA, Mazuki AAM, Yokoyama T (2010) Dynamic properties of pultruded natural fibre reinforced composites using split Hopkinson pressure bar technique. Mater Des 31:4209–4218

Peng X, Fan M, Hartley J, Al-Zubaidy M (2012) Properties of natural fiber composites made by pultrusion process. J Compos Mater 46:237–246

Safiee S, Mazuki AAM, Akil HM, Bakar AA, Ishak ZAM (2009a) Influence of water absorption on coefficient of the thermal expansion (CTE) of pultruded jute fiber reinforced polyester composite (PJFRPC). In: Proceedings of the 4th international conference of recent advances in materials, minerals& environment and the 2nd Asia symposium on material and processing, 1–3 June, Batu Ferringhi

Safiee S, Akil HM, Mazuki AAM, Bakar AA, Ishak ZAM (2009b) Transverse and longitudinal measurements of pultruded kenaf fibre reinforced unsaturated polyester and their impact on interfacial residual stresses in composites. In: Proceedings of the international conference on kenaf & allied fibres 2009 (ICKAF 2009), 1–3 December, Kuala Lumpur

Safiee S, Akil HM, Mazuki AAM, Ishak ZAM, Bakar AA (2011) Properties of pultruded jute fiber reinforced unsaturated polyester composites. Adv Compos Mater 20:231–244

Sapuan SM, Abdalla HS (1998) A prototype knowledge-based system for the material selection of polymeric-based composites for automotive components. Compos A Appl Sci Manuf 29A: 731–742

Sharma D, Mccarty TA, Roux JA, Vaughan G (1998a) Fluid mechanics analysis of a two-dimensional pultrusion die inlet. Polym Eng Sci 38:1611–1622

Sharma D, Mccarty TA, Vaughan G, Roux JA (1998b) Pultrusion die pressure response to changes in die inlet geometry. Polym Compos 19:180–192

Singh B, Gupta M (2005) Performance of pultruded jute fibre reinforced phenolic composites as building materials for door frame. J Polym Environ 13:127–137

Sundararaman S, Shabeer A, Garg A, Chandrashekhara K (2005) Impact characterization of pultruded soy-based composites. In: Proceedings of the SAMPE conference, 1–5 May, Long Beach, pp 1–10

Van de Velde K, Kiekens P (2001) Thermoplastic pultrusion of natural fibre reinforced composites. Compos Struct 54:355–360

Vuppalapati RR, Menta VGK, Chandrashekhara K (2011) Synthesis and performance evaluation of soy-based aliphatic polyurethane nanocomposites for pultrusion. In: Proceedings of the SAMPE conference, 23–26 May, Long Beach

Vuppalapati RR, Menta VGK, Chandrashekhara K, Schuman T (2014) Manufacturing and impact characterization of soy-based polyurethane pultruded composites. Polym Compos 35:1070–1077

Zamri MH, Akil HM, Shahidan HA, Ishak ZAM, Bakar AA (2011) Pultruded natural fibre reinforced composites: preparation, properties and application. In: Proceedings of the nineteenth annual international conference on composites or nano engineering, 24–30 July, Shanghai

Zamri MH, Akil HM, Bakar AA, Ishak ZAM, Leong WC (2012) Effect of water absorption on pultruded jute/glass fiber-reinforced unsaturated polyester hybrid composites. J Compos Mater 46:51–61

Zamri MH, Akil HM, Ishak ZAM, Bakar AA (2014) Effect of different fiber loadings and sizes on pultruded kenaf fiber reinforced unsaturated polyester composites. Polym Compos. doi:10.1002/pc.23025, (in press)

Zhu J, Castle J, Berring N, Chandrashekhara K, Flanigan KV, Kapila S (2001) Mechanical characterization of soy-based pultruded composites. In: Proceedings of the CFA composites conference, Tampa, paper no. M-203, 3–6 October, pp 1–8

Zhu J, Chandrashekhara K, Flanigan KV, Kapila S (2004) Manufacturing and mechanical properties of soy-based composites using pultrusion. Compos A Appl Sci Manuf 35:95–101

Chapter 12
Bio-nanocomposites from Natural Fibre Derivatives: Manufacturing and Properties

Nurhidayatullaili Muhd Julkapli, Samira Bagheri, and S.M. Sapuan

Abstract Bionano-materials are playing an important role in a number of applications due to their inherent eco-friendly advantages since the last few decades. These materials are being explored as the potential alternatives to traditional synthetic materials for diverse applications with particular emphasis as green reinforcement and offer a number of advantages including considerable toughness, flexibility, easy processing, and recyclability. Indeed, nano-sized materials often display an outstanding equilibrium between toughness and strength and frequently enhanced characteristics of their individual components. The effect of different manufacturing conditions and different surface modification techniques for the bionano-materials as well as their polymer composites is discussed in details. Indeed, bionano-material-reinforced polymer composites are emerging very rapidly as the potential substitute to the metal- or ceramic-based materials in applications including automotive, aerospace, marine, sporting goods, and electronic industries.

Keywords Nanotechnology • Biomass • Biodegradable • Renewable resources

12.1 Introduction

Because of the growing environmental awareness, nano materials derived from natural sources (bionano-materials) are becoming more predominant in use. This, in turn, also increases the application of bionano-materials in the polymeric composite system, gaining preference over carbon fibre and glass fibre (Zheng-Ming et al. 2003;

N.M. Julkapli • S. Bagheri
Nanotechnology & Catalysis Research Centre (NANOCAT), University Malaya, IPS Building, 50603 Kuala Lumpur, Malaysia

S.M. Sapuan (✉)
Department of Mechanical and Manufacturing Engineering, Universiti Putra Malaysia (UPM), 43400 Serdang, Selangor, Malaysia
e-mail: sapuan@upm.edu.my

M.S. Salit et al. (eds.), *Manufacturing of Natural Fibre Reinforced Polymer Composites*, DOI 10.1007/978-3-319-07944-8_12

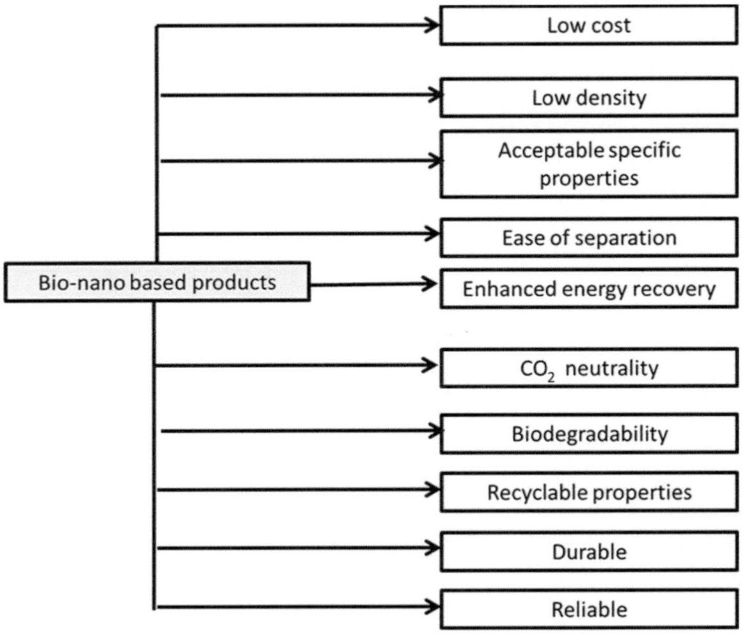

Fig. 12.1 Advantages of bionano-based products

Hao et al. 2002; Ioannis 2005) (Fig. 12.1). Bionano-materials signify a traditional category of renewable material which, nowadays, are experiencing a great renewal (Hao et al. 2002).

12.1.1 Natural Fibre Technology in Composite Materials

The application of natural fibre in composites was expected to be a growing market. Natural fibre only costs as little as USD 0.50 kg^{-1} and can be increased in a few months. Therefore, according to the natural fibre composite market trend and forecast Trend, Forecast, and Opportunity Analysis, the global natural fibre composite market touched USD 289.3 million with a compound annual growth rate of 15 % from 2005 (Farzana et al. 2006). This figure is expected to reach USD 531.3 million with an annual growth rate of 11 % for the next 10 years (Richard and John 2007).

The performance of natural fibres depends on several elements, such as the crystalline cell dimensions, composition, micro-fibrillar angle, structure, defects, physical/mechanical properties, hydrophilicity/hydrophobicity, and surface functionality of natural fibres (Joshia et al. 2004; Sgricciaa et al. 2008). The natural fibres' physical properties are important in the composite system and included dimensions, defects, strength, and structure (O'Donnell et al. 2004). For example, a high aspect ratio

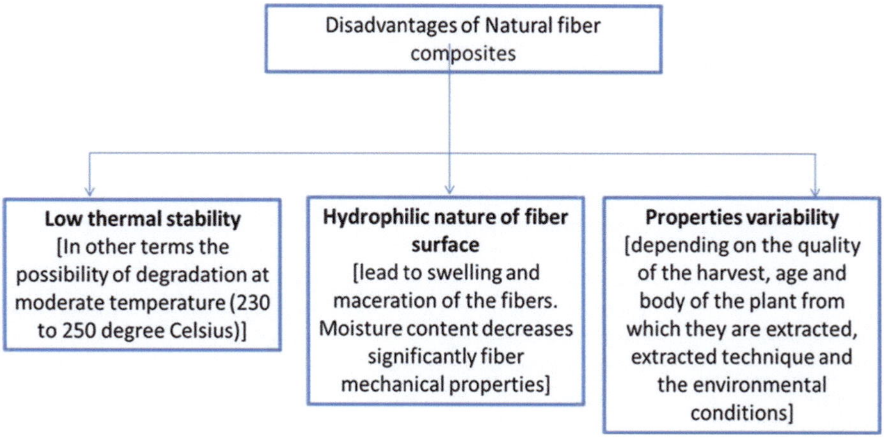

Fig. 12.2 Disadvantages of natural fibre composites

(length/width) which shows the possible strength properties carries a significant importance in natural fibre composites. The natural fibres potentially help in selecting natural fibres as a filler to use composites. Indeed, its dimensions, strength, defects, variability, structure, and crystallinity must also be considered (Muhammad and Mohini 2003).

The primary natural fibres' disadvantages in the particular composites are the poor adaptability of matrix and fibre and their comparatively high moisture adsorption (Fig. 12.2). Consequently, modifications of natural fibre are considered and result in a change in properties of fibre surface to enhance their adhesion with different matrices (Jayamol et al. 2001; Ming et al. 2005). A prototypical stiffness and strength could be attained with a strong interface which is very brittle in nature with easy crack propagation over the fibre and matrix. A huge number of researches have been done, displaying the significance of the interface and the effect of several kinds of surface modifications on the physical and mechanical properties of the natural fibre-reinforced composite system (Table 12.1).

12.1.2 Nanotechnology in Composite Materials

Replacement of traditional microcomposites of nanocomposite materials grew quickly in the last 20 years to overwhelm the restrictions of the micrometer scale, synthesizing novel structures and materials having extraordinary flexibility, enhanced physical performances, and noteworthy industry impact (Michael and Philippe 2000; Ramanathan et al. 2008). Generally, inorganic nanofillers utilized to generate nanocomposites, which compare to organic one, their biodegradability, processability, and biocompatibility, are highly restricted (Peter et al. 1999).

Table 12.1 Surface modification of fibre-reinforced composite system

Surface modification	Examples	Advantages	References
Chemical modification	Mercerization, oxidation, cross-link, grafting, alkaline, benzoylation, silane, maleated coupling, acetylation, acrylation and acrylonitrile grafting, permanganate, peroxide and isocyanate	• Create more sites for chemical links • Significant performance enhancement of low-fibre content • Reduction in water absorption	Xue et al. (2007)
Physical modification	Thermo-treatment, calendaring, stretching, corona and plasma treatment	• Increasing the strength of the fibres	Bledzki et al. (2009)
Biological modification	Enzymatic functionalization	• Increase the compatibility and sensitivity	Maya and Rajesh (2008)
Nanotechnology modification	Mechanical and chemical reduction process	• Increase surface area and functionality	István and David (2010)

All these necessities are connected by nanofibres as one category of bionanomaterial. On Earth, nanofibres are the most ample polymers due to its biodegradability property and high-strength nanocomposites generated using its structure (Mohammad and Karen 2006). An ample and increasing literature dedicated to nanofibres is attained either by a disintegration shearing action to gain micro-fibrillated fibres or by chemical acid hydrolysis treatment. In general, the nanofibres with elongated rodlike features and characteristics contain high-purity single fibre formed from different plant origins under controlled conditions (Xue et al. 2007). The nature of plant origin and hydrolysis conditions plays an important role on nanofibre dimensions. This allows a complete dispersion of nanofibres and maximizing the available matrix–filler interphase optimizing the organic–inorganic interactions, which are responsible for improving the final properties of the nanocomposites. The quality on the dispersion of nanocomposites components given by macroscopic processing factors, including equipment design, mixing velocity, residence time with very limited possibilities for process optimization (O'Donnell et al. 2004). Therefore, depending on the conditions of the procedure and affinity between matrices and nanofibres, various dispersion states can be attained, including the phase separated with agglomerate microcomposites, inserted where layers are uniformly dispersed, but remain orderly and exfoliated when nanofibres are fully and homogeneously dispersed and entirely disoriented (Richard et al. 1996). Exfoliated structures are continually the ultimate aim in nanocomposite compounding because the best performance, especially concerning barrier properties, has to be attained.

It is clear that polymer and hybrid inorganic–organic nanocomposites are focusing on filler functionalization to promote better polymer–filler interfaces. In particular, the processing approach provides a valid technological route to maximize the final properties of nanocomposites.

12.2 General Manufacturing and Fabrication Method of Bio-nanocomposite Materials

Generally, bio-nanocomposites are generated by utilizing conventional manufacturing methods (designed for traditional fibre-reinforced polymer composites and thermoplastic). The manufacturing methods contain compounding, mixing, extrusion, injection molding, and compression molding, which are appropriate for bionano-material-reinforced thermoplastic, thermoset, and elastomer composites. The resin transfer molding, compression molding, and pultrusion are generally implemented with the thermoset matrix. Mentioned manufacturing methods have well accumulated, and developed experience has confirmed their potential success for the production of bio-nanocomposites with controllable quality until now.

12.2.1 Powder Injection Molding

Powder injection molding is the combination of injection molding and powder metallurgy, which is suitable in the fabrication of small and complex parts in large quantities of bionano-composites (Chow et al. 2003; Won-Jun et al. 2006) (Fig. 12.3). In the powder injection molding process, combining two or more powders with different properties, including powder size, type of powder, and thermal expansion coefficient, is challenging (Kyriaki et al. 2007). For example, a difference in particle size increases the critical powder loading. In a high powder loading condition, a small quantity of binder is necessary. The binder is important in determining the success of the injection stage (Hezhou et al. 2008; Chow et al. 2004). Moreover, differences in thermal expansion coefficients induce residual stress and crack during and after the heating process.

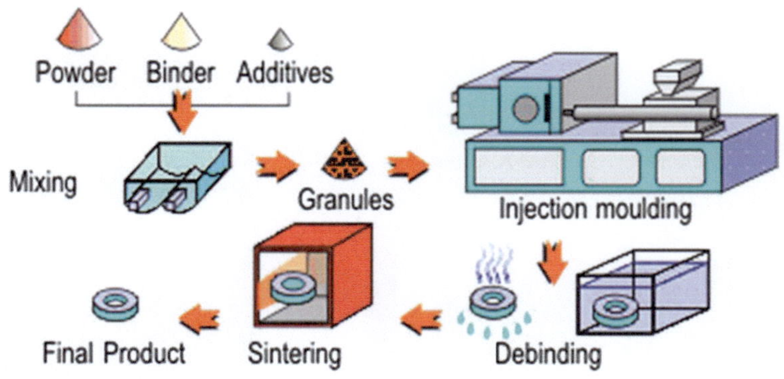

Fig. 12.3 Powder injection molding process

12.2.2 Micro-Cellular Injection Molding

The micro-cellular injection molding is an emerging process with several privileges, able to produce foamed portions. Applying N_2 or CO_2 in the supercritical fluid phase as a physical blowing and plasticizer agent generates lightweight foamed parts with reduced pressure, material, cycle time, and energy (Mingjun et al. 2004; Hrishikesh et al. 2003; Gong et al. 2005) (Fig. 12.4). The micro-cellular injection molding generates components with improved dimensional stability and less thermal degradation and material saving in comparison with the conventional injection molding method (Hrishikesh et al. 2003).

This will occur by exposing the gas/polymer solution to the thermodynamic instability by fast reduction of the solution pressure and increasing its temperature (Mingjun et al. 2005). The actual result is often a drop within the equilibrium solubility of the gas in the composite polymer, leading to the nucleation of frequent microvoids. Other than material formulation and processing parameters, nanofiller and polymer blends also have a robust effect on cell nucleation and growth (Pilla et al. 2007).

12.2.3 Liquid Composite Molding

Using liquid composite molding technique with nanofibre reinforcements offers additional economical, technological, environmental, and ecological advantages. Wide ranges of composite manufacturing processes were covered with the liquid composite molding, including resin transfer molding, resin infusion, and injection

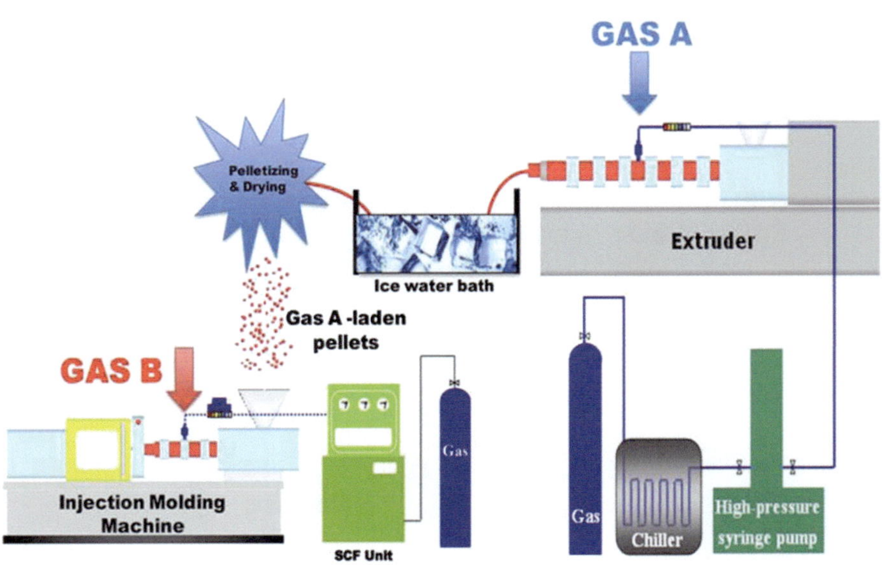

Fig. 12.4 Micro-cellular injection molding

compression molding (Mahrholz et al. 2009; Hammami et al. 1998; Pavel and Suresh 2004). In general, all liquid composite molding techniques involve placement of nanofibrous reinforcing material into some form of closed mold. Based on the manufacturing setup, different types of molds are applied for each liquid composite molding (Baoxing and Tsu-Wei 1999). It shows resistance to flow (quantified by permeability) which significantly influences the required time to fill mold and resin flow patterns. After filling the mold, the polymer matrix will be cured, and on sufficient rigidity, the portion removal from the mold occurs. Permeability and compaction behavior of reinforcement are two essential factors during mold-filling process modeling (Hammami et al. 1998).

In order to understand the through-thickness stress–strain behavior of the reinforcement, accurate general liquid composite molding filling and forces exerted on molds should be modeled (François et al. 2006). Due to compress technique apply towards the nanofibre reinforcement the clamping forces required, which supplied by both internal resin pressure and the stress. Some researchers reported stress relaxation behavior during compaction experiments on nanofibre reinforcements in the absence of a resin. Existence of fluid leads to the apparent lubricant effect, which highly decreases the mold clamping forces (Wang et al. 1994; Lin et al. 1993).

12.2.4 Electrospun Photon Exchange

Electrospinning is a method which produces bio-nanocomposites, using the electricity to drive the spinning process. Therefore, it is a promising alternative to traditional bio-nanocomposite fabrication. Unlike traditional spinning techniques that produce fibre composites in the range of the micrometer (5–50 μm), electrospun photon exchange produces bio-nanocomposites with the range of nanometer size(10–1,000 nm) (Xiaobo et al. 2012; Biao et al. 2008; Dan and Younan 2003) (Fig. 12.5). The bio-nanocomposite synthesis, via a top-down nano-manufacturing process, is a money-saving process which leads to continuous nanofibres that are relatively easy to align, assemble, and process into applications (Biao et al. 2008). Therefore, this technique provides extraordinary properties such as small-sized dimension associated with large specific surface areas, a high degree of structural perfection, and the resultant extreme mechanical properties (Amit et al. 2005).

To date, two types of electrospun photon exchange process are fabricated, which are an uncharged polymer surrounding a three-dimensional interconnected mat of ionomer nanofibres and an ionomer surrounding an electrospun mat of uncharged polymer nanofibres (Thavasi et al. 2008). Electrospinning can fix a great degree of compositional and morphological control of volume fraction and the diameter of the charged or uncharged polymer nanofibres independently. Furthermore, fabrication of bio-nanocomposites via electrospun process occurs with solid-state "forced assembly" of nanofibre on a collector surface, in contrast with self-assembly in the solution, avoiding phase separation and poor dispersion problems causing large-sized domains of the minor composites components (Jih-Jen et al. 2010).

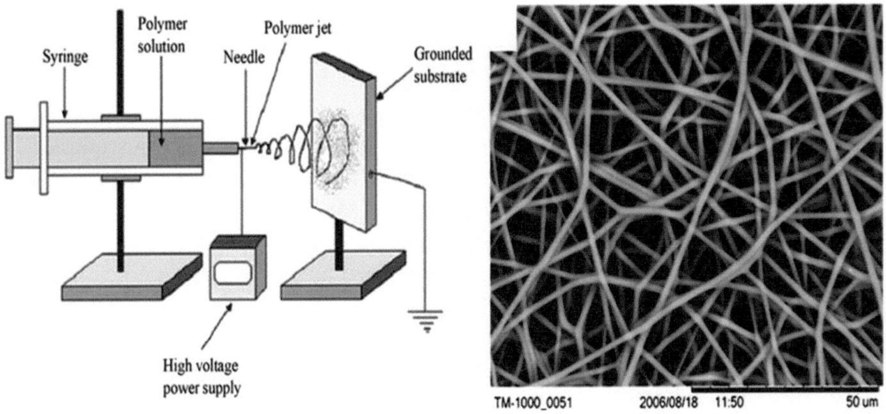

Fig. 12.5 Electrospun photon exchange (Xiaobo et al. 2012)

In general, bio-nanocomposites fabricated through the electrospun method exhibit impressive proton conductivity, water swelling, and mechanical properties.

In general, in the case of the electrospun photon exchange process, the biocompatibility of nanofibre advances with reducing the diameter (it is noteworthy that the electrospun polymer bio-nanocomposites have diameters one to three orders of magnitude smaller than their conventional counterparts). Therefore, electrospinning has a high potential to produce bio-nanocomposites in nanofibrous forms of medical materials, including hard tissue applications and composites of inorganic and polymer phases due to its mechanical and biological properties, which can be tuned appropriately for cellular function and bone formation (Priya et al. 2008; Kim et al. 2008). In addition, the combination of electrospinning with initiated chemical vapor deposition is an effective approach to preparing superhydrophobic bio-nanocomposites. However, this modification process requires the use of special equipments, including chemical vapor deposition reactor and plasma equipment (Amit et al. 2005).

In the meantime, dual electrospun photon exchange process has been introduced whereby the bio-nanocomposites are made by separate and simultaneous electrospinning of two polymers. To construct the first structure, a series of steps is followed, including mechanical mat compaction, solvent vapor exposure, thermal annealing, and acid/water washing (Chi-Ching et al. 2007; Seeram et al. 2006; Sujuan et al. 2010). Designed for the other structure, different series of steps are followed, including mat compaction with heating, thermal annealing, and acid/water washing. This dual process proposes numerous benefits compared to the conventional electrospun fabrication, namely, the elimination of a separate impregnation step (where complete void space filling is often problematic), the choice of making two distinct structures from the same initial dual nanofibre mat, the possibility of expanding the number of polymers that are spun simultaneously to three or more in order to add other properties to the bio-nanocomposites, and the ability to control the spatial distribution of polymers in the bio-nanocomposite thickness direction by changing the relative flow rates of polymers during electrospinning (Sujuan et al. 2010). Therefore, the dual nanofibre electrospinning technique is considered as a robust platform for the fabrication of a wide variety of bio-nanocomposite structure through different processing approaches.

12.2.5 Spray-Up Method

The spray drying method is a direct technique to encapsulate the nanofibre into the polymer matrix homogenously (Yanjun et al. 2010). The preheated spray drier was filled with mixture. The preheating temperature is higher than the boiling point of the solvent (Fig. 12.6). This technique is limited to the high molecular weight and chemically stable polymer, while for those with small molecular weights, it tends to boil off in large quantities at the processing temperature (Kerang et al. 2000; Varun et al. 2006).

12.3 General Properties of Bio-nanocomposite Materials

It has been suggested that bio-nanocomposites compared to their neat polymer and traditional composites will enhance barrier properties and mechanical strength and improve heat resistance. Bio-nanocomposites are predicted to survive the thermal processing stress, transportation, and storage (Michael and Philippe 2000). Furthermore, bio-nanocomposites may allow down gauging due to improvement in their mechanical properties and reduce source material consequently.

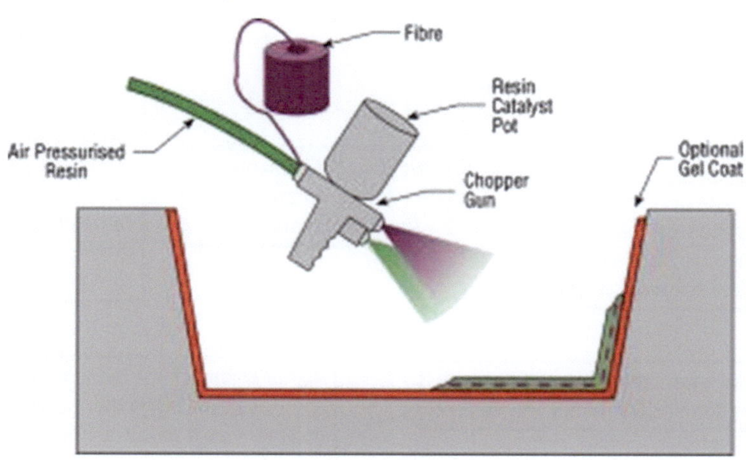

Fig. 12.6 Spray-up method

Indeed, one of the significant properties bio-nanocomposites have is their biodegradability capability. It may lead to fragmentation and loss of mechanical properties or may sometimes be followed by degradation via microorganisms (bacteria, fungi, and algae). This is a multistage procedure which applies enzymes and oxidation to the hydrolysis catalyzed (Mohammad and Karen 2006). The high hydrophilicity property of nanofibres helps easier water penetration into the polymer matrix and activation of the hydrolytic degradation process which leads to the biodegradation process (Richard et al. 1996). Meanwhile biodegradability is one of the main reasons to use nanofibres for the preparation of the bio-nanocomposite material, and the biodegradability of the produced bio-nanocomposites should not be sacrificed fabricating them.

12.4 Thermoset Bio-nanocomposites: Manufacturing and Properties

The application of thermoset polymer composites has been studied widespread by researchers all around the globe due to their enormous advantages (Fig. 12.7). In many applications, which necessitate high performance, including the automotive, nautical, and aeronautical constructions, in addition to those required in the civil and industrial sector, the need to satisfy resistance, stiffness, lifetime, and reliability requirements involves the use of the thermoset polymeric matrix reinforced with bionano-materials (Janis et al. 2000; Omid et al. 2012; Karger-Kocsis and Wu 2004). Therefore, in the case of thermoset bio-nanocomposites, it is important to

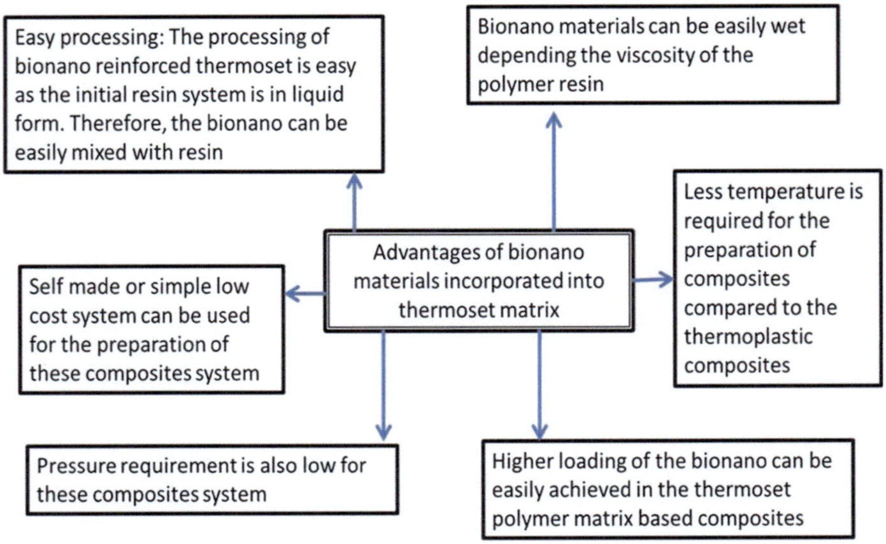

Fig. 12.7 Advantages of bionano-materials incorporated into the thermoset matrix

emphasize the types of nanofibres (including nano-hemp, nano-flax, nano-kenaf, etc.) which are common reinforcements in the thermoset composites for civil and industrial application (Carsten et al. 1999). The possibility to use thermoset bio-nanocomposites with lamination technique with lower temperature and low-pressure technologies significantly broadens their exploitation. Therefore, nanofibre-reinforced thermosets are processed by a relatively simple processing technique, including hand layup and spraying and the liquid composite molding techniques (Karger-Kocsis and Wu 2004). In addition, a variety of other techniques, including centrifugal casting, cold press molding, filament winding, pultrusion, reinforced reaction injection molding, rotational molding, and vacuum forming, were tried, but usage of nanofibre composites was seldom reported. Indeed, thermoset resin also allows the use of advanced close mold processing method including resin mold transfer, which involves a lower impact in the working environment and yet high reinforcement content and performance (Tia and David 2002; Matos Ruiz et al. 2000).

12.4.1 Epoxy Bio-nanocomposites

Epoxy matrix-based composites have been studied by a number of researchers for different applications using different types of nanofibres including nano-bacterial cellulose, nano-electrospun cellulose fibres, nano-recycled cellulose fibres, nano-cellulose whiskers, and nanocrystal cellulose (Rostamiyan et al. 2015; Yu et al. 2014; Oh et al. 2014). Nanofibre-based epoxy composites have been reported to be novel materials for different applications (Yu et al. 2014). Different types of surface functionalization of nanofibres have been carried out for further enhancing their use in epoxy-based polymer composites (Rafiq et al. 2014). For example, the nano palm tree fibre-reinforced epoxy was prepared, and it has been observed that at low frequency, the dielectric constant of the composites increases with the increase in temperature (Abdul Khalil et al. 2013a, b). It was summarized that in addition to the relaxation associated with the glass transition (T_g) of the epoxy resin and ionic relaxation, the nano palm tree fibres also give rise to the other relaxation associated with the interfacial polarization of nano palm tree fibres and epoxy resin (Abdul Khalil et al. 2013b). Besides, the fracture toughness of the epoxy increased simultaneously with the addition of the nanofibre. This is due to the reinforcing effect that strongly depends on the aspect ratio of the nanofibre and its uniform dispersion.

Short-date nano palm tree fibre-reinforced polyepoxy composites were also prepared and studied in the different mechanical property analyses using resin transfer molding techniques (Nik Salleh et al. 2013). The interaction between nanofibres and epoxy matrix is enhanced by treating the nanofibres with chemical modification. The modification was carried out using TEMPO (2,2,6,6-tetramethylpiperidine-1-oxy radical) catalyst mediation (Abdul Khalil et al. 2013b). It is demonstrated that the surface modification assisted the fabrication of epoxy composites using resin transfer molding and increased the mechanical properties.

Nanofibre-reinforced epoxy composites/honeycomb core sandwich panels were also studied in detail for creep properties (Reis et al. 2013). It was observed that at ambient conditions (with relative humidity at 20–50 %), sandwich panels showed linear viscoelastic properties (stress level less than 30 % of their failure stress) (Yeh and Chiu 2011). However, higher relative humidity (65 %) produced a significant acceleration of creep strain. The test result of nano-flax fibre-reinforced epoxy resin produced by the resin transfer molding method was absorbed, and at the nanofibre volume fraction of 21–47 %, the nanofibre direction elastic modulus was 15–39 GPa, and the tensile strength was recorded at 132–280 MPa (Tsai and Chang 2011). Another study tested the similar nano-flax fibre/epoxy composites with 50 % of nanofibre volume fraction and presented 32 GPa as its nanofibre direction elastic modulus and 268 MPa as its tensile strength. Moreover, surface treatment to remove the moisture of nanofibre affects the bending strength in which after the surface treatment of 1–3 % NaOH, the surface treatment improved the strength of overall composites (Yeh and Lin 2009).

To test the real capability of nanofibre reinforced with epoxy composites, the water adsorption behavior of the carpet waste jute yarn reinforced with polymer matrix was studied. Along with the water absorption behavior, mechanical properties including flexural properties of nanofibre-reinforced composites and their impact were also studied after aging in distilled water (Wang et al. 2012; Zou et al. 2006). With the purpose of improving the nanofibre adhesion with the epoxy matrix, alkali treatment of the nanofibres was carried out and indicated that alkali treatment increases the mechanical properties and decreases the water adsorption of the resulting composites. This behavior was attributed to the strong interface improvement between the reinforcement and the nanofibres (Zou et al. 2006).

Hybrid nanofibre using nano-jute and nano-bagasse as the reinforcement was prepared (Jochen and Andrzej 1997, 1999; Bhosale et al. 2014). Both nanofibres were treated with furfuryl alcohol to increase the adhesion between the composite components. It was confirmed through different techniques that the surface modification decreased the composites' hydrophilicity and significantly increased the overall mechanical properties of the composites (Mishra and Biswas 2014).

12.4.2 Polyester Bio-nanocomposites

Polyester resin is commonly used as composite matrix because of its reasonable price, ease of handling, and balanced mechanical, electrical, and chemical properties as well as fire resistance properties (Godina et al. 2004; Navin et al. 2000; El-Tayeb 2008). However, polyester resin is unstable and degradable in the environment. Therefore, some studies focused on incorporating nanofibres in order to increase the modulus both below and above the glass transition temperature of polyester (Sami and Ridha 2007). It is generally accepted that comprising toughness would lead to the increase in stiffness. It has also been reported that the toughness of nanofibre/polyester composites is hooked on the morphology and composition of nanofibre in polyester matrix. At low concentration, proper dispersion improves the toughness (Navin et al. 2000).

Nano-Luffa fibre-reinforced polyester composites have been used to tailor the polyester properties (Tanobe et al. 2014; Seki et al. 2012; Ghali et al. 2011). To improve the nanofibre–polyester interface, nano-Luffa fibre was alkali-treated at two temperatures, with varying alkali concentrations. It was observed that similar to other nano-natural fibres (Boynard et al. 2003), the nanofibres' external surface was completely removed and nanofibres with 5 % NaOH treatment exhibited the best flexural mechanical properties that attributed to mechanical interlock due to the increase of nanofibre roughness and the increase of the nanofibre contact area (Boynard and D'Almeida 2000).

Nano-sugarcane bagasse-reinforced polyester composites were also prepared and examined for tribological applications (Loh et al. 2013; Rodrigues et al. 2011). For that, compression molding along with hand layup techniques was used to prepare the composites with chopped 1.5 and 10 mm nanofibres randomly distributed and unidirectional mat nanofibres (Wu 2011). The SEM micrographs of composites showed a fairly good bonding between nano-sugarcane bagasse and matrix which can be served as the replacement to nano-glass fibre-reinforced composites (Brugnago et al. 2011; Lee and Mariatti 2008).

The effect of chemical treatments on mechanical properties and water adsorption potential of nano-bamboo fibre-reinforced polyester also gained a lot of interest (Monteiro et al. 2014; Rassiah et al. 2014a). The chemical modification of nanofibres in mat form was carried out with maleic anhydride, permanganate, benzoyl chloride, benzyl chloride, and pre-impregnation (Rassiah et al. 2014b). It was summarized that the mentioned pretreatment significantly improved the mechanical properties and water adsorption potential of the resulting nanocomposites.

Nano-Brazilian sisal fibre-reinforced polyester composites were also prepared and studied for mechanical properties and pullout analysis (Ornaghi et al. 2014; Oladele et al. 2014). The surface functionalization of these nanofibres with NaOH and *N*-isopropyl acrylamide was subsequently carried out to get the polyester composites with optimum mechanical properties (Belaadi et al. 2014). Lignin content and density of nanofibres reduced with the chemical treatment and *N*-isopropyl acrylamide treatment caused a significant reduction in moisture absorption. The optimum result was achieved by treating the nanofibres with 2 % *N*-isopropyl acrylamide (Pereira et al. 2014).

Nano-kapok fibre-based polyester composites were prepared and subjected to different mechanical properties (Venkata Reddy et al. 2008, 2009). The surface modification of nano-Kapok fibres was carried out using alkali treatment, before being hybridized with glass and nano-sisal fabrics in polyester matrix (Mwaikambo and Bisanda 1999). It was recorded that the impact properties of composites were significantly increased with the loading of nanofibres. It enhanced the mechanical properties due to the excellent dispersion of the reinforcement in the polyester composites along with better load distribution (Venkata Reddy et al. 2008).

Currently, some research works paid attention to the fabrication and properties of nano-bacterial cellulose-reinforced polyester composites. The composites were prepared using resin transfer molding, and to improve the interface, silane functionalization of nanofibres was carried out. It was revealed that silane-modified nanofibre-reinforced composites exhibit greater nanofibre–resin adhesion strength in comparison with untreated composites.

12.4.2.1 Epoxy/Polyester Bio-nanocomposites

Nano palm fibre-reinforced polyester and epoxy matrices were prepared using the short fibre form. Due to its hydrophobic matrices, epoxy/polyester owns repulsive forces leading to aggregation and poor interfacial contact of the nanofibre (Amor et al. 2014). Moreover, the greater presence of surface OH groups on reactive cellulosic surfaces with respect to nanofibres may also permit further modification to alter the hydrophilicity (Ibrahim et al. 2013; Shinoj et al. 2011). Thus, surface modification of nanofibres was carried out to improve interface adhesion between the nanofibre and matrix (Amor et al. 2014). As an example, in the case of nanocellulose-reinforced epoxy/polyester bio-nanocomposites, sulfate ester groups were randomly introduced on the surface during H_2SO_4 hydrolysis due to the formation of sulfate ester with high surface acidity, while HCl hydrolysis results in hydroxylated surfaces (Kaddami et al. 2006).

12.4.3 Polyurethane Bio-nanocomposites

Polyurethane has commercial benefits as it can be molded, injected, extruded, and recycled (Jae et al. 2005; Khalid et al. 2007; Cynthia et al. 2005). One of the intended uses of polyurethane is the incorporation of bionano-material, including a nanofibre that is biocompatible and nontoxic. Polyurethane bio-nanocomposites are generally produced either using the solgel technique or the intercalation technique (Khalid et al. 2007).

In general process of polyurethane composites, the achieved reinforcement of nanofibre into polyurethane depending on the nanofibre intrinsic characteristics, including aspect ratio, shape, surface area, and rigidity (Saralegi et al. 2014; Pant et al. 2014). The dispersion and distribution state of nanofibres can be altered by the processing conditions, while compatibility/affinity with polyurethane can be improved by surface modification of nanofibres (Xiao et al. 2012; Auad et al. 2012).

12.4.4 Phenolic Bio-nanocomposites

Phenol formaldehyde is the most commonly used resin systems, mostly thanks to its extreme temperature and fire resistance (Seena et al. 2002). The polymerization reaction between phenols and aldehydes (phenol and formaldehyde in specific) generates phenolic resins (Sreekala et al. 2000; Richa et al. 2000). This polymer can be molded, and cure (cross-linking) may be affected by the temperature changes (Richa et al. 2000).

In phenolic composites, phenol formaldehyde is the most commonly used resin system. Lignocellulosic *Saccharum cilliare* nanofibre-reinforced polymer biocomposites were prepared using the compression molding technique employing phenol formaldehyde as a novel polymer matrix (Maiti et al. 2010; Kaith et al. 2010). It was found that different mechanical properties including tensile strength, flexural strength, compressive strength, and wear resistance increase up to 30 %, and beyond this loading, the properties decreased (Maiti et al. 2010). Surface modification of this bionano-material was carried out using silane treatment through aminopropyl triethoxysilane to evaluate the morphological and physicochemical properties (Singha and Thakur 2009).

Another type of bionano-material, which is nano-sisal fibre, is studied at different loading on phenolic composites. To further enhance the adhesion between nano-sisal fibres with the phenolic matrix, these bionano-materials were subsequently hydroxymethylated (Zeng et al. 2014; Bu et al. 2014). In other studies, lignin was extracted from nano-pinus-type wood and sugarcane bagasse (Wei et al. 2013). The lignin was further modified via a methylation process to produce methylated lignin. Another study has utilized chlorine dioxide as an oxidative to modify the nano-sisal fibre surface and subsequently the furfuryl alcohol and poly(furfuryl alcohol) treatment before introducing it to phenolic matrix (Li et al. 2014). Changes in the surface morphology of the nanofibres were confirmed by SEM analysis and demonstrated that nanofibre modification with methylated lignin induced significant changes in the morphology of the nanofibre surface. The modified nanofibre-reinforced composite also demonstrated greater impact strength due to the strong bonding between the composite components and was confirmed under SEM micrograph. Furthermore, the surface-modified nanofibres reinforced with phenolic matrix also exhibit low water absorption, and phenolic resin penetrated very well into the modified nanofibre (Da Silva et al. 2012; Zeng et al. 2010). This process resulted in the blocking of water passage through their channel.

12.4.5 Polyvinyl Ester Bio-nanocomposites

Along with other esters used in polymer composite preparation, polyvinyl ester was also considerably investigated (Everson et al. 2006; Mallarino et al. 2005). Nano-bast fibres and various other types of nano-natural fibre-reinforced vinyl ester polymer composites have been prepared using resin transfer molding specifically for the automotive applications (Mallarino et al. 2005). For example, vinyl ester composites reinforced with natural nanofibres from *Luffa cylindrica* fruit were prepared after modification treatment. Different extraction treatment along with mercerization and esterification with dianhydride compounds was used to modify these nanofibres (Bazle et al. 2001; Sullivan et al. 1984). At the same time, nano-pulp fibre-reinforced polyvinyl ester composites were studied to evaluate their different mechanical and thermal properties. It was observed that modification of nanofibre resulted in more compatibility with the polymer matrix, which indicated that the incorporation of nano-pulp fibres considerably improved the composite storage moduli compared to pristine polyvinyl ester (Thielemans et al. 2002; Liliana et al. 2009).

Another study on nano-flax fibre-reinforced polyvinyl ester composites manufactured using vacuum-assisted resin transfer molding method recorded that, at a nanofibre fraction of 25 %, the tensile strength was recorded at 122.4 MPa (Hongyan et al. 2008). In the case of nano-flax fabric composites, it was found that nanofibre volume fraction and the tensile strength are 25 % and 62 MPa, respectively (Roberts et al. 2003). Nano-flax fibre was selected to be incorporated into the polyvinyl ester due to its greater strength and better mechanical behavior than other nano-natural fibres, and the polyvinyl ester was selected as a resin due to its cheaper price, easier procurement, and better treatment for the resin injection (Goutianos et al. 2006).

Hybrid polymer composites of nano-oil palm empty fruit bunch fibre along with glass fibre were prepared using polyvinyl ester as polymer matrix. It was recorded that mechanical properties of the composites were significantly better than the mechanical and chemical board.

12.5 Thermoplastic Bio-nanocomposites: Manufacturing and Properties

The different thermoplastic composite systems based on the different types of thermoplastic reinforced with the number of nanofibres (nano-kenaf, nano-jute, nano-flax, nanocellulose) have been produced and characterized. In order to take full advantage of the nano-sized nanofibre reinforcement, a good dispersion in the thermoplastic matrix was required to enhance the reinforcement of hydrophilic phase and nonpolar matrix compatibility (Denault et al. 2014; Gates and Sun 1991).

The processing of nanofibre/thermoplastic composites most commonly involves extrusion of ingredients above the melt temperature of the matrix, followed by shaping of the operation including injection molding and thermoforming (Carneiro and Maia 2000).

12.5.1 Polypropylene Bio-nanocomposites

Some research works focused on developing the nano-kenaf fibre-reinforced polypropylene composites and concluded that it could be thermoformed for a wide variety of applications with properties that are comparable to existing synthetic composites (Pang and Ismail 2014; Hao et al. 2014; Sallih et al. 2014). Furthermore, the presence of other bio-fibres including garlic nano-powder or nisin usually incorporated into polypropylene as antimicrobial agents leads to detriments in the tensile strength and elastic modulus values and leads to increase of the strain at break (Jirukkakul 2013). Another study found that a combination of nanocellulose derived from empty fruit bunch nanofibres and polypropylene matrix improved mechanical properties of polypropylene (Zainudin et al. 2014; Rivai et al. 2014; Ramli et al. 2013). The bonding effects of regenerated nanocellulose natural fibre were studied using a Raman technique, and the bonding was found to be dependent on the nanocellulose concentration, pH value, degree of crystallinity, and ionic strength of the oxidized nanocellulose aqueous suspension (Chee et al. 2013).

12.5.2 *Polyethylene Bio-nanocomposites*

Polyethylene is a hydrophobic thermoplastic composite, widely used in everyday life. Due to polyethylene's cheap price, availability, and good processability, it is a promising matrix material for bio-nanocomposites (Kuruvilla et al. 1996; Tony et al. 2005). This cheap price plastic constitutes about 70 % of the whole amount of thermoplastic consumed by the plastic industries (Abdelmouleha et al. 2007). Therefore, utilizing nanofibres significantly reinforces the high-density polyethylene matrix and the bio-nanocomposites applied in lots of value-added products instead of those conventional applications (Kuruvilla et al. 1996).

Nanocellulose derived from cotton reinforced polyethylene biocomposites fabricated through an extrusion process (Fukuya et al. 2014; Iwamoto et al. 2014). Before the fabrication process, two separate methods were applied to overcome the nanocellulose dispersion challenge. Firstly, "the physical pretreatment method" was used to preblend polyethylene in a nanocellulose water suspension before extruding. Secondly, the mixture of the first step was dehydrated and freeze-dried before being compounded and extruded by HAAKE minilab (Abdelmouleha et al. 2007). This is a good technique to save time and the environment due to its dispersing agent-free nature. The "dispersion agent method" is another technique which uses polyoxyethylene as a dispersing agent. After applying polyoxyethylene dispersion as a coating agent, the nanofibres were mixed into the HAAKE minilab with polyethylene granules before being extruded. In order to increase nanocellulose and polyethylene matrix compatibility, maleic anhydride-grafted polyethylene was applied as a coupling agent (Pan et al. 2014; Lin and Dufresne 2013).

12.5.3 *Polyvinyl Acetate Bio-nanocomposites*

There are some interests to study the impact on addition of nanocrystalline cellulose into polyvinyl acetate. For example, it reported the effectiveness of nanocrystalline cellulose extracted from *Phormium tenax* leaf natural fibre by acid hydrolysis as reinforcement phase in the polyvinyl acetate matrix (Kaboorani et al. 2012). The effect of nanocrystalline cellulose types and the content of polyvinyl acetate structural properties were investigated deeply, and their effects on the water adsorption capacity have been evaluated in the case of the potential application in food packaging material (Nancy et al. 2006; Zhang and Zhu 2000).

12.5.4 Polycaprolactone Bio-nanocomposites

The efficiency of nano *L. cylindrica* fibre was investigated as a reinforcement agent for the polycaprolactone matrix (Siang et al. 2012). In general, nano *L. cylindrica* fibre was well known in the field of polymer composites because of its high cellulose content and low density (0.82–0.92 g cm^{-3}), which is lower than the common nano-natural fibres including nano-sisal (1.26 g cm^{-3}), nano-hemp (1.48 g cm^{-3}), nano-ramie (1.5 g cm^{-3}), and nano-cotton (1.51 g cm^{-3}) (Libin et al. 2013; Gilberto et al. 2013). It was observed that polycaprolactone matrix mechanical and thermal properties enhance incorporate with nano *L. cylindrica.*

To exploit further on the potential of nano *L. cylindrica* fibre/polycaprolactone composites, its water barrier properties with varying content of nanofibres have been studied through complementary approaches including water sorption and water permeation kinetics (Arantzazu et al. 2014).

12.5.5 Polyvinyl Alcohol Bio-nanocomposites

Polyvinyl alcohol is the chief hydrophilic water-soluble polymer synthesis up to now. As a result of its outstanding chemical and physical properties, easy processing technique, low cytotoxicity, and biocompatible synthetic polymer, it has been widely used in different fields of biotechnology and biomedicine (Dutta et al. 2002). Polyvinyl alcohol is a highly versatile polymer offering a wide spectrum of property profiles, which opens the way of using it in a broad field of applications, including as a matrix for biodegradable composites (Lindsey and Street 1984). The nanocomposite approach has emerged in the last two decades as an efficient strategy to upgrade the structural and functional properties of polyvinyl alcohol (Mirmohsenia and Wallace 2003).

A new class of engineering-designed polyvinyl bio-nanocomposites was introduced by incorporating cellulose nanocrystal to overcome the limited biological performance and to enhance the mechanical properties of polyvinyl alcohol (Zhi-Gang et al. 2002; Li et al. 2013a, b, 2014). Indeed, the use of cellulose nanocrystal as bionano-materials in polyvinyl alcohol is to increase the load-bearing constituent in developing new and inexpensive biodegradable materials due to their high aspect ratio (Pereda et al. 2011). In this case, cellulose nanocrystal is extracted from a wide variety of natural sources available throughout the world, including ramie, potato, starch, cotton, wood, and others. The addition of cellulose nanocrystal was found to produce 100 % improvement in the tensile modulus of certain polyvinyl alcohol composites (Baheti and Militky 2013; Lee et al. 2009).

12.6 Elastomer Bio-nanocomposites: Manufacturing and Properties

Using nanofibres as reinforcing materials in elastomer bio-nanocomposites is considered as an innovative environment-saving application of natural materials by reducing CO_2 emissions in the atmosphere during their production, processing, and use (O'Connor 1977; Anongnuch et al. 1999; Coran et al. 1974). Nanofibre-reinforced elastomers have mechanical advantages over conventional ones, being 30 % lighter and 3–4 times stronger than the currently used material. They are also less susceptible to damage from heat, chemicals, and spilled gasoline (Tayaa et al. 1998). However, the strength of elastomer-reinforced nanofibres remains far below the theoretical properties. This is due to the limited compatibility between hydrophilic nanofibres and hydrophobic elastomer matrices (George 1971). Poor dispersion of nanofibres and the insufficient adhesion between the nanofibres and elastomer matrix also induce unsatisfactory mechanical properties of the bio-nanocomposites. To overcome the compatibility difficulties, nanofibres have been subjected to physical treatment (plasma), chemical modifications (esterification, etherification, oxidation, silylation, and polymer grafting by click chemistry), and direct silanol nanofibre condensation with prior hydrolysis treatment (Anongnuch et al. 1999). To solve the dispersion problem, the usage of an organic medium, including surfactant, is the most common way.

12.6.1 Natural Rubber Latex Bio-nanocomposites

Natural rubber latex is a colloidal dispersion of *cis*-1,4 Polyisoprene in a form of milky sap which is obtained from the incised bark of *Hevea brasiliensis* trees (Siby and Karger-Kocsic 2004; Kalaprasad and Alain 2003; Julien et al. 2010; Srisuwan et al. 2014). There are some limitations of natural rubber latex due to its hardly degraded solid waste upon disposal (Kalaprasad and Alain 2003). This is due to the presence of inorganic materials, including antioxidants, and sulfur cross-links that make natural rubber latex resistant to microbial infections. Even though its polyisoprene backbone chain degradation can occur in the natural environment, the process is time-consuming. Thus, the combination of natural rubber latex with nanofibre-based filler has been studied to accelerate the process (Phummor et al. 2014). This is used to make the latex-based products decomposable by microorganism this is brought by the several characteristics of nanofibre-based filler, including its cheapness, natural abundance, high susceptibility to biodegradation, and being a renewable resource. On the other hand, the incorporation of nanofibres largely increased the tensile strength, tensile modulus, and abrasion resistance of the natural rubber latex (Wongsorat et al. 2014; Yong and Mustafa 2014). This consequence resulted in a homogenous and well-dispersed bio-nanofibres in the natural rubber latex matrix (Fig. 12.8).

Fig. 12.8 Scanning electron micrographs of (**a**) natural rubber latex and (**b**) natural rubber latex/cellulose whisker nanocomposites containing 10 % bagasse cellulose whiskers (Julien et al. 2010)

Initially, most of the nanofibre has no reinforcement ability of the natural rubber latex compound due to its hydrophilic characteristics. In order to achieve superior reinforcement, several techniques including the addition of coupling agent, ball milling, surface modification, acid hydrolysis, enzyme hydrolysis, and high-pressure homogenization have been applied (Tzounis et al. 2014; Manaila et al. 2014; Supri et al. 2014; Wisittanawat et al. 2014). For example, N_2-pyrolyzed rice bran nanofibre at 900 °C has been used to improve the tribological, friction, thermal, vulcanization, and mechanical properties of the natural rubber latex (Supri et al. 2014). This surface modification of rice bran nanofibre exhibits various fascinating characteristics, including porous structure, high hardness, low density, low fraction, and high wear resistance (Xu and Kong 2014).

12.6.2 Hydrogenated Nitrile Butadiene Rubber Bio-nanocomposites

Hydrogenated nitrile butadiene rubber shows the resistance against oil, chemical, heat, pressure, and ozone (Ding et al. 2015). They are broadly used in the automobile, aviation, oil field, and other industries for the high-performance rubber fittings. However, the crystallization of its ethylene group occurred at a temperature below −30 °C and deteriorated other physical properties (Rajesh et al. 2008, 2013; Lu et al. 2008). By incorporating nanofibre onto its rubber chains, the composite rubber possesses the balanced properties of high strength, elongation, and good oil resistance to a wide range of temperatures. This is brought by the formation of inter H_2 bonds between the rubber and nanofibres (Rajesh et al. 2008). However, careful analysis discovers that it is not beneficial to the elasticity of rubber by adjusting the amount of nanofibres at a wide range, because high strain hysteresis (high loss factor) always brings slow response of its recovery. Therefore, maintaining the low content of nanofibres may be the key to tune the properties of produced composites (Konar et al. 2013; Threepopnatkul et al. 2011). Furthermore, it has been reported that the amount of nanofibres has important effects on the microstructure and nanofibre morphologies of the composites which may connect directly with the reinforcements of the bulk. It is believed that tuning the properties of hydrogenated nitrile butadiene rubber by nanofibres with H_2 bonds might be important to develop modification technologies in the sealing applications (Konar et al. 2013).

12.7 Biopolymer Bio-nanocomposites

12.7.1 Starch Bio-nanocomposites

As a result of its renewability, abundance, and biodegradability, starch has been widely studied as a structural or functional polymer, while its poor mechanical properties and its sensitivity to water are two important drawbacks that should be taken into account (Cristian et al. 2009; Nancy et al. 2009). Starch is the suitable candidate to be embedded in a matrix composed of the hydrophobic polymer when the moisture sensitivity has to be decreased. However, the addition of plasticizer decreases brittleness in the case of mechanical properties, and its mechanical strength degrades accordingly. Furthermore, to improve the low mechanical tensile properties, a classical strategy consists of including rigid nano-sized filler (Deeptangshu 2008). One possibility and factor in contributing the mechanical performance of starch based composites is utilizing nano sized filler (Zhanjun et al. 2011; Nayaka 2010).

Reinforcement of waxy maize nanofibres with filler content between 0 and 5 % into the starch matrix not only showed improvement of mechanical properties but also reduced the water vapor permeability by 40 % (Hélène et al. 2006; Anglès and Alain 2000).

12.8 Potential Applications of Bio-nanocomposites

12.8.1 Automobile

Automotive related industries area be the chief market for the bionano-composites since numerous automotive parts are produced using nanofibre at the present including flax, hemp, kenaf, and sisal. But nowadays, abaca, pineapple leaf, coir, oil palm,

Table 12.2 Advantages of bionano-composites in automotive industries

Justification	Reasons	References
Raw material sources	Bionano-composites made with easily available renewable sources	Hartmut and Ulf (2003)
Properties	Lighter weight, low energy consumption, and low-cost product	Chae et al. (2005)
Volatility in oil prices	Impact substitute material markets and nano-natural fibre cost less	Jin-Hae et al. (2004)
Environmental advantages	Government policy	Marion et al. (2008)

bagasse, bamboo, wheat straw, curaua, and rice husk fibres are gaining interest and importance in both research and application in the automotive field (My Ahmed et al. 2005; James et al. 2011). The advantage of using bionano-composites in automotive industries could be money-saving, weight reduction, recycling, and marketing incentives rather than technical demand (Table 12.2).

12.8.2 Food Packaging

The main function of food packaging is to maintain the quality and safety of food products during storage and transportation and to extend the shelf life of food products. In order to achieve those benefits, unfavorable factors or conditions, including spoilage microorganisms, chemical contaminants, O_2, moisture, light, and external force, should be controlled and prevented (Valapa et al. 2015; Reddy and Rhim 2014). Therefore, the good food package should have all these properties in addition to the common standard properties of packaging materials including mechanical, optical, and thermal (Sun et al. 2014).

With this respect, polymer bio-nanocomposites offer an ideal biodegradable food packaging material (Cunha et al. 2014; Ghaderi et al. 2014). These concerns include degradation rates under various conditions, changes in mechanical properties during storage, potential for microbial growth, and release of harmful compounds into packaged food products. In addition, polymer bio-nanocomposites are an excellent medium for incorporating a wide variety of additives including antioxidants, antifungal agents, antimicrobials, colors, and other nutrients (Khan et al. 2014).

12.8.3 Structural Applications

Structural parts are usually utilized in civil structures, and construction used to be extremely large. Due to the effect of the shape, size, and dimension of large structures on the polymer flow and wettability in bionano-composite production, controlling these elements seems very important (Aboshio et al. 2015; Fasahat et al. 2015;

Jölly et al. 2014). Nanofibre is a hydroscopic material, which brings the curing reaction of the resin for structural application into challenge. In order to overcome this issue, all of the moisture that could polymerize by reacting with the resin should be eliminated by treating the nanofibre (Sprenger et al. 2014).

12.8.4 Tissue Engineering

Bionano-filled polymer composites were utilized as matrices for tissue repairing and regenerating and drug delivery in a hydrogel form in the case of tissue engineering and regenerative medical field application (Kucinska-Lipka et al. 2015; Zhang et al. 2014; Kim et al. 2014). For example, crystal nanocellulose/polyvinyl alcohol composites accurately improve targeted controlled drug delivery to the preferred site of action (Zhang et al. 2014). The possibility to incorporate crystal nanocellulose into polyvinyl alcohol has numerous advantages, including the isolation of drug, slower drug release rates, and achievement of different drug release profiles, as well as the incorporation of multiple drugs in different microsphere populations (Bhaarathy et al. 2014; Shi et al. 2014).

12.9 Conclusion

In the last decade, bionano-composites consisting of polymeric matrix materials and nanofibres have attracted scientific and industrial interest due to their improved properties. Their usages cover a varied range of applications such as high-barrier packaging for food and electronics and automotive and aerospace applications. At low nanofibre content as compared with the common micro and macro or neat counterparts, they reveal high property improvements in terms of lightweight, mechanical, corrosion resistance, lower density, lower weight components, thermal gas permeability resistance, and fire-retardant properties. Other advantages of bio-nanocomposites over traditional polymeric composites are economic viability, reduced tool wear during machining operation, improved energy recovery, reduced dermal and respiratory irritation, and biodegradability. Numerous life cycle assessment studies conducted with bio-nanocomposites prove these advantages.

References

Abdelmouleha M, Boufia S, Belgacemb MN, Dufresne A (2007) Short natural-fibre reinforced polyethylene and natural rubber composites: effect of silane coupling agents and fibres loading. Compos Sci Technol 67(7):1627–1639

Abdul Khalil HPS, Firoozian P, Jawaid M, Akil HM, Hassan A (2013a) Preparation of activated carbon filled epoxy nanocomposites: morphological and thermal properties. J Therm Anal Calorim 113(2):623–631

Abdul Khalil HPS, Fizree HM, Bhat AH, Jawaid M, Abdullah CK (2013b) Development and characterization of epoxy nanocomposites based on nano-structured oil palm ash. Composites B 53:324–333

Aboshio A, Green S, Ye J (2015) Experimental investigation of the mechanical properties of neoprene coated nylon woven reinforced composites. Compos Struct 120:386–393

Amit B, Dan L, Younan X, Samson AJ (2005) Electrospun nanofibers of blends of conjugated polymers: morphology, optical properties, and field-effect transistors. Macromolecules 38(11):4705–4711

Amor IB, Arous M, Kallel A (2014) Interfacial polarization phenomena in palm tree fiber-reinforced epoxy and polyester composites. J Compos Mater 48(21):2631–2638

Anglès MN, Alain D (2000) Plasticized starch/tunicin whiskers nanocomposites. 1. Structural analysis. Macromolecules 33(22):8344–8353

Anongnuch C, Chakrit S, Taweechai A, Sauvarop B-L, Wiriya M (1999) Improvement of interfacial adhesion of poly(m-phenylene isophthalamide) short fiber–thermoplastic elastomer (SEBS) composites by N-alkylation on fiber surface. J Appl Polym Sci 74(10):2414–2422

Arantzazu VG, Marina RS, Ana BS, del María CGS (2014) Characterization and degradation characteristics of poly(ε-caprolactone)-based composites reinforced with almond skin residues. Polym Degrad Stab 108:269–279

Auad ML, Richardson T, Hicks M, Mosiewicki MA, Aranguren MI, Marcovich NE (2012) Shape memory segmented polyurethanes: dependence of behavior on nanocellulose addition and testing conditions. Polym Int 61(2):321–327

Baheti V, Militky J (2013) Reinforcement of wet milled jute nano/micro particles in polyvinyl alcohol films. Fibers Polym 14(1):133–137

Baoxing C, Tsu-Wei C (1999) Compaction of woven-fabric preforms in liquid composite molding processes: single-layer deformation. Compos Sci Technol 59(10):1519–1526

Bazle AG, John WG Jr, Hassan M, Roshan PR, Anwarul H, Shaik J, Travis AB, Bruce KF (2001) High strain-rate behavior of plain-weave S-2 glass/vinyl ester composites. J Compos Mater 35(13):1201–1228

Belaadi A, Bezazi A, Maache M, Scarpa F (2014) Fatigue in sisal fiber reinforced polyester composites: hysteresis and energy dissipation. Procedia Eng 74:325–328

Bhaarathy V, Venugopal J, Gandhimathi C, Ponpandian N, Mangalaraj D, Ramakrishna S (2014) Biologically improved nanofibrous scaffolds for cardiac tissue engineering. Mater Sci Eng C 44:268–277

Bhosale SH, Singh VV, Rangasai MC, Bandyopadhyay-Ghosh S, Ghosh SB (2014) Integration of biobased functionalized feedstock and plastisol in epoxy resin matrix toward developing structural jute biocomposites with enhanced impact strength and moisture resistance properties. Polym Compos. doi:10.1002/pc 23192

Biao D, Hongwei S, Hongquan Y, Hui Z, Ruifei Q, Xue B, Guohui P, Shaozhe L, Fang W, Libo F, Qilin D (2008) Upconversion properties of Ln^{3+} Doped $NaYF_4$/polymer composite fibers prepared by electrospinning. J Phys Chem C 112(5):1435–1440

Bledzki AK, Reihmane S, Gassan J (2009) Properties and modification methods for vegetable fibers for natural fiber composites. J Appl Polym Sci 59(8):1329–1336

Boynard CA, D'Almeida JRM (2000) Morphological characterization and mechanical behavior of sponge gourd (*Luffa cylindrica*)-polyester composite materials. Polym Plast Technol Eng 39(3):489–499

Boynard CA, Monteiro SN, D'Almeida JRM (2003) Aspects of alkali treatment of sponge gourd (*Luffa cylindrica*) fibers on the flexural properties of polyester matrix composites. J Appl Polym Sci 87(12):1927–1932

Brugnago RJ, Satyanarayana KG, Wypych F, Ramos LP (2011) The effect of steam explosion on the production of sugarcane bagasse/polyester composites. Composites A 42(4):364–370

Bu Z, Hu J, Li B (2014) Novel silicon-modified phenolic novolac resins: non-isothermal curing kinetics, and mechanical and thermal properties of their biofiber-reinforced composites. Thermochim Acta 575:244–253

Carneiro OS, Maia JM (2000) Rheological behavior of (short) carbon fiber/thermoplastic composites. Part I: The influence of fiber type, processing conditions and level of incorporation. Polym Compos 21(6):960–969

Carsten Z, Rolf M, Jürgen F (1999) Morphology and toughness/stiffness balance of nanocomposites based upon anhydride-cured epoxy resins and layered silicates. Macromol Chem Phys 200(3):661–670

Chae HH, Young BL, Jin WB, Jae YJ, Byeong UN, Tae WH (2005) Preparation and mechanical properties of polypropylene/clay nanocomposites for automotive parts application. J Appl Polym Sci 98(1):427–433

Chee CY, Yong GK, Abdullah LC, Nadarajah K (2013) Effect of nanosilica and titania on thermal stability of polypropylene/oil palm empty fruit fibre composite. J Biobased Mater Bioenergy 7(1):169–174

Chi-Ching K, Chia-Hung L, Wen-Chang C (2007) Morphology and photophysical properties of light-emitting electrospun nanofibers prepared from poly(fluorene) derivative/PMMA blends. Macromolecules 40(19):6959–6966

Chow WS, Mohd Ishak ZA, Karger-Kocsis J, Apostolov AA, Ishiakud US (2003) Compatibilizing effect of maleated polypropylene on the mechanical properties and morphology of injection molded polyamide 6/polypropylene/organoclay nanocomposites. Polymer 44(24):7427–7440

Chow WS, Ishak ZAM, Ishiaku US, Karger-Kocsis J, Apostolov AA (2004) The effect of organoclay on the mechanical properties and morphology of injection-molded polyamide 6/polypropylene nanocomposites. J Appl Polym Sci 91(1):175–189

Coran AY, Boustany K, Hamed P (1974) Short-fiber—rubber composites: the properties of oriented cellulose-fiber—elastomer composites. Rubber Chem Technol 47(2):396–410

Cristian JG, Fernando GT, Clara MG, Omar PT, Josep C-F, Juan M-P (2009) Development of self-assembled bacterial cellulose-starch nanocomposites. Mater Sci Eng C 29(4):1098–1104

Cunha M, Berthet M-A, Pereira R, Covas JA, Vicente AA, Hilliou L (2014) Development of polyhydroxyalkanoate/ beer spent grain fibers composites for film blowing applications. Polym Compos. doi:10.1002/pc23093

Cynthia RL, Sibylle G, Katarzyna G, Sylwester G, Andreas G, Mauro A (2005) Fibrin–Polyurethane composites for articular cartilage tissue engineering: a preliminary analysis. Tissue Eng 11(9):1562–1573

Da Silva CG, Oliveira F, Ramires EC, Castellan A, Frollini E (2012) Composites from a forest biorefinery byproduct and agrofibers: lignosulfonate-phenolic type matrices reinforced with sisal fibers. Tappi J 11(9):41–49

Li D, Xia Y (2003) Fabrication of titania nanofibers by electrospinning. Nano Lett 3(4):555–560

Deeptangshu SC (2008) Understanding amylose crystallinity in starch-clay nanocomposites. J Polym Sci B 46(10):979–987

Denault J, Vu-Khanh T, Foster B (2014) Tensile properties of injection molded long fiber thermoplastic composites. Polym Compos 10(5):313–321

Ding X, Wang J, Zhang S, Wang J, Li S (2015) Composites based on CB/CF/Ag filled EPDM/NBR rubber blends with high conductivity. J Appl Polym Sci 132(4). Article number 41357

Dutta P, Biswas S, Subodh KD (2002) Dielectric relaxation in polyaniline–polyvinyl alcohol composites. Mater Res Bull 37(1):193–200

El-Tayeb NSM (2008) A study on the potential of sugarcane fibers/polyester composite for tribological applications. Wear 265(1):223–235

Everson K, Grace C, Dongyan W, Charles AW, Jeanne MH (2006) Probing synergism, antagonism, and additive effects in poly(vinyl ester) (PVE) composites with fire retardants. Polym Degrad Stab 91(6):1209–1218

Farzana H, Mehdi H, Masami O, Russell EG (2006) Review article: polymer-matrix nanocomposites, processing, manufacturing, and application: an overview. J Comp Mater 40(17):1511–1575

Fasahat F, Dastjerdi R, Mojtahedi MRM, Hoseini P (2015) Wear properties of high speed spun multi-component PA6 nanocomposite fabrics; abrasion resistance mechanism of nanocomposites. Wear 322:117–125

François T, Edu R, Vincent A, Sofiane S (2006) Advanced numerical simulation of liquid composite molding for process analysis and optimization. Composites A 37(6):890–902

Fukuya MN, Senoo K, Kotera M, Yoshimoto M, Sakata O (2014) Controlling of crystallite orientation for poly(ethylene oxide) thin films with cellulose single nano-fibers. Polymer 55(16): 4401–4404

Gates TS, Sun CT (1991) Elastic/viscoplastic constitutive model for fiber reinforced thermoplastic composites. AIAA J 29(3):457–463

George CD (1971) Short fiber-elastomer composites. J Elastomers Plast 3(4):230–248

Ghaderi M, Mousavi M, Yousefi H, Labbafi M (2014) All-cellulose nanocomposite film made from bagasse cellulose nanofibers for food packaging application. Carbohydr Polym 104(1): 59–65

Ghali L, Aloui M, Zidi M, Bendaly H, M'sahli S, Sakli F (2011) Effect of chemical modification of *Luffa cylindrica* fibers on the mechanical and hygrothermal behaviours of polyester/luffa composites. BioResources 6(4):3836–3849

Gilberto S, Julien B, Nadège F, Sabrina B, Stéphane M, Alain D (2013) Thermal and mechanical properties of bio-nanocomposites reinforced by *Luffa cylindrica* cellulose nanocrystals. Carbohydr Polym 91(2):711–717

Godina N, Hugueta S, Gaertnera R, Salmonb L (2004) Clustering of acoustic emission signals collected during tensile tests on unidirectional glass/polyester composite using supervised and unsupervised classifiers. NDT E Int 37(4):253–264

Gong S, Yuan M, Chandra A, Kharbas H, Osorio A, Kharbas A, Turng LS (2005) Microcellular injection molding. Int Polym Process 20(2):202–214

Goutianos S, Peijs T, Nystrom B, Skrifvars M (2006) Development of flax fibre based textile reinforcements for composite applications. Appl Compos Mater 13(4):199–215

Hammami A, Gauvin R, Trochu F (1998) Modeling the edge effect in liquid composites molding. Composites A 29(5):603–609

Hao F, Weidong L, Chyi-Shan W, Richard AV (2002) Generation of electrospun fibers of nylon 6 and nylon 6-montmorillonite nanocomposite. Polymer 43(3):775–780

Hao A, Chen Y, Chen JY (2014) Creep and recovery behavior of kenaf/polypropylene nonwoven composites. J Appl Polym Sci 131(17):8864–8874

Hartmut P, Ulf K (2003) Future nanotechnology developments for automotive applications. Mater Sci Eng C 23(6):737–741

Hélène A, Sonia M-B, Patrice D, Alain D (2006) Thermoplastic starch-waxy maize starch nanocrystals nanocomposites. Biomacromolecules 7(2):531–539

Hezhou Y, Xing Yang L, Hanping H (2008) Fabrication of metal matrix composites by metal injection molding—a review. J Mater Process Technol 200(1):12–24

Hongyan C, Menghe M, Xin D (2008) Influence of moisture absorption on the interfacial strength of bamboo/vinyl ester composites. Composites Part A App 40(12):2013–2019

Hrishikesh K, Paul N, Mingjun Y, Shaoqin G, Lih-Sheng T, Rick S (2003) Effects of nano-fillers and process conditions on the microstructure and mechanical properties of microcellular injection molded polyamide nanocomposites. Polym Compos 24(6):655–671

Ibrahim AH, Leman Z, Mohd Sapuan S (2013) Tensile properties of impregnated sugar palm (*Arenga pinnata*) fibre composite filled thermosetting polymer composites. Adv Mater Res 701:8–11

Ioannis SC (2005) Novel nanocomposites and nanoceramics based on polymer nanofibers using electrospinning process—a review. J Mater Process Technol 167(2):283–293

István S, David P (2010) Microfibrillated cellulose and new nanocomposite materials: a review. Cellulose 17(3):459–494

Iwamoto S, Yamamoto S, Lee S-H, Endo T (2014) Mechanical properties of polypropylene composites reinforced by surface-coated microfibrillated cellulose. Composites A 59:26–29

Jae WC, Jeong WK, Yong CJ, Nam SG (2005) Electroactive shape-memory polyurethane composites incorporating carbon nanotubes. Macromol Rapid Commun 26(5):412–416

James N, Paul W, Krzysztof P, Kambiz K (2011) Natural fibre-reinforced polymer composites and nanocomposites for automotive applications. Cellulose Fibers Bio Nano Polym Compos 2011: 661–700

Janis MB, David C, Richard AV (2000) Thermoset-layered silicate nanocomposites. Quaternary ammonium montmorillonite with primary diamine cured epoxies. Chem Mater 12(11): 3376–3384

Jayamol G, Sreekala MS, Sabu T (2001) A review on interface modification and characterization of natural fiber reinforced plastic composites. Polymer Eng Sci 41(9):1471–1485

Jih-Jen W, Yan-Ru C, Wen-Pin L, Chun-Te W, Chuh-Yung C (2010) Construction of nanocrystalline film on nanowire array via swelling electrospun polyvinylpyrrolidone-hosted nanofibers for use in dye-sensitized solar cells. ACS Nano 4(10):5679–5684

Jin-Hae C, Sung JK, Yong LJ, Seungsoon I (2004) Poly(ethylene terephthalate) nanocomposites by in situ interlayer polymerization: the thermo-mechanical properties and morphology of thehybrid fibers. Polymer 45(3):919–926

Jirukkakul N (2013) A study of Mu Yor sausage wraps using chitosan films incorporating garlic oil, lemon grass oil and galangal oil. Int Food Res J 20(3):1199–1204

Jochen G, Andrzej KB (1997) Effect of moisture content on the properties of silanized jute-epoxy composites. Polym Compos 18(2):179–184

Jochen G, Andrzej KB (1999) Possibilities for improving the mechanical properties of jute/epoxy composites by alkali treatment of fibres. Compos Sci Technol 59(9):1303–1309

Jölly I, Schlögl S, Wolfahrt M, Pinter G, Fleischmann M, Kern W (2014) Chemical functionalization of composite surfaces for improved structural bonded repairs. Composites B 69:296–303

Joshia SV, Drzalb LT, Mohanty AK, Arorac S (2004) Are natural fiber composites environmentally superior to glass fiber reinforced composites? Composites A 35(3):371–376

Julien B, Mohammad LH, Cecile B, Enas AH, Nahla AE-W, Alain D (2010) Mechanical, barrier, and biodegradability properties of bagasse cellulose whiskers reinforced natural rubber nanocomposites. Ind Crops Prod 32(3):627–633

Kaboorani A, Riedl B, Blanchet P, Fellin M, Hosseinaei O, Wang S (2012) Nanocrystalline cellulose (NCC): a renewable nano-material for polyvinyl acetate (PVA) adhesive. Eur Polym J 48(11):1829–1837

Kaddami H, Dufresne A, Khelifi B, Bendahou A, Taourirte M, Raihane M, Issartel N, Sautereau H, Gérard J-F, Sami N (2006) Short palm tree fibers – thermoset matrices composites. Composites A 37(9):1413–1422

Kaith BS, Jindal R, Jana AK, Maiti M (2010) Development of corn starch based green composites reinforced with *Saccharum spontaneum* L fiber and graft copolymers – evaluation of thermal, physico-chemical and mechanical properties. Biores Technol 101(17):6843–6851

Kalaprasad GN, Alain D (2003) Crab shell chitin whiskers reinforced natural rubber nanocomposites. 3. Effect of chemical modification of chitin whiskers. Biomacromolecules 4(6): 1835–1842

Karger-Kocsis J, Wu C-M (2004) Thermoset rubber/layered silicate nanocomposites. Status and future trends. Polym Eng Sci 44(6):1083–1093

Kerang H, Shunliang J, Chuck Z, Ben W (2000) Flow modeling and simulation of SCRIMP for composites manufacturing. Composites A 31(1):79–86

Khalid MZ, Haq NB, Ijaz AB (2007) Methods for polyurethane and polyurethane composites, recycling and recovery: a review. React Funct Polym 67(8):675–692

Khan A, Huq T, Khan RA, Riedl B, Lacroix M (2014) Nanocellulose-based composites and bioactive agents for food packaging. Crit Rev Food Sci Nutr 54(2):163–174

Kim G-M, Lee S-M, Michler GH, Roggendorf H, Gösele U, Knez M (2008) Nanostructured pure anatase titania tubes replicated from electrospun polymer fiber templates by atomic layer deposition. Chem Mater 20(9):3085–3091

Kim JI, Pant HR, Sim H-J, Lee KM, Kim CS (2014) Electrospun propolis/polyurethane composite nanofibers for biomedical applications. Mater Sci Eng C 44:52–57

Konar SK, Gu R, Sain M (2013) Preparation and characterization of baked nitrile latex foam reinforced with biomasses. Ind Crops Prod 42(1):261–267

Kucinska-Lipka J, Gubanska I, Janik H, Sienkiewicz M (2015) Fabrication of polyurethane and polyurethane based composite fibres by the electrospinning technique for soft tissue engineering of cardiovascular system. Mater Sci Eng C 46:166–176

Kuruvilla J, Sabu T, Pavithran C (1996) Effect of chemical treatment on the tensile properties of short sisal fibre-reinforced polyethylene composites. Polymer 37(23):5139–5149

Kyriaki K, Hiroyuki F, Lawrence TD (2007) A new compounding method for exfoliated graphite-polypropylene nanocomposites with enhanced flexural properties and lower percolation threshold. Compos Sci Technol 67(10):2045–2051

Lee SC, Mariatti M (2008) The effect of bagasse fibers obtained (from rind and pith component) on the properties of unsaturated polyester composites. Mater Lett 62(15):2253–2256

Lee S-Y, Mohan DJ, Kang I-A, Doh G-H, Lee S, Han SO (2009) Nanocellulose reinforced PVA composite films: effects of acid treatment and filler loading. Fibers Polym 10(1):77–82

Li C, Fan H, Wang D-Y, Hu J, Wan J, Li B (2013a) Novel silicon-modified phenolic novolacs and their biofiber-reinforced composites: preparation, characterization and performance. Compos Sci Technol 87:189–195

Li W, Zhao X, Huang Z, Liu S (2013b) Nanocellulose fibrils isolated from BHKP using ultrasonication and their reinforcing properties in transparent poly (vinyl alcohol) films. J Polym Res 20(8). Article number 210

Li W, Wu Q, Zhao X, Huang Z, Cao J, Li J, Liu S (2014) Enhanced thermal and mechanical properties of PVA composites formed with filamentous nanocellulose fibrils. Carbohydr Polym 113:403–410

Libin S, Rongjia J, Yu Z, Yali J (2013) Electrospun cellulose nanocrystals/polycaprolactone nanocomposite fiber mats. J MacromolSci B 53(5):1–10

Liliana R, Constantin NC, Dan R (2009) Effect of UV radiation on some polymeric networks based on vinyl ester resin and modified lignin. Polym Test 28(3):296–300

Lin N, Dufresne A (2013) Physical and/or chemical compatibilization of extruded cellulose nanocrystal reinforced polystyrene nanocomposites. Macromolecules 46(14):5570–5583

Lin RJ, James LL, Ming JL (1993) Mold filling and curing analysis in liquid composite molding. Polym Compos 14(1):71–81

Lindsey SE, Street GB (1984) Conductive composites from polyvinyl alcohol and polypyrrole. Synth Met 10(1):67–69

Loh YR, Sujan D, Rahman ME, Das CA (2013) Review sugarcane bagasse – the future composite material: a literature review. Resour Conserv Recycl 75:14–22

Lu L, Zhai Y, Zhang Y, Ong C, Guo S (2008) Reinforcement of hydrogenated carboxylated nitrile-butadiene rubber by multi-walled carbon nanotubes. Appl Surf Sci 255(5):2162–2166 (PART 1)

Mahrholz T, Stängle J, Sinapiusa M (2009) Quantitation of the reinforcement effect of silica nanoparticles in epoxy resins used in liquid composite moulding processes. Composites A 40(3):235–243

Maiti M, Kaith BS, Jindal R, Jana AK (2010) Synthesis and characterization of corn starch based green composites reinforced with *Saccharum spontaneum* L graft copolymers prepared under micro-wave and their effect on thermal, physio-chemical and mechanical properties. Polym Degrad Stab 95(9):1694–1703

Mallarino S, Chailan JF, Vernet JL (2005) Glass fibre sizing effect on dynamic mechanical properties of cyanate ester composites I. Single frequency investigations. Eur Polym J 41(8):1804–1811

Manaila E, Stelescu MD, Craciun G, Surdu L (2014) Effects of benzoyl peroxide on some properties of composites based on hemp and natural rubber. Polym Bull 71(8):2001–2022

Marion P, Julasak J, Jerry YYH, Athanasios M, Adam FL, Karen W, Gerhard K, Milo SPS, Alexander B (2008) Surface modification of natural fibers using bacteria: depositing bacterial cellulose onto natural fibers to create hierarchical fiber reinforced nanocomposites. Biomacromolecules 9(6):1643–1651

Matos Ruiz M, Cavaillé JY, Dufresne A, Gérard JF, Graillat C (2000) Processing and characterization of new thermoset nanocomposites based on cellulose whiskers. Compos Interfaces 7(2):117–131

Maya JJ, Rajesh DA (2008) Recent developments in chemical modification and characterization of natural fiber-reinforced composites. Polym Compos 29(2):187–207

Michael A, Philippe D (2000) Polymer-layered silicate nanocomposites: preparation, properties and uses of a new class of materials. Mater Sci Eng R 28(1):1–63

Ming QZ, Min ZR, Xun L (2005) Fully biodegradable natural fiber composites from renewable resources: all-plant fiber composites. Compos Sci Technol 65(15):2514–2525

Mingjun Y, Lih-Sheng T, Shaoqin G, Daniel C, Chris H, Rick S (2004) Study of injection molded microcellular polyamide-6 nanocomposites. Polym Eng Sci 44(4):673–686

Mingjun Y, Andreas W, Shaoqin G, Lih-Sheng T (2005) Effects of nano- and micro-fillers and processing parameters on injection-molded microcellular composites. Polym Eng Sci 45(6): 773–788

Mirmohsenia A, Wallace CG (2003) Preparation and characterization of processable electroactive polyaniline–polyvinyl alcohol composite. Polymer 44(12):3523–3528

Mishra V, Biswas S (2014) Three-body abrasive wear behavior of short jute fiber reinforced epoxy composites. Polym Compos. doi:10.1002/pc.23178

Mohammad M, Karen IW (2006) Polymer nanocomposites containing carbon nanotubes. Macromolecules 39(16):5194–5205

Monteiro SN, Margem FM, de Souza Martins LB, Loiola RL, Oliveira MP (2014) Tensile strength of polyester matrix composites reinforced with giant bamboo (*Dendrocalamus giganteus*) fibers. Mater Sci Forum 775–776:308–313

Muhammad P, Mohini MS (2003) Carbon storage potential in natural fiber composites. Resour Conserv Recycl 39(4):325–340

Mwaikambo LY, Bisanda ETN (1999) Performance of cotton-kapok fabric-polyester composites. Polym Test 18(3):181–198

My Ahmed AS, Fannie A, Alain D (2005) Review of recent research into cellulosic whiskers, their properties and their application in nanocomposite field. Biomacromolecules 6(2):612–626

Nancy LGR, Wim T, Alain D (2006) Sisal cellulose whiskers reinforced polyvinyl acetate nanocomposites. Cellulose 13(3):261–270

Nancy LG, Laura R, Alain D, Mirta IA, Silvia G (2009) Physico-mechanical properties of biodegradable starch nanocomposites. Macromol Mater Eng 294(3):169–177

Navin C, Ajay N, Somit N (2000) Three-body abrasive wear of short glass fibre polyester composite. Wear 242(1):38–46

Nayaka SK (2010) Biodegradable PBAT/starch nanocomposites. Polym Plast Technol Eng 49(14):1–10

Nik Salleh NG, Sofian Alias M, Gläsel H-J, Mehnert R (2013) High performance radiation curable hybrid coatings. Radiat Phys Chem 84:70–73

O'Connor JE (1977) Short-fiber-reinforced elastomer composites. Rubber Chem Technol 50(5): 945–958

O'Donnell A, Dweib MA, Wool RP (2004) Natural fiber composites with plant oil-based resin. Compos Sci Technol 64(9):1135–1145

Oh HJ, Kim HY, Kim SS (2014) Effect of the core/shell-structured meta-aramid/epoxy nanofiber on the mechanical and thermal properties in epoxy adhesive composites by electrospinning. J Adhes 90(9):787–801

Oladele IO, Daramola OO, Fasooto S (2014) Effect of chemical treatment on the mechanical properties of sisal fibre reinforced polyester composites. Leonardo Elect J Pract Technol 13(24):1–12

Omid Z, Aminreza K, Mostafavic SM (2012) Preparation, optimization and thermal characterization of a novel conductive thermoset nanocomposite containing polythiophene nanoparticles using dynamic thermal analysis. Polym Degrad Stab 97(1):3–13

Ornaghi HL, Zattera AJ, Amico SC (2014) Dynamic mechanical properties and correlation with dynamic fragility of sisal reinforced composites. Polym Comp 36(1): 1–10

Pan M, Mei C, Li G, Du J (2014) Ammonium polyphosphate improving physicochemical properties of rice straw-high density polyethylene composites. Trans Chin Soc Agric Eng 30(16):328–333

Pang AL, Ismail H (2014) Influence of kenaf form and loading on the properties of kenaf-filled polypropylene/waste tire dust composites: a comparison study. J Appl Polym Sci 131(19). Article number 40877

Pant HR, Kim HJ, Joshi MK, Pant B, Park CH, Kim JI, Hui KS, Kim CS (2014) One-step fabrication of multifunctional composite polyurethane spider-web-like nanofibrous membrane for water purification. J Hazard Mater 264:25–33

Pavel S, Suresh GA (2004) Desirable features in mold filling simulations for liquid composite molding processes. Polym Comp 25(4):355–367

Pereda M, Amica G, Rácz I, Marcovich NE (2011) Structure and properties of nanocomposite films based on sodium caseinate and nanocellulose fibers. J Food Eng 103(1):76–83

Pereira AC, Margem FM, Monteiro SN, Faria R Jr, Cordeiro T (2014) Characterization of thermal properties of polyester matrix reinforced with sisal fibers by photoacoustic technique. TMS Annual Meeting, 2014, pp 449–456

Peter CLB, Zhen W, Thomas JP (1999) Polymer-layered silicate nanocomposites: an overview. Appl Clay Sci 15(1):11–29

Phummor P, Riyajan S-A, Tangboriboonrat P (2014) Physical properties of polymer blend: natural rubber glove waste/polystyrene foam waste/cellulose fiber via brabender. Adv Mater Res 844:57–60

Pilla S, Kramschuster A, Gong S, Chandra A, Turng L-S (2007) Solid and microcellular polylactide-carbon nanotube nanocomposites. Int Polym Process 22(5):418–428

Priya ARS, Subramania A, Young-Sam J, Kang-Jin K (2008) High-performance quasi-solid-state dye-sensitized solar cell based on an electrospun PVdF–HFP membrane electrolyte. Langmuir 24(17):9816–9819

Rafiq A, Al-Qadhi M, Merah N, Ali Y (2014) Mechanical behavior of hybrid glass fibre/epoxy clay nanocomposites. Adv Mater Res 894:336–341

Rajesh C, Manoj KC, Unnikrishnan G, Purushothaman E (2013) Dielectric properties of short nylon-6 fiber-reinforced NBR composites. Adv Polym Technol 32(Suppl 1):E90–E102

Rajesh C, Unnikrishnan G, Purushothaman E (2008) Investigation of interfacial adhesion in nylon-6 fibre/NBR composites through restricted equilibrium swelling technique. Compos Interf 15(5):527–548

Ramanathan T, Abdala AA, Stankovich S, Dikin DA, Herrera-Alonso M, Piner RD, Adamson DH, Schniepp HC, Chen X, Ruoff RS, Nguyen ST, Aksay IA, Prud'Homme RK, Brinson LC (2008) Functionalized graphene sheets for polymer nanocomposites, letter abstract. Nat Nanotechnol 3:327–331

Ramli R, Yunus RM, Beg MDH (2013) Thermal gravimetric/Fourier transform infrared spectroscopy analyses of oil palm fibre reinforced polypropylene composites. J Thermoplast Compos Mater 26(4):497–512

Rassiah K, Megat Ahmad MMH, Ali A, Sihombing H (2014a) Valorization of an industrial organosolv-sugarcane bagasse lignin: characterization and use as a matrix in biobased composites reinforced with sisal fibers. Adv Environ Biol 8(8):2632–2639

Rassiah K, Megat Ahmad MMH, Ali A (2014b) Mechanical properties of laminated bamboo strips from *Gigantochloa scortechinii*/polyester composites. Mater Des 57:551–559

Reddy JP, Rhim J-W (2014) Characterization of bionanocomposite films prepared with agar and paper-mulberry pulp nanocellulose. Carbohydr Polym 110:480–488

Reis PNB, Santos P, Ferreira JAM, Richardson MOW (2013) Impact response of sandwich composites with nano-enhanced epoxy resin. J Reinf Plast Compos 32(12):898–906

Richa A, Saxena NS, Sharma KB, Thomas S, Sreekala MS (2000) Activation energy and crystallization kinetics of untreated and treated oil palm fibre reinforced phenol formaldehyde composites. Mater Sci Eng A 277(1):77–82

Richard AV, John FM (2007) Polymer nanocomposites with prescribed morphology: going beyond nanoparticle-filled polymers. Chem Mater 19(11):2736–2751

Richard AV, Klaus DJ, Edward JK, Emmanuel PG (1996) Microstructural evolution of melt intercalated polymer–organically modified layered silicates nanocomposites. Chem Mater 8(11):2628–2635

Rivai M, Gupta A, Islam MR, Beg MDH (2014) Characterization of oil palm empty fruit bunch and glass fibre reinforced recycled polypropylene hybrid composites. Fibers Polym 15(7): 1523–1530

Roberts J, Jānis A, Lennart W (2003) Strength and adhesion characteristics of elementary flax fibres with different surface treatments. Composites A 34(7):603–612
Rodrigues EF, Maia TF, Mulinari DR (2011) Tensile strength of polyester resin reinforced sugarcane bagasse fibers modified by esterification. Procedia Eng 10:2348–2352
Rostamiyan Y, Fereidoon A, Mashhadzadeh AH, Ashtiyani MR, Salmankhani A (2015) Using response surface methodology for modeling and optimizing tensile and impact strength properties of fiber orientated quaternary hybrid nano composite. Composites B 69:304–316
Sallih N, Lescher P, Bhattacharyya D (2014) Factorial study of material and process parameters on the mechanical properties of extruded kenaf fibre/polypropylene composite sheets. Composites A 61:91–107
Sami BB, Ridha BC (2007) Influence of fibre orientation and volume fraction on the tensile properties of unidirectional alfa-polyester composite. Compos Sci Technol 67(1):140–147
Saralegi A, Gonzalez ML, Valea A, Eceiza A, Corcuera MA (2014) The role of cellulose nanocrystals in the improvement of the shape-memory properties of castor oil-based segmented thermoplastic polyurethanes. Compos Sci Technol 92:27–33
Seena J, Sreekalab MS, Oommena Z, Koshyc P, Sabu T (2002) A comparison of the mechanical properties of phenol formaldehyde composites reinforced with banana fibres and glass fibres. Compos Sci Technol 62(14):857–1868
Seeram R, Kazutoshi F, Wee-Eong T, Thomas Y, Zuwei M, Ramakrishna R (2006) Electrospun nanofibers: solving global issues. Mater Today 9(3):40–50
Seki Y, Sever K, Erden S, Sarikanat M, Neser G, Ozes C (2012) Characterization of *Luffa cylindrica* fibers and the effect of water aging on the mechanical properties of its composite with polyester. J Appl Polym Sci 123(4):2330–2337
Sgricciaa N, Hawleya MC, Misrab M (2008) Characterization of natural fiber surfaces and natural fiber composites. Composites A 39(10):1632–1637
Shi Y, Han H, Quan H, Zang Y, Wang N, Ren G, Xing M, Wu Q (2014) Activated carbon fibers/poly(lactic-co-glycolic) acid composite scaffolds: preparation and characterizations. Mater Sci Eng C 43:102–108
Shinoj S, Visvanathan R, Panigrahi S, Kochubabu M (2011) Oil palm fiber (OPF) and its composites: a review. Ind Crops Prod 33(1):7–22
Siang YC, Hanafi I, Qumrul A (2012) Effect of maleic anhydride on kenaf dust filled polycaprolactone/thermoplastic sago starch composites. Bioresources 7(2):1–10
Siby V, Karger-Kocsic J (2004) Melt-compounded natural rubber nanocomposites with pristine and organophilic layered silicates of natural and synthetic origin. J Appl Polym Sci 91(2):813–819
Singha AS, Thakur VK (2009) Fabrication and characterization of S. cilliare fibre reinforced polymer composites. Bull Mater Sci 32(1):49–58
Sprenger S, Kothmann MH, Altstaedt V (2014) Carbon fiber-reinforced composites using an epoxy resin matrix modified with reactive liquid rubber and silica nanoparticles. Compos Sci Technol 105:86–95
Sreekala MS, Kumaran MG, Seena J, Maya J, Sabu T (2000) Oil palm fibre reinforced phenol formaldehyde composites: influence of fibre surface modifications on the mechanical performance. Appl Compos Mater 7(5):295–329
Srisuwan L, Jarukumjorn K, Suppakarn N (2014) Effect of bis (triethoxysilylpropyl) tetrasulfide on physical properties of rice husk fiber/natural rubber composites. Adv Mater Res 844:263–266
Sujuan W, Qidong T, Feng Y (2010) Hybrid photovoltaic devices based on poly (3-hexylthiophene) and ordered electrospun ZnO nanofibers. J Phys Chem C 114(13):6197–6200
Sullivan JL, Kao BG, Oene HV (1984) Shear properties and a stress analysis obtained from vinyl-ester iosipescu specimens. Exp Mech 24(3):223–232
Sun H, Wang W, Li W, Zhang X (2014) Preparation technique of *Eichhornia crassipes*-based composite packaging materials. Trans Chin Soc Agric Eng 30(12):258–264

Supri AG, Tan SJ, Ismail H, Teh PL (2014) Properties of (Low-density polyethylene)/(natural rubber)/(water hyacinth fiber) composites: the effect of polyaniline. J Vinyl Addit Technol 20(2):122–130
Tanobe VOA, Flores-Sahagun THS, Amico SC, Muniz GIB, Satyanarayana KG (2014) Sponge gourd (*Luffa cylindrica*) reinforced polyester composites: preparation and properties. Def Sci J 64(3):273–280
Tayaa M, Kima WJ, Onob K (1998) Piezoresistivity of a short fiber/elastomer matrix composite. Mech Mater 28(1):53–59
Thavasi V, Singh G, Ramakrishna S (2008) Electrospun nanofibers in energy and environmental applications. Energy Environ Sci 1:205–221
Thielemans W, Can E, Morye SS, Wool RP (2002) Novel applications of lignin in composite materials. J Appl Polym Sci 83(2):323–331
Threepopnatkul P, Teppinta W, Sombatsompop N (2011) Effect of co-monomer ratio in ABS and wood content on processing and properties in wood/ABS composites. Fibers Polym 12(8): 1007–1013
Tia BT, David PA (2002) Morphology development in layered silicate thermoset nanocomposites. Compos Sci Technol 62(7):1033–1041
Tony M, Petra P, Peter H, Michael M, Darren M, Steven EJB, Gerard PB, Daniel B, Patrick L, John PQ (2005) Polyethylene multiwalled carbon nanotube composites. Polymer 46(19): 8222–8232
Tsai J-L, Chang N-R (2011) Investigating damping properties of nanocomposites and sandwich structures with nanocomposites as core materials. J Comp Mater 45(21):2157–2164
Tzounis L, Debnath S, Rooj S, Fischer D, Mäder E, Das A, Stamm M, Heinrich G (2014) High performance natural rubber composites with a hierarchical reinforcement structure of carbon nanotube modified natural fibers. Mater Des 58:1–11
Valapa RB, Pugazhenthi G, Katiyar V (2015) Fabrication and characterization of sucrose palmitate reinforced poly(lactic acid) bionanocomposite films. J Appl Polym Sci 132(3). Article number 41320
Varun S, Megan MMB, Blake RP, Ayusman S (2006) Silver bromide nanoparticle/polymer composites: dual action tunable antimicrobial materials. J Am Chem Soc 128(30):9798–9808
Venkata Reddy G, Venkata Naidu S, Shobha Rani T (2008) Kapok/glass polyester hybrid composites: tensile and hardness properties. J Reinf Plast Compos 27(16):1775–1787
Venkata Reddy G, Venkata Naidu S, Shobha Rani T (2009) A study on hardness and flexural properties of kapok/sisal composites. J Reinf Plast Compos 28(16):2035–2044
Wang TJ, Wu CH, James LL (1994) In-plane permeability measurement and analysis in liquid composite molding. Polym Compos 15(4):278–288
Wang J, Feng L, Feng Y, Yan A, Ma X (2012) Preparation and properties of organic rectorite/epoxy resin nano-composites. Polym Plast Technol Eng 51(15):1583–1588
Wei C, Zeng S, Huang S, Liu H, Lv J, Yu C (2013) Preparation and properties of sisal fiber cellulose microcrystal reinforced phenol-formaldehyde resins composites. Acta Mater Compos Sinica 30:306–310
Wisittanawat U, Thanawan S, Amornsakchai T (2014) Mechanical properties of highly aligned short pineapple leaf fiber reinforced – nitrile rubber composite: effect of fiber content and bonding agent. Polym Test 35:20–27
Wongsorat W, Suppakarn N, Jarukumjorn K (2014) Effects of compatibilizer type and fiber loading on mechanical properties and cure characteristics of sisal fiber/natural rubber composites. J Comp Mater 48(19):2401–2411
Won-Jun L, Sang-Eui L, Chun-Gon K (2006) The mechanical properties of MWNT/PMMA nanocomposites fabricated by modified injection molding. Compos Struct 76(4):406–410
Wu C-S (2011) Performance and biodegradability of a maleated polyester bioplastic/recycled sugarcane bagasse system. J Appl Polym Sci 121(1):427–435
Xiao S, Xu W, Ma H (2012) Fabrication and characterization of mechano-modulated PET/BPU nanofibrous mats as potential vascular grafts materials. Fibers Polym 13(5):18–625

Xiaobo X, Huaqing Y, Debin Z, Jiapeng Z, Huang C, Wei C, Jiye C (2012) Subwavelength and nanometer diameter optical polymer fibers as building blocks for miniaturized photonics integration. In: N Das (ed) Optical communication. Published October 3, 2012 under CC BY 3.0 license. ISBN 978-953-51-0784-2

Xu ZH, Kong ZN (2014) Mechanical and thermal properties of short-coirfiber-reinforced natural rubber/polyethylene composites. Mech Compos Mater 50(3):353–358

Xue L, Lope GT, Satyanarayan P (2007) Chemical treatments of natural fiber for use in natural fiber-reinforced composites: a review. J Polym Environ 15(1):25–33

Yanjun X, Callum ASH, Zefang X, Holger M, Carsten M (2010) Silane coupling agents used for natural fiber/polymer composites: a review. Composites A 41(7):806–819

Yeh M-K, Chiu Y-W (2011) Finite element analysis of centrally-debonded composite sandwich beam under four point bending. Adv Mater Res 335–336:351–354

Yeh M-K, Lin C-M (2009) Bending strength of sandwich beams with nanocomposites core. Adv Mater Res 79–82:577–580

Yong KC, Mustafa A (2014) Natural rubber-rubberwood fibre laminated composites with enhanced stab resistance properties. J Rubber Res 17(1):1–12

Yu L, Kang H, Lim Y-S, Lee CS, Shin K, Park JS, Han JH (2014) Electrically conductive epoxy nanocomposites with expanded graphite/carbon nanotube hybrid fillers prepared by direct hybridization. J Nanosci Nanotechnol 14(12):9139–9142

Zainudin ES, Yan LH, Haniffah WH, Jawaid M, Alothman OY (2014) Effect of coir fiber loading on mechanical and morphological properties of oil palm fibers reinforced polypropylene composites. Polym Compos 35(7):1418–1425

Zeng M, Wei C, Xiong X-M, Yu C-B (2010) Dynamic mechanical and thermal properties of sisal fiber/nanometer-particles phenol formaldehyde resin composites. Polym Mater Sci Eng 26(9):37–40

Zeng SH, Wei C, Zeng M, Xiong XM, Liu HX, Lv J, Qin AM (2014) Friction properties of sisal fiber reinforced nano-SiO_2 phenol formaldehyde resin brake composites. Appl Mech Mater 490:64–68

Zhang Z, Zhu S (2000) Microvoids in unsaturated polyester resins containing poly(vinyl acetate) and composites with calcium carbonate and glass fibers. Polymer 41(10):3861–3870

Zhang J, Qiu K, Sun B, Fang J, Zhang K, Ei-Hamshary H, Al-Deyab SS, Mo X (2014) The aligned core-sheath nanofibers with electrical conductivity for neural tissue engineering. J Mater Chem B 2(45):7945–7954

Zhanjun L, Lei Z, Minnan C, Jiugao Y (2011) Effect of carboxylate multi-walled carbon nanotubes on the performance of thermoplastic starch nanocomposites. Carbohydr Polym 83(2): 447–451

Zheng-Ming H, Zhangb Y-Z, Kotakic M, Ramakrishnab S (2003) A review on polymer nanofibers by electrospinning and their applications in nanocomposites. Compos Sci Technol 63(15):2223–2253

Zhi-Gang S, Wang X, Hsing IM (2002) Composite nafion/polyvinyl alcohol membranes for the direct methanol fuel cell. J Memb Sci 210(1):147–153

Zou C, Fu M, Fothergill JC, Rowe SW (2006) Influence of absorbed water on the dielectric properties and glass-transition temperature of silica-filled epoxy nanocomposites. Annual Report – conference on electrical insulation and dielectric phenomena, CEIDP. Article number 4105435, 321–324

Chapter 13
The Manufacturing of Natural Fibre-Reinforced Composites by Resin-Transfer Molding Process

Deepak Verma

Abstract Natural fibre composites became a most important and versatile theme of research. This area/theme not only utilizes the agricultural wastes such as rice husk, bagasse fibres, banana fibres, etc. but also provides a new material made from these wastes which are light in weight, have low cost, and have high mechanical strength. Nowadays, there are so many methods available for the processing or fabrication of polymer composites. Some of these methods are hand layup method, injection molding method, compression molding, spray-up method, resin-transfer molding (RTM), etc. In this chapter, we are specifically discussing about the RTM process and their important parameters for manufacturing of the polymer composites. Various mechanical properties and characterization of composite materials made by RTM have also been discussed and reported here.

Keywords Polymer composites • Resin-transfer molding • Mechanical properties • Morphological study

13.1 Introduction

The resin-transfer molding (RTM) is one of the most important technologies available for composite making. The capability of RTM process is seen in making of large complex three-dimensional part having great dimensional variation and good surface finish. Better scheme of RTM process fabricates 3-D complicated parts and offered fabrication of low value structural parts in average amounts. The RTM process is generally known as cost-effective manufacturing method for the fabrication of composites. In RTM method, a catalyzed thermosetting resin is injected into an

D. Verma (✉)
Department of Mechanical Engineering,
Graphic Era Hill University,
Dehradun, Uttarakhand 248001, India
e-mail: dverma.mech@gmail.com

M.S. Salit et al. (eds.), *Manufacturing of Natural Fibre Reinforced Polymer Composites*, DOI 10.1007/978-3-319-07944-8_13

enclosed metal mold which contains a reinforcement preform. The preform is compressed to the defined fibre volume fraction when the matched mold is closed. The mold is filled with resin, and fibres become wet at that time and curing takes place inside the mold. There are some benefits of RTM on various available composite making processes; some of these are autoclave molding, compression molding, etc. RTM process is generally considered as a closed mold process which ultimately lowers the worker's subjection to detrimental volatiles, like styrene, and relates room temperature processing resins. Nevertheless, RTM process is a costly process and the tool design becomes arduous when manufacturing large and complicated shaped parts.

RTM manufactured continuous fibre-reinforced composite. RTM distinctly reduces the manufacturing cycle times as compared to other available composite making methods, and it is a feasible method for mass production of composites parts.

13.2 Detailed Study of RTM Process

The RTM process generally used a stiff closed mold. Figure 13.1 shows the RTM process sequence by considering easy case. A half mold is used in this process to arrange the preform. After this, close the mold and compress the preform. Then, introduce the resin into the mold by using positive gradient pressure which ultimately replaces the air entrapped into the preform. Vacuum is also utilized at some exclusive vents and removes the entrapped air from the mold. The preform is impregnated by the resin into the mold. After that, the curing of composite is considered to start. The next and last step is to open the mold and remove the part from it. The mold closing sequence is distinguished by the compression of the reinforced fibre which permits to have the appropriate thickness. The compression changes the microstructure of the preform, developing substantial distortions and nonlinear viscoelastic effects. These effects show changes in energy of the material and generated stresses (residual or leftover) just because of the viscoelastic response of the fibres. Still, stresses released throughout the impregnation phase. The introduction of resin assured the entire impregnation of the preform, whereas in general poor

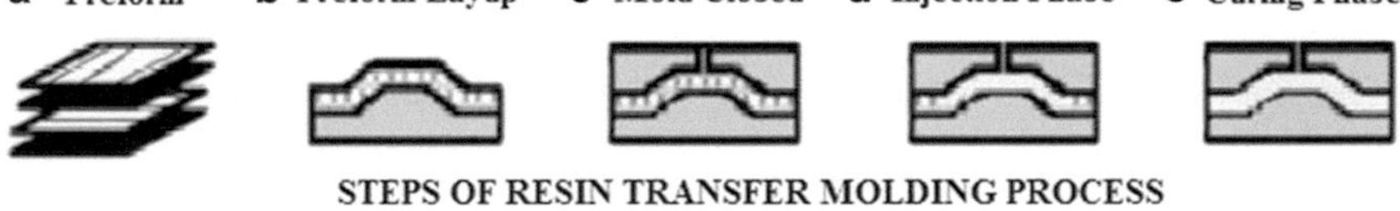

Fig. 13.1 Sequence of the main steps of RTM process. (**a**) realization of the preform, (**b**) deposition and draping of the preform in the mold cavity, (**c**) closing mold, (**d**) injection phase, and (**e**) curing phase

impregnation of the fibres gives dry spots with disordered adherence between the layers which ultimately make the surface rough and irregular (Laurenzi and Marchetti 2012).

13.2.1 Important Components of RTM Process

There are five important components in the RTM process.

1. Resin and catalyst container
2. Pumping unit
3. Mixing chamber
4. Resin injector
5. Molding unit

Two separate containers are to be taken for the resin and catalyst. The resin container should be bigger in contrast to the catalyst container. Both of the containers pass through the pumping unit having separate outlets which goes to the mixing chamber. The purpose of the pumping unit is to transfer and mix both of the resin and catalyst in the mixing chamber. This mixture is introduced into the mold cavity by resin injectors. The molding unit consists of upper half and lower half molds. Generally, heating systems are attached with the molding unit. Some vents are provided into the mold cavity to release the entrapped gases during clamping.

13.3 Components Made by RTM Process

The RTM process generally uses two main materials for the fabrication of composites; these are reinforcement and resin. The RTM process used a vast range of various reinforcement and resin systems for the manufacturing of a composite material with a wide range of properties. The important parameter generally considered for the designing of a composite product is the good understanding of the raw material properties as it has been considered that each fibre and resin system has their individual contribution for making the composites. The main aim of the reinforcement is to provide good mechanical properties like stiffness, tensile strength, and impact strength, whereas on the other side the resin system provides physical properties to the composite in addition to the resistance to corrosion, fire, etc. and also gives some rigidity which is important from the solid composite point of view.

The performance and the process of composite design are affected by the reinforcing materials such as the choice of the fibres (for reinforcement) and fibre structure. Fibre selection depends upon the performance like modulus, strength, durability, compressibility, and adequate volume fractions for the proper placement of the reinforced material during preforming operations.

13.4 Selection of Important Parameters for RTM Process

There are some variables and parameters available in RTM process which are dependent on each other. The combination of both of the variables and parameters influences not only the process but also the quality and finishing of the product. Therefore, they should be thoroughly determined. Pressure, temperature, viscosity, and volume fraction are some of the significant parameters which cannot be omitted in design of the composite material. The other parameters that should be taken into account are the angle of strike of the nozzle, the fibres placement, the flow route and shear rates, etc. It has been observed that the resin flows swiftly in the direction of the fibres; thus, the flow movement is generally based upon the kind of fabric used and the number of overlapped layers. For the attainment of the homogeneous distribution of the resin, it is essential to have an accurate number of skins.

13.4.1 Thickness

The thickness is also another parameter. If the thickness is not considered under specified limit, then it may cause the formation of the voids and dry spots on the composite and also influence the flow progress and impregnation of the fibres. The thickness becomes a crucial design requirement specifically in the instance of the addition of reinforcements and ribs.

13.4.2 Injection Pressure

The injection pressure is another important parameter which not only evaluates the velocity of the resin which enters in to the mold but also the hydraulic pressure of the mold. Generally, the entering velocity elucidates the pouring time. The pouring time should not be less. If it is less, then it results not only in an inadequate impregnation of the fibres but also in the similar time untimely gelation of the resin occurred. The injection pressure generally related to the dispersion of the resin on the preform. If the dispersion of resin is not uniform, then it results in the formation of air bubbles in the matrix which ultimately affects the surface quality and the mechanical properties of the composite. This parameter also has some importance with the viscosity term called "fibre wash." The fibres can move directly under the injection pressure if the binder mixed too rapidly in contact with the resin.

13.4.3 Temperature

The temperature is also an exceptionally significant parameter and factually associated with the injection pressure and the resin viscosity. The increase in temperature results in the decrease in pouring time which ultimately lowers the pressure.

Similarly, a decrease in temperature results in an increase in the viscosity which results in an increase in the pressure to confirm the transport of the resin itself.

13.5 Influence of RTM Process on Final Product

The quality and the performance of the product mainly depend on the process used for the fabrication. For example, if the composite has a good surface finish, then it ultimately gives the better mechanical properties. A good surface finish averts the inclusion of the elements like dust which produces and intensifies the micro-cracks inside the structure resulting in lower mechanical properties. Figure 13.2 demonstrates the steps of the process that affect the response of material and define the connection among the process and conclusive implementation. The two stages, viz., compaction and impregnation stages, control the imperfections due to voids and dry spots. These can be defined as follows.

13.5.1 Compaction Stage

For the fabrication of the advanced composite materials, the compaction stage plays an important role. This stage generally appears by the use of an external pressure, which makes disposition of the fibres and alteration of microstructure. In RTM, the compaction phase takes place by closing the mold which results in the attainment of the fibre volume fraction by the preform. The large numbers of distortions in the preform are generally observed in the compaction phase. This only happened by a variation in energy observed in the material. For example, when a compaction of multilayer preform is taking place, the fibres move and flatten with other fibres, thus producing interlayer nesting phenomena which depend on the failure mode and the

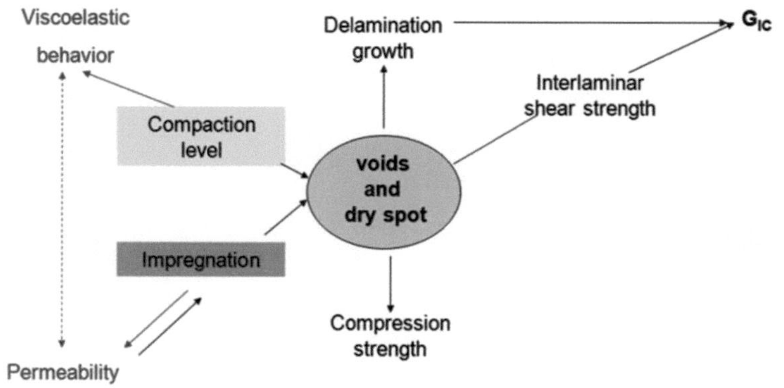

Fig. 13.2 Relations between manufacturing phases and material behavior (Laurenzi and Marchetti 2012)

fibres' structure. After removal of the compression load, fibres do not return to their original position inside the preform.

If there will be any changes like mechanical and spatial changes in the preform, which modifies the mechanical properties of the composite materials, then it generally refers to the phenomenon of nesting. Specifically, the nesting affects the rigidity and resistance of the piece. Certainly, the stiffness is based on the placing of the layers. If the layer is set without any reduction in the thickness, then there is no significant change noticed in any mechanical properties.

13.5.2 Impregnation Phase

The voids formation and dry spots are the defects induced by the RTM process and become the major problems which affect the production quality of the product. The voids cause the failure of the matrix material. The mechanism of crack formation is generally seen in composites by the emergence of micro-voids among fibres which ultimately results in the delamination of the composites. The two resistance levels by preform are observed in the resin flow, i.e., between the fibre bundles and inside the fibre bundles. This ultimately can be understood by the two different permeabilities, i.e., the permeability between the fibre bundles and inside the fibre bundles. The kind of flow scale intent the void formation in composites. The phenomenon of emergence of macro-voids in the macroscale can be observed when the displaced air remains trapped in the resin system or, on the other hand, when inadequate pressure overcomes the resistance of the preform. In case of microscale, micro-voids generally result or give rise to porosity. The general mechanism of emergence of voids in fibre bundles can be shown in Fig. 13.3. The mechanism states that the flow front flows around the tows and continues to impregnate the tow after the passage of the front. If the trapped air (in tow) is not deported during the start of the pouring process, then it ultimately results in the generation of the micro-voids at the center. Macroscopic pressure drop is another reason of the formation of micro-voids. The

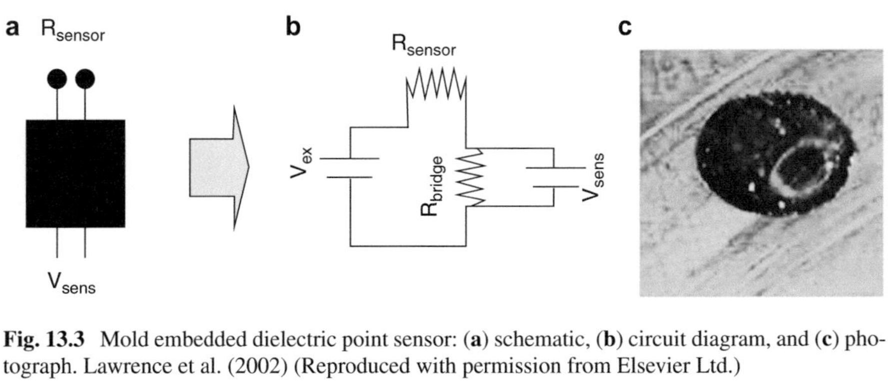

Fig. 13.3 Mold embedded dielectric point sensor: (**a**) schematic, (**b**) circuit diagram, and (**c**) photograph. Lawrence et al. (2002) (Reproduced with permission from Elsevier Ltd.)

pressure drop offers an observable change in the permeability along the preform and can be justified by the sink effect. In this, the fibre bundles generally considered as fluid sinks. In this, the fibres of the bundle are placed much closer than that existing between two bundles. In the same time, the flow of resin takes place between the fibre bundles. This ultimately results in the impregnation of the bundle (Laurenzi and Marchetti 2012).

13.6 RTM Composites: A Past Research

Lawrence et al. (2002) developed a design and control methodology for mold filling by simulation by using some sensors and actuators. This recognizes the distraction of flow in the mold filling process and directs the resin to successfully flow into the mold without formation of voids. This technique is applied and approved for a mold geometry which generally has the complicated features such as tapered regions and thick regions. The position of the sensors in the mold can be observed by the application of software tools. These tools noticed the expected distraction and recommend flow control actions for additional actuators to redirect the flow. Laboratory hardware is used for the automation of the pouring process. These sensors control the flow of the resin and impregnate all the fibres completely against the distraction in the process.

The point sensor senses the resin in significant locations. From a manufacturing point of view, the sensors are the combined part of the tooling. The data acquisition system was used to connect the sensors. The small current flow was used to determine the presence of resin when it wets the two poles of the sensor. The half bridge setup was used to detect the current excited by a 10 VDC power supply incorporated in the data acquisition system as shown in Fig. 13.3.

Leclerc and Ruiz (2008) carried out an experimental analysis on various kinds of fibrous reinforcements for the optimal analysis of the impregnation conditions in relation to the content of macro-/micro-voids to the local capillary number. The effect of voids on the mechanical properties of the laminates was also studied. From Fig. 13.4, it is noticed that the macro-voids generally evolved at less speed during fibre impregnation. This is only because of the capillary action, and they can overreach 15 % in volume. The reason behind the micro-void formation pertains to the high velocity of the resin, but at a very less rate (approx. 2 % in volume). From this, it can be observed that the resin impregnation velocity has some critical value, and it should not fall below this. Figure 13.4 shows an optimum impregnation velocity which ultimately results in minimum void formation.

Hsiao and Gangireddy (2008) used a vacuum-assisted resin-transfer molding (VARTM) process which infuses a little quantity of carbon nanofibres with resin into the glass fibre preform which ultimately controls and decreases the spring-in angle. From these results, it can be observed that the carbon nanofibres totally restrained the spring-in angle of L-shaped composites. These spring-in angles were measured only after being de-molded. Figure 13.5 represents and differentiates the

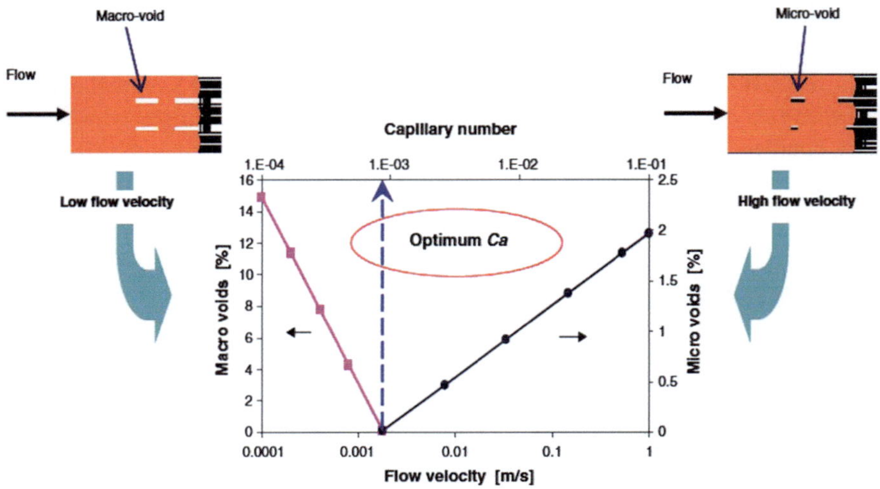

Fig. 13.4 Macroscopic and microscopic void formation during fibre impregnation related to the flow velocity and capillary number (Ca). An optimum Ca corresponds to minimum void [5]. Leclerc and Ruiz (2008) (Reproduced with permission from Elsevier Ltd.)

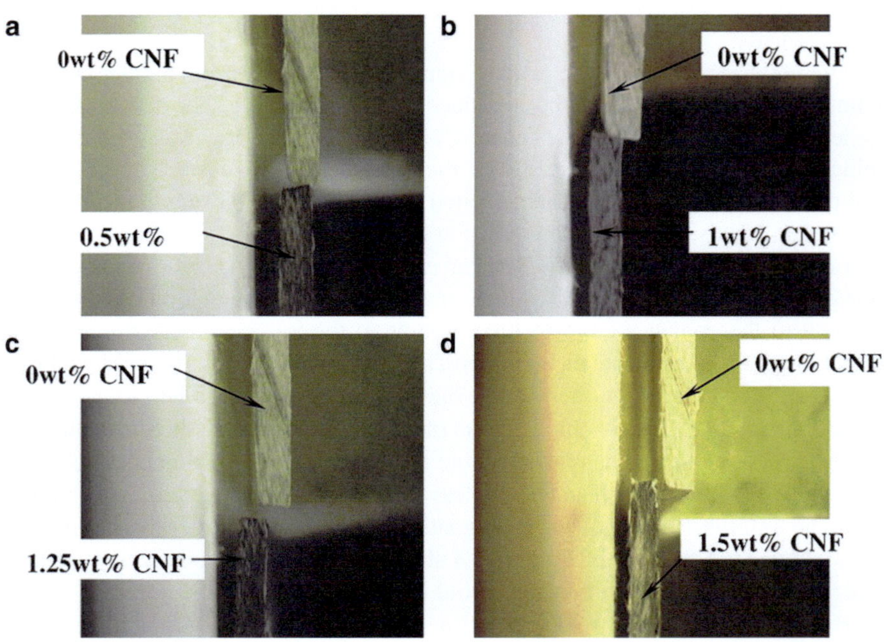

Fig. 13.5 Top-view comparison of the spring-in between the control sample and the samples with various CNF weight fractions. (**a**) 0 and 0.5 wt% CNF specimens. (**b**) 0 and 1 wt% CNF specimens. (**c**) 0 and 1.25 wt% CNF specimens. (**d**) 0 and 1.5 wt% CNF specimens. Hsiao et al. (Reproduced with permission from Elsevier Ltd.)

Table 13.1 Experimental data for the L-shaped composite specimens (Hsiao and Gangireddy 2008) (Reproduced with permission from Elsevier Ltd.)

	0 wt%	0.5 wt%	1 wt%	1.25 wt%	1.5 wt%
Room temperature	20.1	21.4	21.0	21.5	21.1
ΔT_{top} (°C)	−10.2	−8.0	−6.4	−4.8	−6.4
ΔT_{mid} (°C)	−10.8	−8.3	−6.0	−5.6	−6.4
ΔT_{bottom} (°C)	−11.6	−8.1	−6.4	−5.6	−5.9
$\Delta t_{average}$ (°C)	−10.9	−8.1	−6.4	−5.6	−5.9
Spring-in angle (°)	2.340±0.318	1.830±0.318	1.330±0.318	0.972±0.318	0.634±0.318

spring-in effects of the various L-shaped composites of 0, 0.5, 1, 1.25, and 1.5 wt% carbon nanofibre weight fractions.

The L-shaped samples were observed from the top. These samples were placed in a 90° bracket with their base sides fixed. Figure 13.5 represents that the 0 wt% carbon nanofibre part has the highest spring-in distortion as compared to the composite parts having some fractions of CNFs in the matrix. The composite having CNF's percentage of 1.5 wt% has the lowest spring-in distortion in contrast to the other available specimens. Table 13.1 shows the measured spring-in angles for the L-shaped parts. The temperature information is also showed in Table 13.1 during the cure process. It has been observed from Table 13.1 and Fig. 13.5 that the spring-in angles of the L-shaped composite part decrease noticeably by increasing the wt% fraction of carbon nanofibres in the matrix.

Sreekumar et al. (2012) studied the electrical properties of sisal fibre-reinforced polyester composites manufactured by RTM process by considering fibre loading, frequency, and temperature. From the study, it has been found that the dielectric constant (ε′), loss factor (ε″), dissipation factor (tan δ), and conductivity increase by increasing the fibre content for all range of frequencies. From these results, it has been noticed that the values are found maximum for composites with fibre content of 50 vol.%. The volume resistivity also varied with fibre loading at smaller frequency. The increase in temperature ultimately improves the dielectric constant values by decreasing the glass transition temperature. This variation is only because of the fibre content.

Figure 13.6 shows the variation of the dielectric constant of sisal/polyester composites as a function of frequency. From Fig. 13.6, it has been observed that the dielectric constant increases by increasing fibre content on all range of frequencies such as polyester<R20<R30<R40<R50. The ε′ values primarily based upon the effects of interfacial, orientation, atomic, and electronic polarizations in the material. The interfacial polarization occurs just because of the differences in conductivities. The orientation and interfacial polarizations also depend on the concentration of fillers. The available polar groups in the natural fibres have resulted in an important increment in ε′ value by increasing the fibre loading. This can be understood by Fig. 13.7 showed the variation of the ratio between dielectric constant of composites and matrix behaving as a function of fibre content. In case of polyester, the

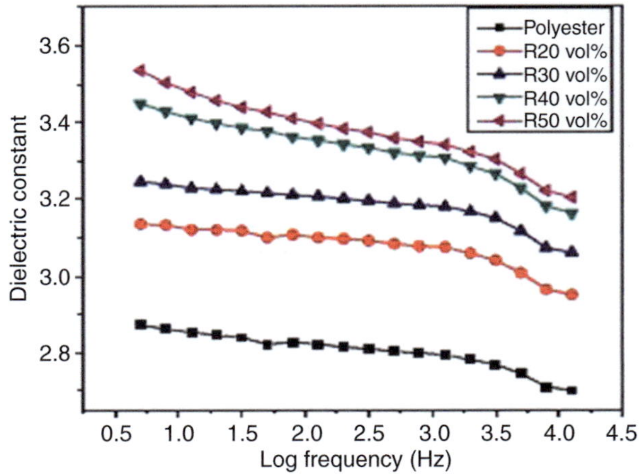

Fig. 13.6 Variation of dielectric constant of sisal/polyester composites as function of fibre content and frequency at 30 °C (Reproduced with permission from Elsevier Ltd.)

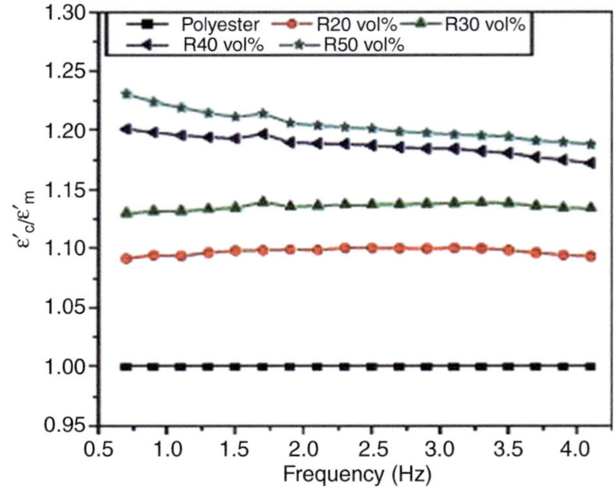

Fig. 13.7 Plot of ratio of dielectric constant of composites with that of matrix as a function of fibre content at 30 °C (Reproduced with permission from Elsevier Ltd.)

ε' value is low because it has only momentary atomic and electronic polarizations. On the other hand, it is noticed that for a stated fibre loading, the ε' gives maximum values at low frequencies. This is because of the reduction in the orientation polarization with increased frequency. Loss factor (ε'') is the average power factor over a given period of time which describes the losses in transmission and distribution. The variation of ε'' for the composites is shown in Fig. 13.8. It can be noticed that ε'' is high in lower frequency region and is found maximum for the composites with a fibre content of 50 vol.%. By increasing the frequency, the value of ε'' decreases rapidly. The dissipation factor (function of frequency) at various fibre loadings is shown in Fig. 13.9. This dissipation factor decreases by increasing the frequency. It has been noticed that increasing the fibre content in the low frequency region represents a significant variation, i.e., it is minimum for polyester and maximum for R50

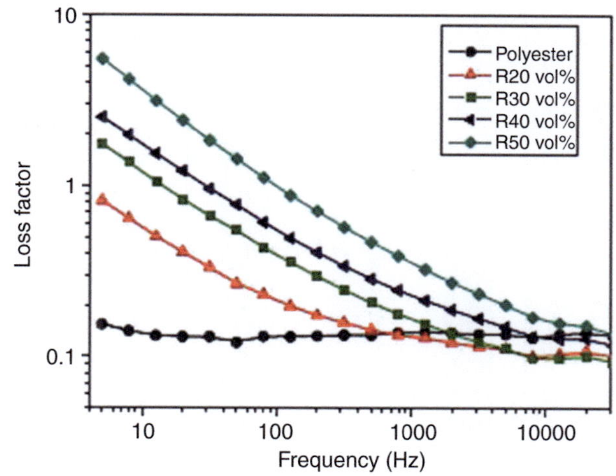

Fig. 13.8 Variation of loss factor of sisal/polyester composites as function of frequency at 30 °C (Reproduced with permission from Elsevier Ltd.)

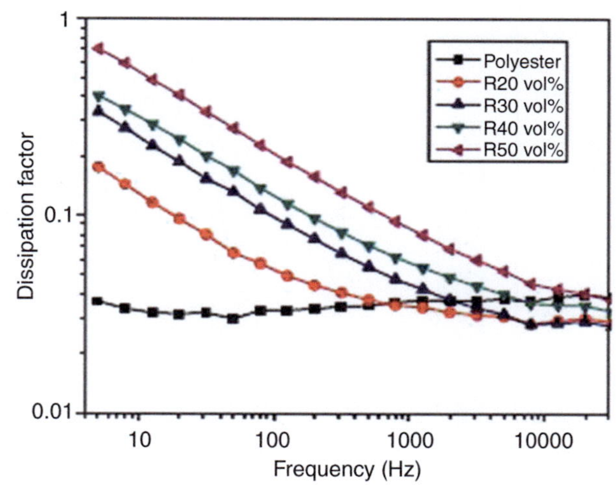

Fig. 13.9 Variation of dissipation factor of sisal/polyester composites as a function of frequency at 30 °C (from paper Sreekumar et al. 2012) (Reproduced with permission from Elsevier Ltd.)

for a given range of frequency. On the other hand, in higher frequency range, tan δ curves come nearer as seen in the loss factor. Also increasing the fibre content results in an increase in the number of polar groups which again increases the orientation polarization which results in an improvement in the dissipation factor.

Pearce et al. (1998) fabricated different types of carbon/epoxy plates by RTM. The plates were fabricated by using the same fabrics, and then it sends for microstructural image analysis and interlaminar shear strength (ILSS) testing according to CRAG standards. Quantimet 570 automatic image analyzer was used to measure some relationships between measured permeabilities and finished microstructures.

Schmachtenberg et al. (2005) used and applied ultrasonic techniques for online monitoring and controlling of RTM. In this study, a master mold was fabricated to

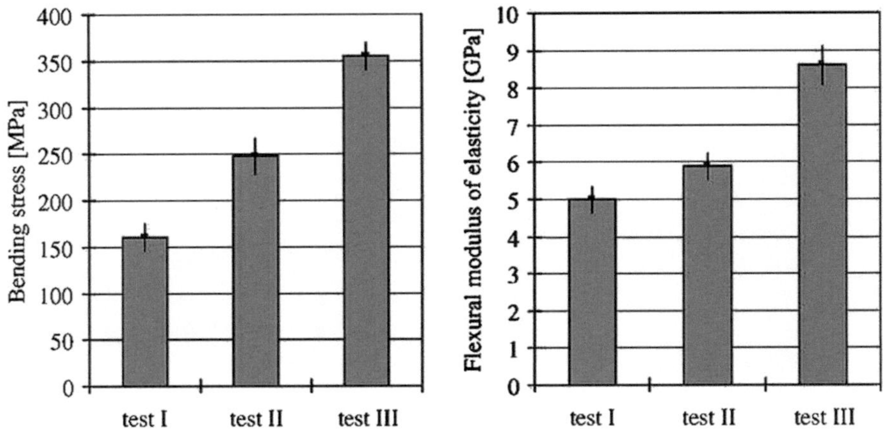

Fig. 13.10 Three-point bending properties of the RTM samples, preform: continuous random mat (Schmachtenberg et al. 2005) (Reproduced with permission from Elsevier Ltd.)

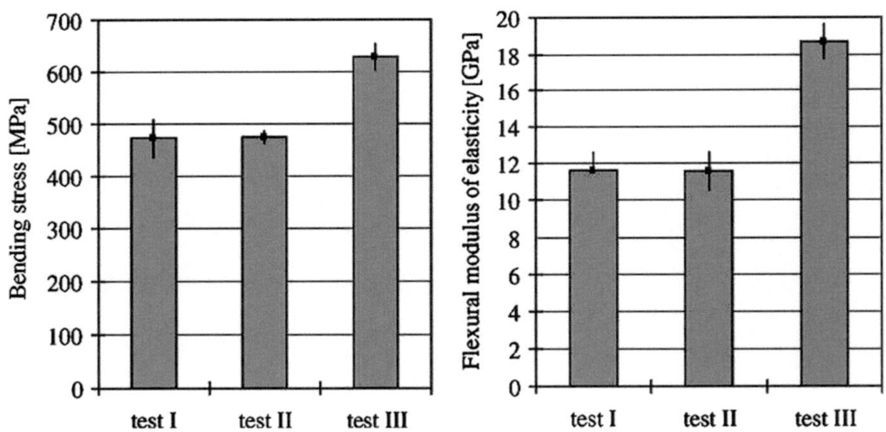

Fig. 13.11 Three-point bending properties of the RTM samples, preform: multiaxial layers (Schmachtenberg et al. 2005) (Reproduced with permission from Elsevier Ltd.)

examine the ultrasonic technique as compared to pressure transducers and dielectric sensors. The main objective of this study was to constrain the flow front during injection and to observe the curing behavior of the resin.

The RTM process is completed with this mold by direct transmission sensors of 8 mm diameter and a frequency of 4 MHz. Firstly, the investigation can be carried out by applying measurement device to the RTM process, and then different types of resin systems were distinguished. After this study, a correlation between different process parameters, ultrasonic signals, and part properties was observed. Three-point bending tests were carried out for the estimation of the mechanical properties of the RTM composites. The bending tests were also carried out on composite laminates.

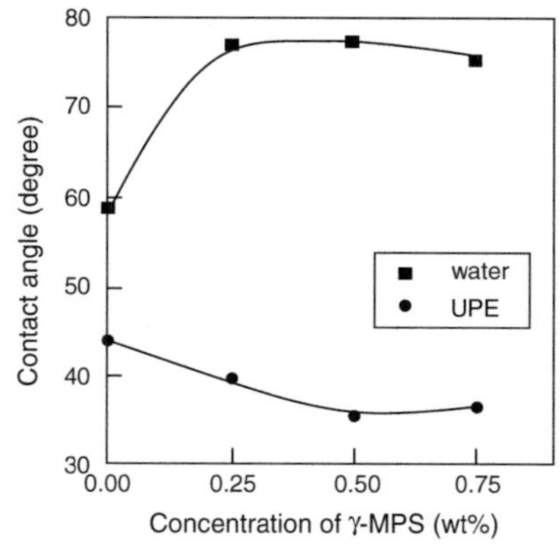

Fig. 13.12 Advancing contact angles of glass plate with the variation of g-MPS treatment concentration. Lee et al. (2002) (Reproduced with permission from Elsevier Ltd.)

The mechanical characteristics of the sample can be shown in Figs. 13.10 and 13.11. From these results, it has been noticed that a correlation between the measured amplitude maxima and the mechanical properties exists. From Fig. 13.10 and 13.11, it can be noticed that by giving variation in the process parameters, the mechanical value of the composite with random mat become doubled.

Lee et al. (2002) investigated the influence of glass fibre surface modification on the flow properties of unsaturated polyester (UPE) resin by the resin-transfer molding process. This g-methacryloxypropyl trimethoxysilane (γ-MPS) was adopted as a glass fibre surface modifier. It has been observed that the γ-MPS treatment decreases the surface energy of glass fibre by advancing the contact-angle measurement. Darcy's law was used for measuring the unsteady state permeability of glass fabric preforms.

Figure 13.12 shows the advancing contact angles of the glass plate in distilled water and UPE resin at various γ-MPS treatment concentrations. In case of water, increasing the concentration of γ-MPS ultimately increases the advancing contact angle and shows a plateau value. On the other hand, the contact angle decreases by dipping the glass plate into UPE resin under the same experimental conditions for water. Figure 13.13 shows the flexural properties of the glass fibre/UPE composites. From Fig. 13.13, it has been noticed that the flexural strength and modulus of γ-MPS-treated glass fibre/UPE composite are higher than those of untreated composites. When the silane coupling agent is used to treat the glass fibre surface, there is a formation of siloxane bond that takes place on the glass fibre surface via a condensation reaction. The improved interfacial adhesion leads to form a chemical bond, at the interface.

Rouison et al. (2004) used RTM process to develop hemp/kenaf fibre/UPE composites. In this research, the fibre mats having a moisture content of 4.3–5 % were dried in the mold under vacuum condition and got the final moisture value of around 1–2 %. The low permeability of the fibre mat ultimately increases the resin injection time at high fibre contents. They keep the temperature of the mold constant to obtain homogeneous curing of the part at fast rate. Figure 13.14 represent the results of

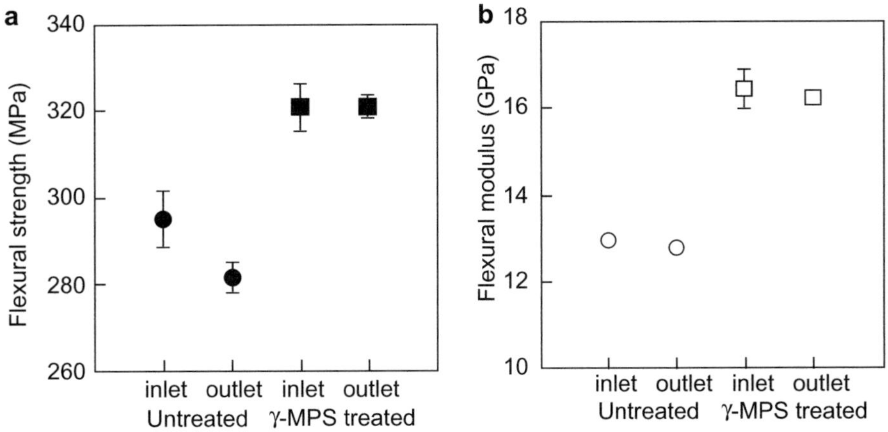

Fig. 13.13 Flexural property of the glass fibre/UPE composites: (**a**) flexural strength and (**b**) flexural modulus (Lee et al. 2002) (Reproduced with permission from Elsevier Ltd.)

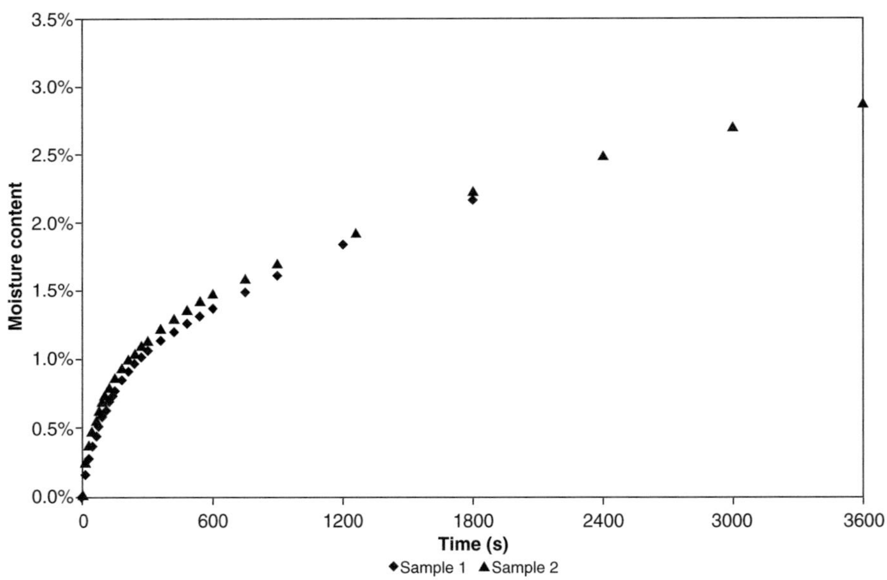

Fig. 13.14 Water absorption rate of the fibres under standard conditions after drying for 2 h at 105 °C (Rouison et al. 2004) (Reproduced with permission from Elsevier Ltd.)

these experiments. From figure, it can be noticed that the moisture content at 50 %RH was only of 4.1 %. The water available in the fibres begins to increase across this point to reach about 12 % at 94 %RH. So for this reason, the fibres were kept in a standard room at 23 °C, 50 % RH permanently because of the variations in

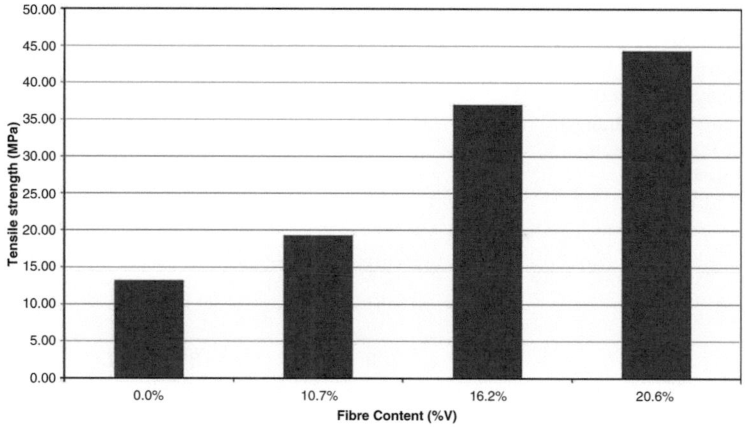

Fig. 13.15 Tensile strength of natural fibre composites (Rouison et al. 2004) (Reproduced with permission from Elsevier Ltd.)

atmospheric conditions. The moisture absorption rate of the fibres after drying at 105 °C for 2 h was also measured.

The tensile strength of the composites is presented in Fig. 13.15. An improvement in the tensile strength of the samples is observed by increasing the fibre content. A considerable improvement was noticed between the pure resin and the sample with 20.6 % fibres, from 13.1 to 44.3 MPa. From these results, it has been found that the tensile strength of composites with 20 % fibres is around 100 MPa, whereas on the other hand the flexural tests of the same composite showed a highest value of 71.4 MPa.

Lin et al. (2006) prepared layered silicate/glass fibre/epoxy hybrid composites by using a VARTM process. In this study, two directions, i.e., parallel and perpendicular to the flow direction of resin, were used, and then the effects of the fibre direction on the clay distribution in the hybrid composites were checked. The morphological study of the composites was investigated by using X-ray diffraction (XRD) and transmission electron microscopy (TEM). The dispersion of clay in the composites was also noticed by using scanning electron microscopy (SEM). The results showed that reinforcing a small amount of organoclay in the glass fibre/epoxy composites ultimately enhances their mechanical and thermal properties.

Three-point bending test was used for the investigation of the flexural properties of the nanocomposites in both directions, i.e., parallel and perpendicular to the fibre alignment. Figure 13.16 shows that reinforcement of clay ultimately increases the transverse flexural modulus of composite. The impact properties of the nanocomposites with various clay loadings are presented in Fig. 13.17. It is observed that the perpendicular impact direction decreases the impact strength with clay loading, while if the impact is along the fibre direction, the impact strength increases by increasing the clay loading.

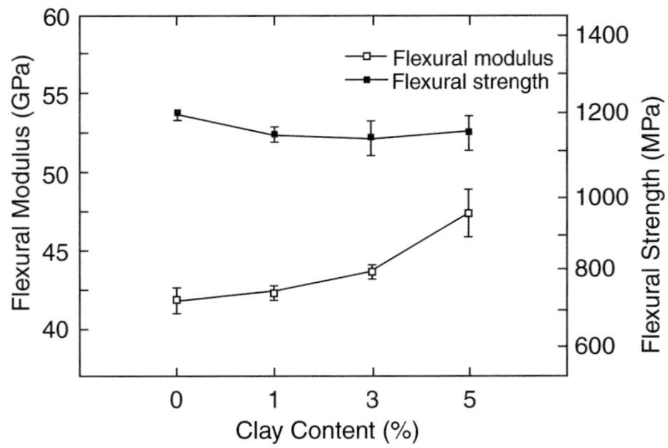

Fig. 13.16 Flexural properties of the nanocomposites with different clay loadings in the fibre direction (Lin et al. 2006) (Reproduced with permission from Elsevier Ltd.)

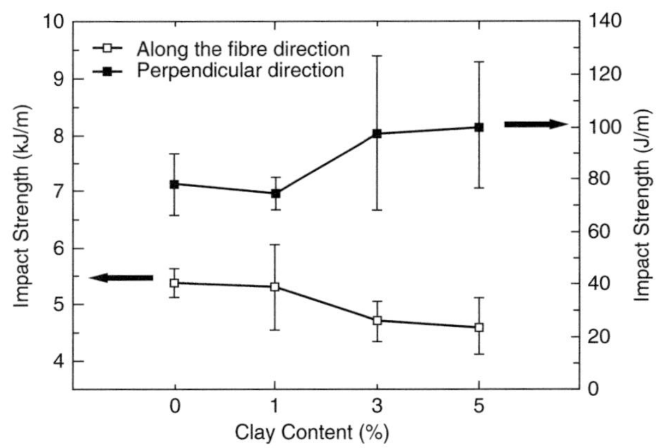

Fig. 13.17 Impact strength of nanocomposites with different clay loadings in two directions (Lin et al. 2006) (Reproduced with permission from Elsevier Ltd.)

Papargyris et al. (2008) incorporated microwave heating in RTM process. About 50 % reductions in cure cycle time were achieved by using microwave heating. The mechanical testing result shows the same values of flexural properties of the two types of composites. Microwave-cured composites observed about 9 % increase of the ILSS. This improvement in ILSS is due to a less resin viscosity at the first stage of the curing process as confirmed by SEM. Again, both kinds of composites yielded less void content (<2 %).

Figure 13.18 shows the viscosity change against time at three different heating rates (2, 5, and 10 °C/min) for the neat LY/HY5052 resin system. By increasing the heating rate, the viscosity of the neat resin decreases considerably from 0.12 to 0.068 Pa s and reached a final value of 0.043 Pa s. Excessive heating rate ultimately results in lesser resin viscosity, which would make the resin to flow and makes resin to completely soak the carbon fibre surface at the starting phase of the curing process. On the other hand, a lesser viscosity enhances the interfacial bonding and

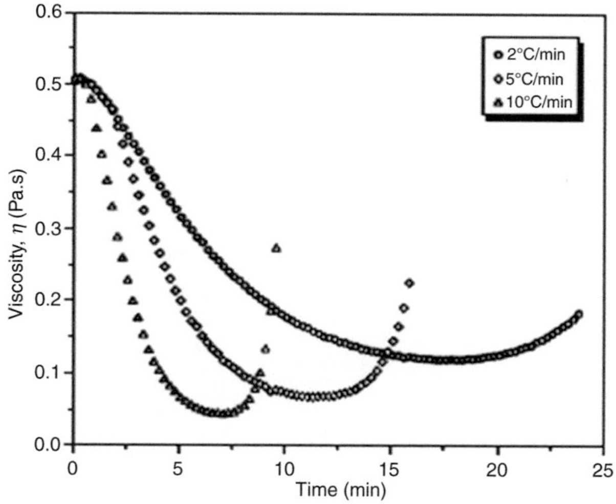

Fig. 13.18 Viscosity versus time at 2, 5, and 10 °C/min for neat LY/HY5052 resin system (Papargyris et al. 2008) (Reproduced with permission from Elsevier Ltd.)

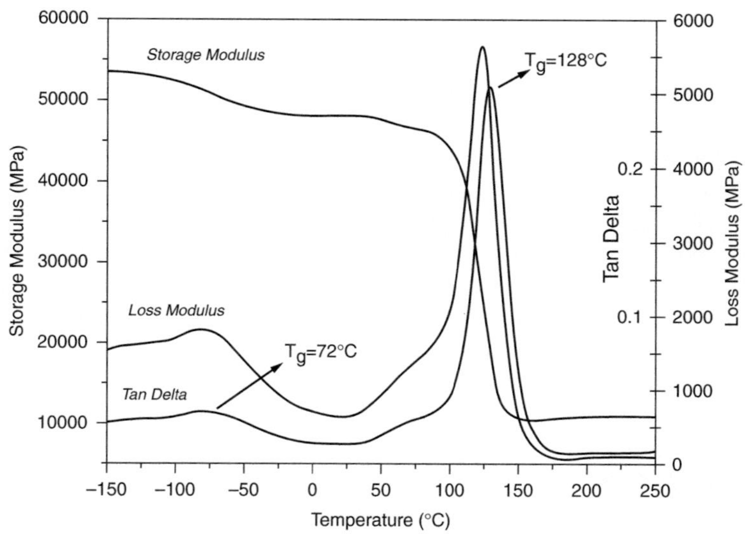

Fig. 13.19 Storage modulus, loss modulus, and tangent delta for the LY/HY5052/carbon composite system cured by conventional thermal RTM (Papargyris et al. 2008) (Reproduced with permission from Elsevier Ltd.)

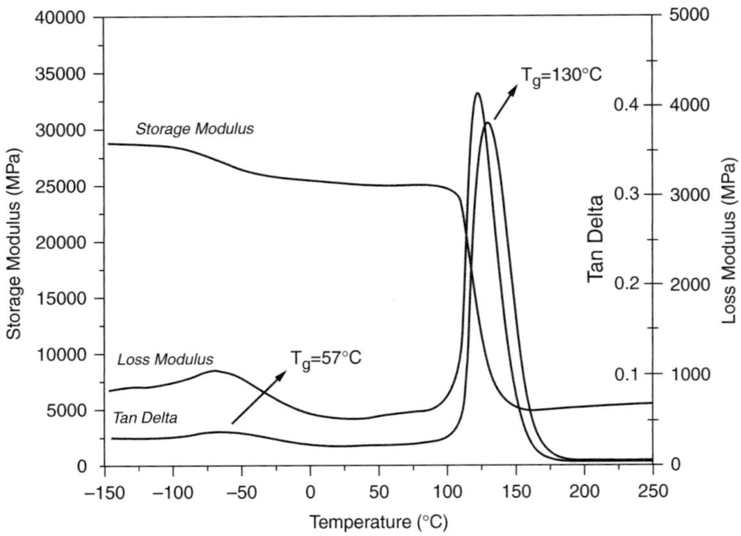

Fig. 13.20 Storage modulus, loss modulus, and tangent delta for the LY/HY5052/carbon composite system cured by microwave RTM (Papargyris et al. 2008) (Reproduced with permission from Elsevier Ltd.)

improves adhesion between the resin and the fibres. Figures 13.19 and 13.20 show the typical DMTA curves at 2 °C/min scan rate and 1 Hz frequency for the thermally and microwave-cured composites. In relation to the average peak values of a-transition, there is no significant difference observed for both types of composites. On the other side, microwave heating shows a higher glass transition temperature (Tg) of 130 ± 1 °C, as opposed to 128 ± 2 °C for those cured thermally. A temperature difference of 15 °C was noticed (average peak values of the lower temperature b-transition) between the two composite types (−72 ± 3 °C and −57 ± 3 °C for the thermally and microwave-cured composites).

Wang et al. (2013) used carboxylic acid functionalized carbon nanotubes (CNTs) to modify epoxy to develop a nanocomposite matrix for hybrid composites. High energy sonication was used to disperse the CNTs in epoxy. RTM process was used to manufacture hybrid multiscale composites specimen which contains less CNTs. The optical microscopy was used to characterize the dispersion quality behavior of CNTs in epoxy. A CNT reinforcement of 0.025 wt% notably increases the glass transition temperatures (Tg) of the hybrid multiscale composites. The fractured surface of failed specimens was examined by SEM.

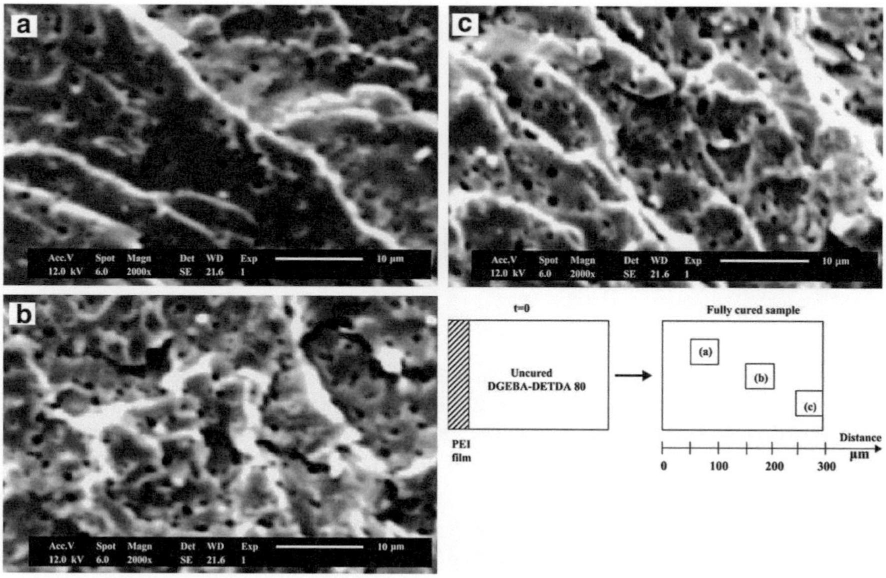

Fig. 13.21 SEM micrographs of fractured specimens of TP/TS blends, obtained from a PEI film embedded in DGEBA–DETDA 80, at different distances from the initial position of the film. Scheme of the position of the observed morphology: (**a**) 0–50 µm, (**b**) 150–200 µm, and (**c**) 250–300 µm (Naffakh et al. 2006) (Reproduced with permission from Elsevier Ltd.)

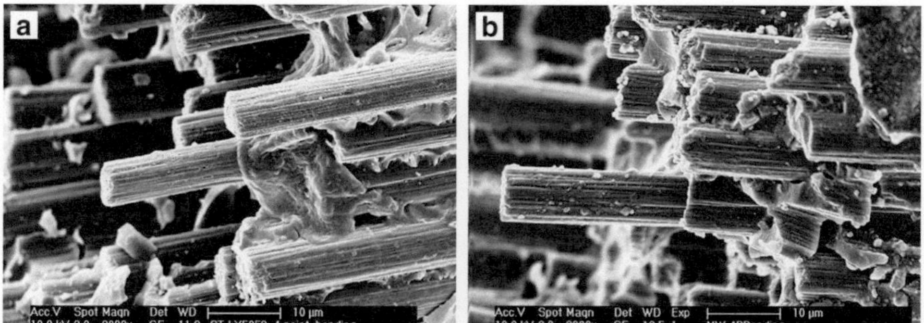

Fig. 13.22 SEM micrographs at the same magnification showing fibre pullout after four-point bending test; (**a**) conventionally cured specimen and (**b**) microwave-cured specimen (Papargyris et al. 2008) (Reproduced with permission from Elsevier Ltd.)

13.7 Morphological Study of RTM Composites

Morphological study can be done by the various available techniques like XRD, SEM, TEM, AFM, etc. SEM can be effectively used for the morphological characterization of natural fibre-reinforced composites. This is the most extensively

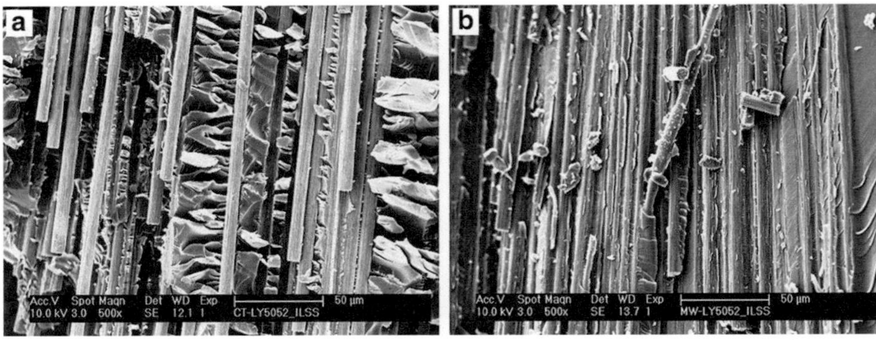

Fig. 13.23 SEM micrographs at the same magnification after interlaminar shear test showing (**a**) clean fibres in conventionally cured specimen and (**b**) fibres coated with resin in microwave-cured specimen (Papargyris et al. 2008) (Reproduced with permission from Elsevier Ltd.)

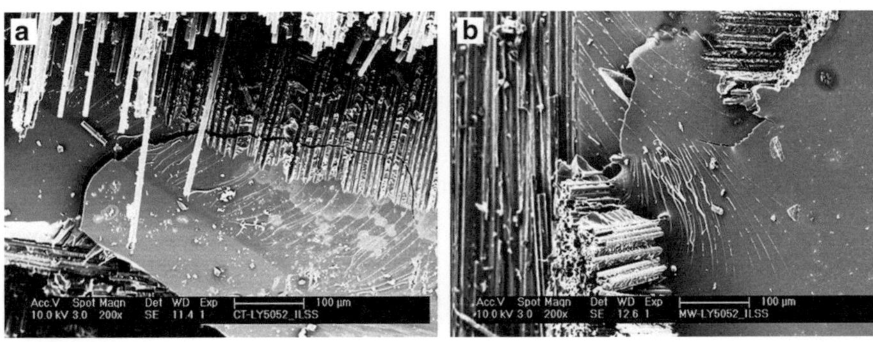

Fig. 13.24 SEM micrographs at the same magnification showing matrix fracture after interlaminar shear test; (**a**) conventionally cured specimen and (**b**) microwave-cured specimen (Papargyris et al. 2008) (Reproduced with permission from Elsevier Ltd.)

used method for the analysis of the surface of composite materials. Figure 13.21 represents the SEM image of "model" blends manufactured from epoxy–amine/thermoplastic films (without glass fibres) noticed at various diffusion lengths. The SEM images showed only one domain size of PEI; this is due to the maximum rate of diffusion of the thermoplastic.

Papargyris et al. (2008) examined the fracture surfaces of composite samples by using SEM. Figures 13.22–13.24 showed the topographic appearance of both types of composites. It can be noticed from Fig. 13.22a, b that the conventional methods show poor interfacial bonding, which can be observed by fracture surface of the composites, and also show broad fibre pullout in contrast to that manufactured by using microwaves. It is clearer in Fig. 13.23a, b, where there is a notable number of clean fibres that appeared in contrast to fibres completely coated with resin. Brittle

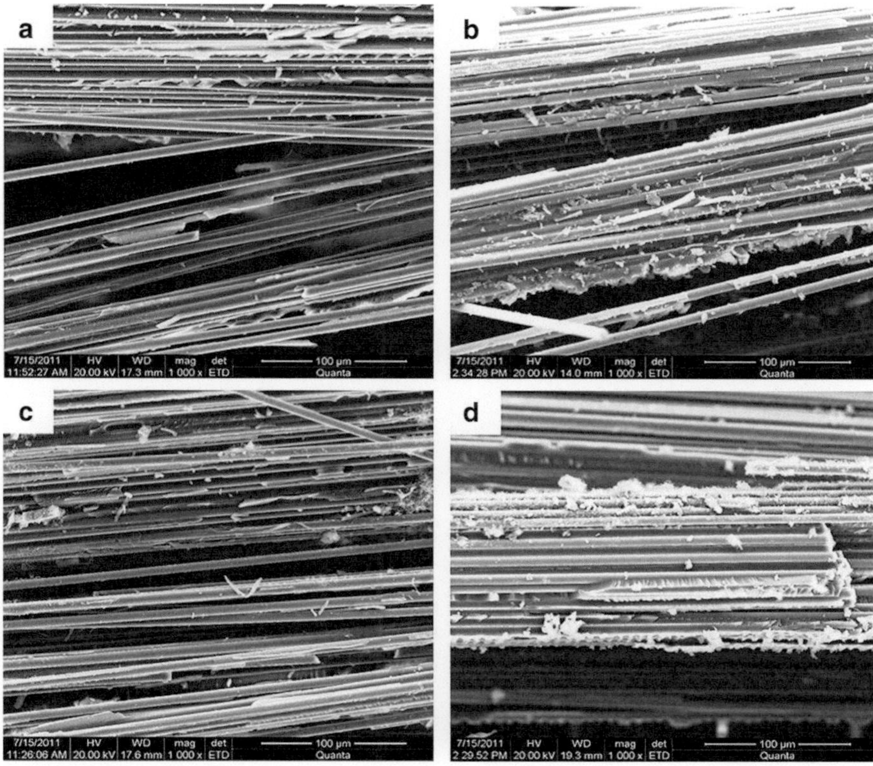

Fig. 13.25 SEM micrographs of fracture surfaces of the hybrid multiscale composites containing various CNTs content: 100 lm scale bar magnification. (**a**) 0 wt%, (**b**) 0.025 wt%, (**c**) 0.075 wt%, and (**d**) 0.1 wt%. (Reproduced with permission from Elsevier Ltd.)

matrix failure sections were also noticed for both types of composites, as presented in Fig. 13.24a, b.

Wang et al. (2013) investigated the fracture surfaces of composites having different amount of CNT content by SEM (Figs. 13.25 and 13.26). Figure 13.25a shows the SEM image of the fracture surface of carbon fibre/epoxy laminates. The fibre pulled out from the matrix surface due to a weak adhesion is confirmed by the smooth fibre surface. Figure 13.26a showed the SEM image of the composites without CNTs. Figure 13.26b showed the composite containing CNTs content of 0.025 wt% which has remarkably different fracture surface. It can be observed from Fig. 13.26b that the continuous reinforcement of carbon fibres was notably smeared with a thin layer of matrix. The combined CNTs were noticed between adjacent carbon fibres as shown in Fig. 13.26b, which ultimately increases the matrix toughness. Figure 13.26c shows the presence of CNTs at the concentration of 0.075 wt%.

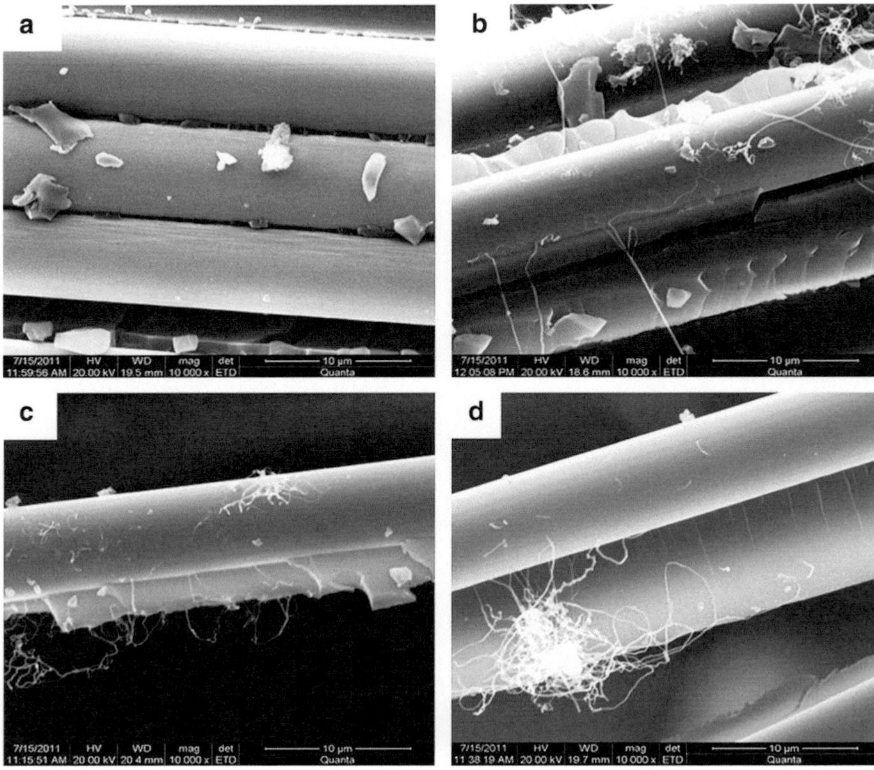

Fig. 13.26 SEM micrographs of fracture surfaces of the hybrid multiscale composites containing various CNTs content: 10 lm scale bar magnification. (**a**) 0 wt%, (**b**) 0.025 wt%, (**c**) 0.075 wt%, and (**d**) 0.1 wt%. (Wang et al. 2013) (Reproduced with permission from Elsevier Ltd.)

13.8 Application of RTM Composites

The applications of the RTM composites generally found in industries are as follows:

(a) In automotive industry, for example, auto body panels, truck air deflector, caravan components, etc.
(b) In consumer products such as dishes antenna, chairs, swim pool panels, bathtub/shower units, solar collectors, etc.
(c) In aerospace and military.
(d) For corrosion resistance applications such as chemical storage tanks, tubes, etc.
(e) Wind turbine blades.
(f) Hollow shapes and complex structures can also be produced by this process.
(g) Big containers and bathtubs are commonly processed through RTM technique.

13.9 Advantages of RTM Composites

The advantages of the RTM composites are as follows:

(a) The part produced by RTM process has good surface finish on both side surfaces of the product.
(b) Any combination or 3D orientation can be achieved by RTM process.
(c) Temperature control device is used for achieving fast cycle time.
(d) The RTM process can be controlled both manually and automatically.
(e) The part made by the RTM process is having uniform thickness throughout.
(f) Low emission achieved during composite fabrication.
(g) Typical dimensional tolerance can be achieved by this process.
(h) The RTM process does not require high injection pressure.
(i) Material wastage is very less as near net shape parts are produced.

13.10 Conclusions

The current chapter focuses on the use of RTM process for the fabrication of the natural fibre-reinforced composite material. The main objective of this chapter is to explore the potential use of RTM process for making polymer composites and also to study the mechanical properties of composites made by the RTM process. Both types of fibres such as chemically treated and untreated can be used for manufacturing of composite materials by considering the various parameters of the said process. The process has good applications in the industry as well, specifically in aerospace industry for making various parts of the aircraft. Wide variety of application of RTM process can also be found for making household equipment. The present chapter also defines the SEM or morphological study of the fractured specimen to understand the fracture mechanism of the composites.

References

Hsiao K-T, Gangireddy S (2008) Investigation on the spring-in phenomenon of carbon nanofiber-glass fiber/polyester composites manufactured with vacuum assisted resin transfer molding. Compos Part A 39:834–842

Laurenzi S, Marchetti M (2012) Advanced composite materials by resin transfer molding for aerospace applications, Compo and Their Prop. Book chapter, Intech open, pp 197–226

Lawrence JM, Hsiao K-T, Don RC, Simacek P, Estrada G, Sozer EM, Stadtfeld HC, Advani SG (2002) An approach to couple mold design and on-line control to manufacture complex composite parts by resin transfer molding. Compos Part A 33:981–990

Leclerc JS, Ruiz E (2008) Porosity reduction using optimized flow velocity in resin transfer molding. Compos Part A 39:1859–1868

Lee G-W, Lee N-J, Jang J, Lee K-J, Nam J-D (2002) Effects of surface modification on the resin-transfer moulding (RTM) of glass-fibre/unsaturated-polyester composites. Compos Sci Technol 62:9–16

Lin L-Y, Lee J-H, Hong C-E, Yoo G-H, Advani SG (2006) Preparation and characterization of layered silicate/glass fiber/epoxy hybrid nanocomposites via vacuum-assisted resin transfer molding (VARTM). Compos Sci Technol 66:2116–2125

Naffakh M, Dumon M, Gerard JF (2006) Study of a reactive epoxy–amine resin enabling in situ dissolution of thermoplastic films during resin transfer moulding for toughening composites. Compos Sci Technol 66:1376–1384

Papargyris DA, Day RJ, Nesbitt A, Bakavos D (2008) Comparison of the mechanical and physical properties of a carbon fibre epoxy composite manufactured by resin transfer moulding using conventional and microwave heating. Compos Sci Technol 68:1854–1861

Pearce NRL, Guild FJ, Summerscales J (1998) An investigation into the effects of fabric architecture on the processing and properties of fibre reinforced composites produced by resin transfer moulding. Compos Part A 29A:19–27

Rouison D, Sain M, Couturier M (2004) Resin transfer molding of natural fiber reinforced composites: cure simulation. Compos Sci Technol 64:629–644

Schmachtenberg E, Schulte zur Heide J, Topker J (2005) Application of ultrasonics for the process control of resin transfer moulding (RTM). Polym Test 24:330–338

Sreekumar PA, Saiter JM, Joseph K, Unnikrishnan G, Thomas S (2012) Electrical properties of short sisal fiber reinforced polyester composites fabricated by resin transfer molding. Compos Part A 43:507–511

Wang B-C, Zhou X, Ma K-M (2013) Fabrication and properties of CNTs/carbon fabric hybrid multiscale composites processed via resin transfer molding technique. Compos Part B 46:123–129

Chapter 14
Manufacturing of Chemically Modified Date Palm Leaf Fibre-Reinforced Polymer Composites

Sarat K. Swain, Gyanaranjan Sahoo, and Niladri Sarkar

Abstract Chemically modified fibre-reinforced polymer composites were manufactured by melt mixing method with several fractions of the modified fibres. Before proceeding to melt mixing method, fibres were made compatible by different chemical modifications. The interaction between date palm leaf fibres and polymeric matrix was examined by Fourier transform infrared (FTIR) spectroscopy, whereas the morphology of chemically modified DPL fibres and DPL fibre-reinforced biocomposites (i.e., polyvinylpyrrolidone (PVP)/DPL and polyvinyl alcohol (PVA)/DPL) was studied by using field emission scanning electron microscopic (FESEM) technique. Because of the strong interfacial interaction between polymers and fibres, mechanical properties were improved. It was found that the tensile strength, elongation at break, and loss modulus of polymer composites were compared with that of virgin polymeric matrix of PVP and PVA. The thermal and conducting properties of composites were studied by using thermogravimetric analysis.

Keywords Biocomposites • Tensile strength • Loss modulus • Date palm leaf fibre

14.1 Introduction

The thought of using natural fibres as reinforcing component in different composite materials is not new, and it has been effectively utilized by various civilizations throughout the world history. As per the earlier literature analysis, a renewed interest on the utilization of natural fibres as binding phase in polymer-based composites has been observed world widely, and this new class of material has been selected as suitable alternative for other fibre-reinforced composites. Fibre-reinforced composites

---

S.K. Swain (✉) • G. Sahoo • N. Sarkar
Department of Chemistry, Veer Surendra Sai University of Technology,
Burla, Sambalpur 768018, Odisha, India
e-mail: swainsk2@yahoo.co.in

M.S. Salit et al. (eds.), *Manufacturing of Natural Fibre Reinforced Polymer Composites*, DOI 10.1007/978-3-319-07944-8_14

have its own importance in the field of civil infrastructure to provide rehabilitees or retrofit existing structure and also boost the life of the built environment. The up-growing attention toward the sustainable built surroundings has challenged engineers to consider environmental factors, social factors, and energy/resource consumption along with the financial constraints and performance criteria. By the passing of time, fibre-reinforced polymer composites are increasingly being considered as an alternative to traditional civil engineering materials, namely, concrete and steel. The lightweight and noncorrosive nature of fibre-reinforced composites along with its high specific strength and stiffness make these composites interesting with the aim to satisfy the performance criteria. The demand of natural fibre-bound polymer composites has grown fast due to the high performance, significant processing, mechanical properties, low cost, low density, renewability, biodegradability and easy cease to surface modification, relative non-abrasiveness, and wide availability. In recent times, a series of investigation have been performed on fibre-based bio-composite materials. Although natural fibres have some better-quality properties as compared to synthetic fibres, they also have some drawbacks due to their surface polarity such as low-dimensional stability, poor moisture resistant, and limited processing temperature. Natural fibres are chemically modified in order to improve the interfacial adhesion between matrices and fibres. Therefore, different techniques of surface modifications on the natural fibres are applied to enhance the interfacial interaction with the polymeric matrix.

The date palm tree, member of the palm tree community with a scientific name *Phoenix dactylifera*, is usually found in the Middle East of India and Pakistan, in Northern Africa (Canary Island), and in California (United States). Traditionally, date palm leaves are utilized in fabricating baskets, mats, and ropes in many areas of the world. Competent utilization of this natural resource in synthesizing natural fibre-reinforced composite would have a high optimistic impact on the environment and improve the economic standard of working-class people (Kaddmi et al. 2006). Now, few works have been adopted by various researchers to develop polymer composite materials using fibres, obtained from different potions of date palm leaf. Mahdavi et al. (2010) have investigated the mechanical and morphological properties of wood plastic composites designed by reinforcing the fibres obtained from trunk, rachis, and petiole of date palm tree in high-density polyethylene (HDPE) matrix. It has been noticed that there is a significant difference between the trunk and petiole on fibre length, but rachis has no considerable difference relative to other parts. The tensile strength of fibres obtained from leaves and stems of date palm tree has been investigated and compared with other natural fibres by Rao et al. (2007). They have observed that the measured mechanical strength of date palm fibre is compatible to that of other fibres such as bamboo, banana, coconut, and sisal. Alawar et al. (2009) have studied the effect of different chemical modifications on date palm fibre surrounding the stems of date palm tree. From their investigation, it has been clearly observed that 1 % of NaOH-treated date palm fibre shows the optimum mechanical properties, whereas HCl-treated date palm leaves show a deterioration in mechanical properties. Two types of resins, i.e., phenol formaldehyde and bisphenol, are used for the synthesis of composite panel (Al Sulaiman 2003), and it was observed that the thermal conductivities of the composites vary with the types of resins (Mohanty et al. 2013). It is also noticed that fibre orientation and size have no reckonable effect on the thermal conductivity.

In the present context, we are going to narrate a brief outline of the manufacturing part of chemically modified date palm leave reinforced polymer composites. For the first case, chemically functionalized leaf fibres of date palm have been introduced in polyvinyl alcohol (PVA) to synthesize biocomposite materials by injection molding technique. By using benzoylation, alkaline, permanganate, peroxide, acrylic acid, and maleic anhydride action on date palm leaf fibre, we have achieved various chemical modifications with good interfacial bonding. Here, different wt% of acrylic acid (10, 20, 30, and 40 %)-modified date palm leaf fibres have been mixed up with PVA polymer to fabricate the desired biocomposite materials. Numerous works have been investigated the preparation of polymer composites reinforced with surface-modified natural fibres (Kang and Kim 2011). However, the work relating to polyvinylpyrrolidone (PVP)–PVA composites with reinforcement of chemically modified fibres was merely mentioned in the literature (Mohanty et al. 2013). So further investigation on mechanical, thermal, and conducting properties of PVA/DPL biocomposite with its manufacturing process is undoubtedly a new work and first reported by our research group. Both the fibre and the prepared biocomposites were characterized by Fourier transform infrared (FTIR), X-ray diffraction (XRD), and field emission scanning electron microscopic (FESEM). It was found that the mechanical and thermal properties of the prepared composites are substantially superior as compared to virgin PVP with a little compromise in conducting properties.

14.2 Experimental

The experimental part of this chapter includes materials and manufacturing part in detail. The main raw material, date palm branches, was collected from different areas of Odisha, India. Different essential chemicals such as benzoylchloride, benzoyl peroxide, sodium hydroxide, potassium permanganate, maleic anhydride, acetone, sodium chloride, benzene, starch, and acrylic acid of analytical grade were used with any further purification. Main matrix components, PVP and PVA, were purchased from Merk (India, Mumbai), and these were used as such. All the solutions were prepared in double-distilled water.

14.3 Preparation of DPL Fibre and Its Chemical Modification

14.3.1 DPL Fibre Preparation

At first, fibres were prepared or made compatible by taking out the strips from date palm leaves and then cutting these into small pieces. After that, these small pieces of fibres were washed with distilled water, followed by drying at 70 °C for 1 day to

Table 14.1 Chemical composition of DPL fibre

Constituents	Composition (%)
Lignin	15.30
Cellulose	54.71
Moisture	6.50
Pectin	1.20
Hemi-cellulose	20.00
Ash	1.75
Wax	0.50

eradicate the moisture completely. Now, these untreated fibres or so-called prepared fibres were processed for chemical modifications. Different chemical treatments such as benzoylation, alkaline, permanganate, peroxide, acrylic acid, maleic anhydride, and sodium hydroxide were performed to improve the adhesiveness which has been explained as follows. Refer to Table 14.1 for the chemical composition of DPL fibres (Pothan et al. 2010).

14.3.2 *Chemical Modifications of DPL Fibres*

14.3.2.1 Alkaline Treatment

In case of alkaline modification, at first the so-called prepared fibres of the date palm leaves were thoroughly washed by sodium hydroxide solution (5 wt% NaOH) for 1 h at 80 °C. After that, the NaOH-treated fibres were cooled down to room temperature and rinsed several times with fresh water to remove unreacted NaOH solution (if any) from the fibre surface, and these fibres were then dried at 60 °C for 24 h. This method was adopted to activate the –OH groups on the lignin and cellulose in the fibre. Actually, the addition of aqueous sodium hydroxide in natural fibres leads to the ionization of the –OH groups to the alkoxide which is given as follows (Lei et al. 2007). These resulting fibres were termed as alkaline pretreated fibres.

$$\text{Fibre}-\text{OH}+\text{NaOH}\rightarrow\text{Fibre}-\text{O}^{-}\text{Na}^{+}+\text{H}_2\text{O} \tag{14.1}$$

14.3.2.2 Benzoylation Treatment

It involves the process where alkaline pretreated fibres were first well suspended in 10 % sodium hydroxide and benzoylchloride solution for 15 min. The organic reaction between the cellulosic –OH groups of date palm leaf fibre and benzoylchloride is revealed in Eq. (14.2) and Agrawal et al. (2000). These fibres were then isolated. These fibres were soaked in ethanol for 1 h to remove the excess benzoylchloride and finally washed with fresh water. At last these fibres were dried at 80 °C for 24 h.

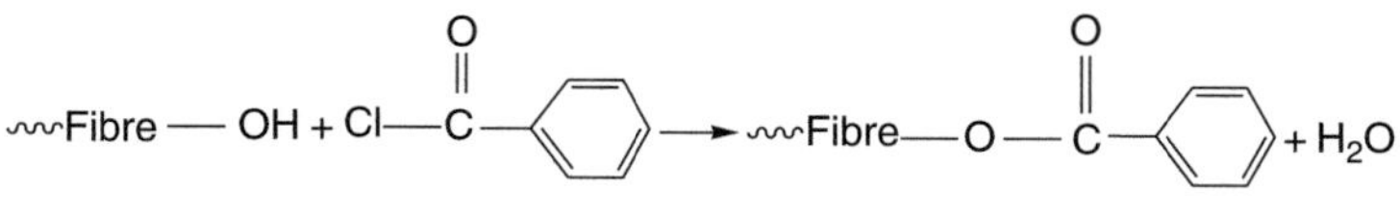

(14.2)

14.3.2.3 Permanganate Treatment

In this process, the alkaline pretreated fibres were immersed in $KMnO_4$ solution (0.003 % in acetone) for 1 min. The graft copolymerization scheme of DPL fibres by highly Mn^{3+} ions during the chemical treatment is given below (Joseph et al. 2000). After that, these fibres were rinsed with fresh water and dried for 24 h at 80 °C.

$$\text{Cellulose} - \text{H} + \text{KMnO}_4 \rightarrow \text{Cellulose} - \text{H} - \text{O} - \text{MnO}_2 - \text{OK}^+ \quad (14.3)$$

$$\text{Cellulose} - \text{H} - \text{O} - \text{MnO}_2 - \text{OK}^+ \rightarrow \text{Cellulose} + \text{H} - \text{O} - \text{MnO}_2 - \text{OK}^+ \quad (14.4)$$

14.3.2.4 Peroxide Treatment

In this method, alkaline pretreated fibres were encrusted with benzoyl peroxide in 6 % acetone solution for half an hour, and then these fibres were washed with distilled water and dried in an oven for 1 day at 80 °C. For a clear understanding, peroxide initiated free radical reaction between DPL cellulose fibres and PVA matrix and is given below (Paul et al. 1997):

$$\text{RO} - \text{OR} \rightarrow 2\text{RO} \quad (14.5)$$

$$\text{RO} + \text{H} - \text{PVA} \rightarrow \text{ROH} + [\text{PVA}]. \quad (14.6)$$

$$\text{H} - \text{Cellulose} + \text{RO} \rightarrow \text{ROH} + [\text{Cellulose}] \quad (14.7)$$

$$[\text{Cellulose}] + [\text{PVA}] \rightarrow \text{PVA} - \text{Cellulose} \quad (14.8)$$

14.3.2.5 Maleic Anhydride Treatment

In this anhydride treatment, alkaline pretreated date palm leaf fibres were mixed up with maleic anhydride solution (5 wt% of in acetone) at 65 °C for 3 h. The organic reaction mechanism of maleic anhydride with DPL fibre and PVA matrix is shown in Eq. (14.9) (Joseph and Thomas 1996). Then, these fibres were washed for several times with distilled water to drive away unreacted chemicals from the fibre surface and then dried at 80 °C for 24 h.

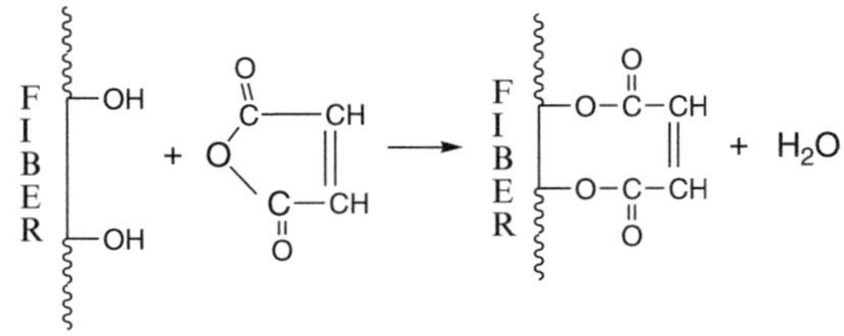

(14.9)

14.3.2.6 Acrylic Acid Treatment

According to this procedure, the alkaline pretreated date palm leaf fibres were first swelled in a binary solvent prepared by benzene and water in a proportion of 45/5 cm^3 for 15 min at 85 °C. The mechanistic pathway of DPL fibre with acrylic acid is given in Eq. (14.10) (Bledzki et al. 1996). Then these modified fibres were washed in several steps with different reagents: firstly, with an alkaline solution of 1 g L^{-1} NaOH, then 6 g L^{-1} NaCl for 15 min, and after that with distilled water for 1 h. Finally, the modified fibres were dried at 80 °C for 24 h in an oven and cooled in a desiccator.

$$\text{Fibre} - \text{OH} + \text{HO} - \text{CO} - \text{CH} = \text{CH}_2 \rightarrow \text{Fibre} - \text{O} - \text{CO} - \text{CH} = \text{CH}_2 + \text{H}_2\text{O} \quad (14.10)$$

14.4 Manufacturing of DPL Fibre-Based Biocomposites

The early manufacturing method for fibre-reinforced composite structural parts used a hand layup technique. Although hand layup is a reliable process, it is by nature very slow and labor intensive. In recent years, particularly due to the interest generated in the automotive industry, there is more emphasis on the development on manufacturing methods that can support mass production. Bag molding, compression molding, and injection compounding molding represent such manufacturing methods.

Molds are made from metal, usually either steel or aluminum, and precision machined to form the features of the desired part. Injection molding is a manufacturing process for producing parts from both thermoplastic and thermosetting plastic materials. The material is feed into a heated barrel, mixed, and forced into a mold cavity where it cools and hardens to the configuration of the cavity. This kind of molding is widely used for manufacturing to form the features of the desired part.

Injection molding machines (shown in Fig. 14.1) consist of a material hopper, an injection ram or screw-type plunger, and a heating unit. They are also known as presses; they hold the molds in which the components are shaped. Presses are rated by tonnage, which expresses the amount of clamping force that the machine can exert. This force keeps the mold closed during the injection process. Tonnage can vary

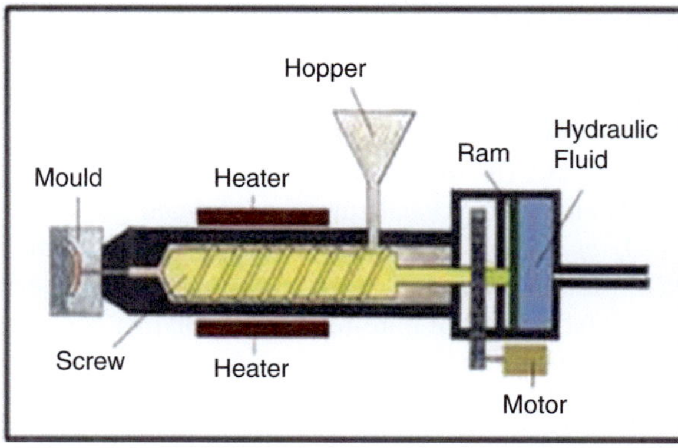

Fig. 14.1 Injection molding machine for processing of treated fibres, PVP and PVP/DPL biocomposites

Fig. 14.2 (**a**) Date palm trees, (**b**) date palm leaves, and (**c**) date palm leaf fibres

from less than 5–6,000 tons, with the higher figures used in comparatively few manufacturing operations. The total clamp force needed is determined by the projected area of the part being molded. This projected area is multiplied by a clamp force from 2 to 8 tons for each square inch of the projected areas. As a rule of thumb, 4 or 5 tons in^{-2} can be used for most products. If the plastic material is very stiff, it will require more injection pressure to fill the mold, thus more clamp tonnage to keep the mold closed. The required force can also be determined by the material used and the size of the part, larger parts require higher clamping force. Advantages of this process include high production rate, design flexibility, relatively low labor, little to no finishing of the parts, minimum scrap losses, and repeatability within tolerances. So it can effectively produce a wide range of materials.

Date palm (*Phoenix dactylifera*) leaf (shown in Fig. 14.2b, c) is one such natural resource whose potential as fibre reinforcement in polymer composite has not been explored till date. The genus *Phoenix* belongs to the Arecaceae family. As a group, there are about 13 different species of *Phoenix*. Some important species of date palm trees (shown in Fig. 14.2a) are found in India. Hence, various measures have been taken to eradicate this species like manual eradication, biological control, chemical methods, etc. But unfortunately, efforts to manage the weed have not been successful. Alternatively, visualizing the luxuriant growth and vigorous survival of this weed, researchers worldwide are trying to find out the potential economic value for its utilization into value-added products and effective method for its management.

The fabrication of fibre-reinforced polymer composites involves a special step to mix all the constituting components (chemically modified date palm leaf fibre, PVP, starch, and polypropylene) in order to prepare the raw material. Thermoplastic starch was prepared by the addition of glycerol and water. Initially, the starch was pre-dried at 80 °C for 1 h in the oven .Then, 70 % of starch, 20 % of glycerol, and 10 % of water were mixed and stirred thoroughly for 30 min. After that, the starch was again pre-dried at 80 °C so that water present in it could be evaporated. In order to incorporate the better compatibility between the main matrix of PVP and date palm leaf fibre, thermoplastic starch and a small amount of polypropylene were added to the mixture. The mixture was kept in a vessel closed with aluminum foil to avoid absorption of moisture and was kept for 24 h under at room temperature. Five samples of different chemically modified fibre volume fractions, e.g., 0, 10, 20, 30, and 40 %, were prepared whose detail compositions are given in Table 14.2.

The preparation of raw material is also the same for date palm leaf fibre-reinforced PVA composites; the only difference is in the matrix phase. Here, PVA serves the purpose of forming the matrix phase, instead of PVP.

Fabrication of final composite specimens was performed by injection molding process. Before fabrication, matrix mixture was prepared by micro-compounder. The DSM Xplore 5 and 15 micro-compounder (shown in Fig. 14.3a) is intended for compounding small quantities of material. The compounder houses the unique vari-batch TM concept, which gives the opportunity to select the batch volume via multiple recirculation channels from 3, 7, and up to 15 ml.

The different batches for micro-compounding are prepared as per Table 14.1. The barrel is heated up to the desired temperature. Before filling the compounder with material, the hopper first needs to be connected. Hopper consists of a funnel, a plunger, and a hollow pipe. Material is poured inside the compounder slowly which moves toward the barrel and allows them to melt before they reach the bottom of the

Table 14.2 Composition of fibre and matrix (PVP + TPS + PP)

Composition of fibre (%)	Weight of fibre (gm)	Weight of PVP + TPS (gm)	Weight of PP (gm)	Total weight (gm)
0	0	13.6	2.4	16
10	1.6	12	2.4	16
20	3.2	10.4	2.4	16
30	4.8	8.8	2.4	16
40	6.4	7.2	2.4	16

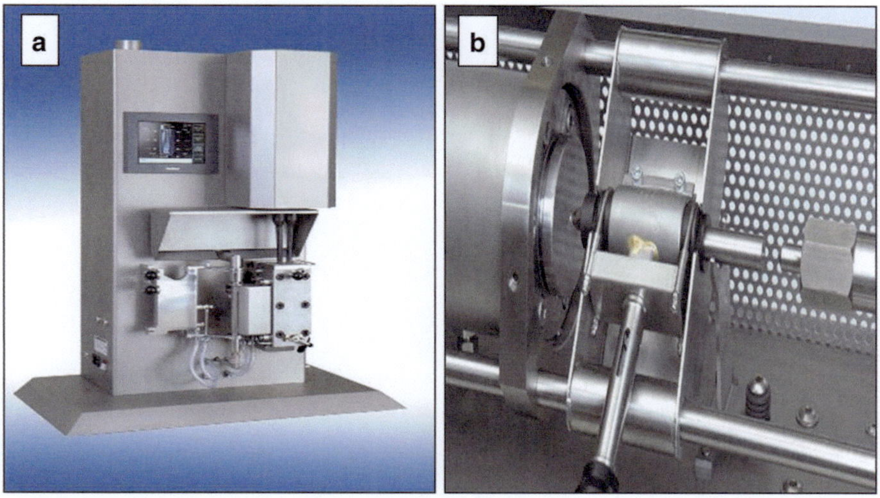

Fig. 14.3 (**a**) Micro-compounder and (**b**) micro-injection molding machine

barrel. Inside the barrel, the mixing screws are rotating at a speed of 100 rpm, i.e., the compounding speed. During this operation, the drain valve is closed. The operation time has been kept at 5 min. In this method, the materials get dissolved into solvents at high temperatures, and as a result, a high viscous solution is prepared. In order to have the material circulate through the compounder during mixing, it must be viscous enough to develop pressure at the end of the barrel. This pressure will push the mixed material through the recirculation channel. At the end of the compounding time, the flow director valve is turned to outflow position (front), and the compounded material is collected from the exit port. After the material has been extruded from the compounder, the motor is stopped and the barrel is cleaned. DSM Micro 10cc injection molding machine is intended for injection molding which is useful for fabrication of small quantities of materials. The process accuracy and reliability of micro-molding are much better than that of conventional molding. Accuracy and reliability mean higher-quality molded parts with fewer problems. Micro-injection molding (shown in Fig. 14.3b) has various parts such as mold and mold holder, nozzle, pressure cylinder, touch screen (control panel), etc.

The molten materials were forced out of micro-compounder and collected inside a cylinder (nozzle). The composite specimens were prepared by forcing hot polymer through compressed air cylinder into a heated mold. The touch screen fully adjusts the time and level of pressure under which the test specimens are made. The molten materials were filled into the nozzle, and the plunger was moved in an outward direction. Once the nozzle is filled up, it is placed inside the carriage at the right angles with the orifice pointing toward the left. Then the handle was pushed downward to lock the nozzle in the carriage. The machine will only work once the protective cover is closed; the pneumatic system stays locked whenever the protective cover is not closed. The machine was started by pressing the "RUN" key on the touch screen.

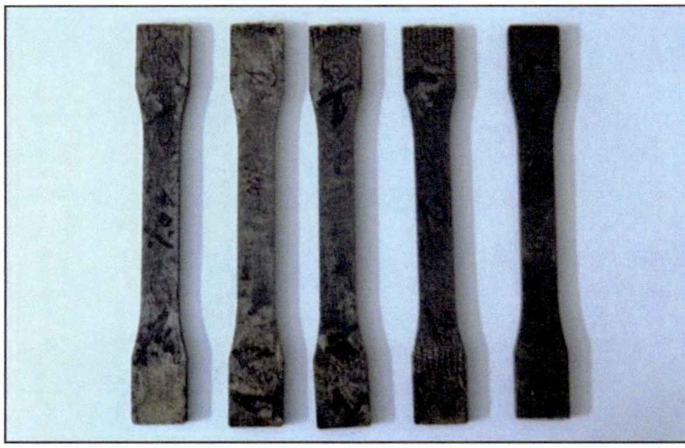

Fig. 14.4 Test specimen

The mold was taken out by the help of ejection pin when the injection had finished. Then the specimens were removed by the help of tongs when the mold was opened (as shown in Fig. 14.4).

14.5 Standard Techniques Used for Properties Investigation

14.5.1 Instrumentation

- FTIR characterization of the prepared samples (in the form of KBr pallet) in the form of spectra was recorded by Shimadzu IR Affinity-1 Fourier infrared spectrophotometer in the range of 4,000–400 cm^{-1}.The surface morphology of the prepared biocomposite was studied with the help of Scanning Electron Microscope from Jeol, Japan (model 5200 with magnification of 30,000). The thermal stability of the prepared samples was performed under the nitrogen purge at a heating rate of 10 °C min^{-1} using a TGA apparatus (model DTG-60, Shimadzu Corporation Japan). XRD patterns of the prepared biocomposites were recorded using an X-ray machine (Rigaku operating at 40 kV and 150 mA).
- Mechanical Property: Tensile tests of the prepared samples were carried out using INSTRON 3382, a servo hydraulic static testing machine having 5 kN load cell. In this case, test specimens were prepared as per ASTM D 638 specification, i.e., 165 mm in length and 3.2 mm in thickness. In order to realize the change in property with fibre content, at each fibre loading, e.g., 0, 10, 20, 30, and 40 %, five samples were tested at a crosshead speed of 2 mm min^{-1} out of which average readings were taken at each fibre loading. A 50 mm extensometer was used to get displacement readings. All tests were performed with controlled humidity of 54 ± 5 % at room temperature (23 ± 2 °C).
- The measurement of flexural strength was done by INSTRON 3382, a servo hydraulic static testing machine, using three-point bending method. Specimens for

flexural test were prepared according to ASTM D790-99 standard with dimensions $127 \times 12.7 \times 3.2$ mm³. In order to realize the change in property with fibre content, at each fibre loading, e.g., 0, 10, 20, 30, and 40 %, five samples were tested at a crosshead speed of 1 mm min^{-1}, and the bending stress was calculated from the measured load as per the equation given below:

$$\delta = WL^3 / 48EI$$

- where W = maximum load (N)

  δ = displacement (mm)
  I = moment of inertia (mm^4)
  E = Young's modulus (Mpa)
  L = distance between the support span (mm)

- For the prepared samples, impact tests were carried out as per ASTM D 256 with the help of pendulum type Izod impact testing machine. Specific hammer velocity and hammer weight 3.46 m s^{-1} and 0.905 kg, respectively, were maintained during experiment. At the time of testing, specimens were clamped vertically as a cantilever beam and then struck by a single swing of the pendulum released from the clamp. Three consistent results were selected for each fibre loading in PVA matrix.
- For dynamic mechanical thermal analysis (DMTA), test the prepared biocomposite sheets which were sliced with average typical dimensions of $32 \times 13 \times 3$ mm^3 according to ASTM D 5026 standards. This measurement was performed with a DMA Q 800 TA instrument at a frequency of 1 Hz with heating from 33 to 160 °C at a heating rate of 2 k min. The instrument was equipped for rectangular samples and working at shear mode. The maximum shear strain was equal to 0.2 %.
- The glass transition temperature which was taken as the onset temperature of the specific heat increment was measured with differential scanning calorimetry (DSC) Q 20 TA calorimeter. This was performed under argon atmosphere from 49 to 260 °C with a heating rate of 10 °C min^{-1}.

14.6 Explanations of Evidenced Results

14.6.1 FTIR Analysis of Polymer Composites

The chemical interaction between the functional groups present in DPL and PVP is characterized by FTIR study. The FTIR results of chemically modified DPL fibre and PVP along with DPL fibre-reinforced PVP biocomposite with 26 wt% of fibre loading are compared and shown in Fig. 14.5a. The observed sharp peak at 2,900 cm^{-1} may be due to the C–H stretching, whereas a week peak at 1,680 cm^{-1} corresponds to the presence of ester group formed after treatment with acrylic acid. However, the FTIR spectrum of treated DPL fibre gives a broad peak at 3,480 cm^{-1}

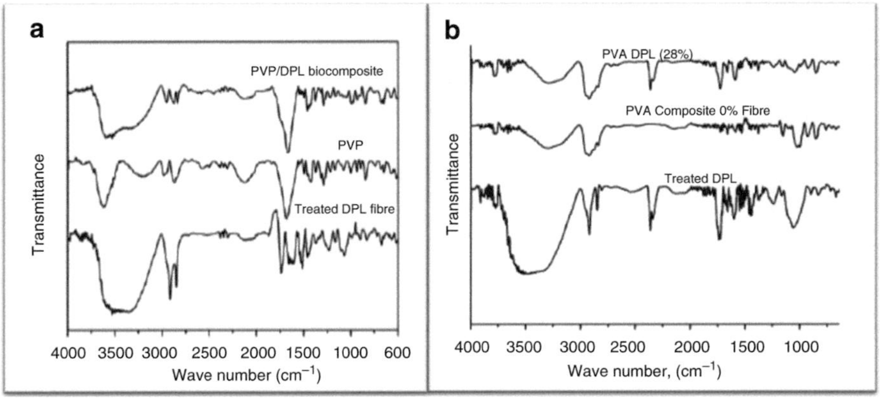

Fig. 14.5 (**a**) FTIR spectra of treated fibres PVP and PVP/DPL biocomposites and (**b**) treated fibres PVA and PVA/DPL biocomposites [*Polym. Compos.* 15:1062–1070 (2013), *Fibres Polym.* 15, 1062–1070 (2014)]

indicating the presence of some unreacted –OH peak of the fibre. The observed peak at 1,750 cm^{-1} may be due to stretching vibration of carbonyl groups (>C=O) of PVP along with its N–H peak at 3,600 cm^{-1}. The FTIR of PVP/DPL biocomposites consists of the carbonyl peak of PVP and the –OH peaks along with the ester peak of DPL fibre which gives the evidence regarding the formation of biocomposites.

Figure 14.5b gives the comparative study of FTIR results of chemically modified DPL fibre, virgin PVA, and DPL fibre-reinforced PVA biocomposite with 28 wt% fibre loading. The FTIR spectrum of treated DPL fibre shows a broad peak at 3,400 cm^{-1} indicating the presence of –OH functional group in the fibre surface, whereas the sharp peak at 2,900 cm^{-1} may be due to the C–H stretching. The peak observed at 1,740 cm^{-1} in the case of DPL fibre may be due to stretching vibration of carbonyl (>C=O) groups that are attached to the fibre backbone. The presence of –OH functional groups of PVA matrix can be realized from the broad peak of FTIR spectrum at about 3,300 cm^{-1}. It is noticed that the FTIR peak of carbonyl group at 1,740 cm^{-1} of DPL fibre is present for PVA/DPL biocomposite and absent for virgin PVA. This gives the supportive information regarding the formation of PVA/DPL biocomposite.

14.6.2 XRD Analysis of Polymer Composites

In order to make a structural analysis, XRD patterns of PVP and PVP/DPL composites have been illustrated in Fig. 14.6a. XRD peak at 2θ of 16° and a broad peak at 22° may be due to the crystalline phase of PVP. For PVP/DPL biocomposites, the left side shift of peaks at 2θ values of 15.5° and 21° is due to the change in crystallinity of PVP in the form of prepared biocomposite. The crystallinity of virgin PVA was also reflected through the observed diffraction peak at 2θ value of 20.5°, and in

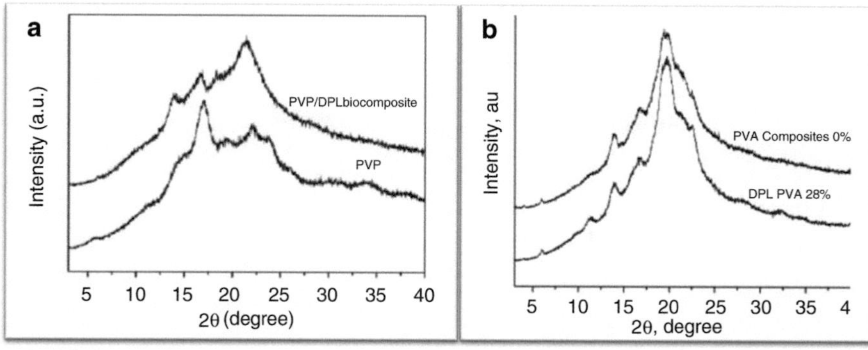

Fig. 14.6 (**a**) Comparative XRD curves of PVP and PVP/DPL biocomposites and (**b**) PVA and PVA/DPL biocomposites [*Polym. Compos.* 15:1062–1070 (2013), *Fibres Polym.* 15, 1062–1070 (2014)]

a comparative study, it is also found in the same place for PVA/DPL biocomposites (Fig. 14.6b). However, the most interesting fact is that the crystallinity of PVA is remained unaltered even after formation of biocomposites. The final conclusion, drawn from XRD study, is the same as for FTIR results.

14.6.3 Morphological Analysis Polymer Composites

The surface texture of the prepared biocomposites was captured through FESEM images of PVP/DPL biocomposite with 26 wt% treated fibre (shown in Fig. 14.7). Change in surface morphology with the incorporation of fibres was compared from FESEM images. For the first biocomposite, i.e., PVP/DPL biocomposite, chemically modified fibres are found to be debundled and distinct with fibre dimension of about 1 μm, represented in Fig. 14.7c. On the other hand, FESEM images of DPL fibre-reinforced PVP biocomposites represent almost a uniform distribution of DPL fibres within PVP matrix forming different flexes of biocomposites (Fig. 14.7d). The average width of the flex is found at around 4–5 μm, and FESEM images of biocomposite also indicate compatible interfacial adhesion between DPL filler and PVP matrix. Figure 14.7a presents the FESEM images of both chemically modified fibre and 28 wt% fibre-treated DPL fibre-reinforced PVA biocomposite. The surface morphology of the second biocomposite, i.e., PVA/DPL biocomposite, is interesting. By the close vision, it is also noticed that DPL fibres are not pulled out from PVP matrix, creating a polymer coating on the surface of the DPL fibres. However, some sort of local agglomeration within PVA polymeric matrix was observed as black patches in FESEM images of the prepared DPL fibre-reinforced PVA biocomposite. Better compatibility of DPL/PVA biocomposite can be well explained by the uniform dispersion of chemically modified DPL fibre within the PVA matrix. Furthermore, DPL fibres were not pulled out from the thermoplastic matrix as the first biocomposites and were covered by polymers.

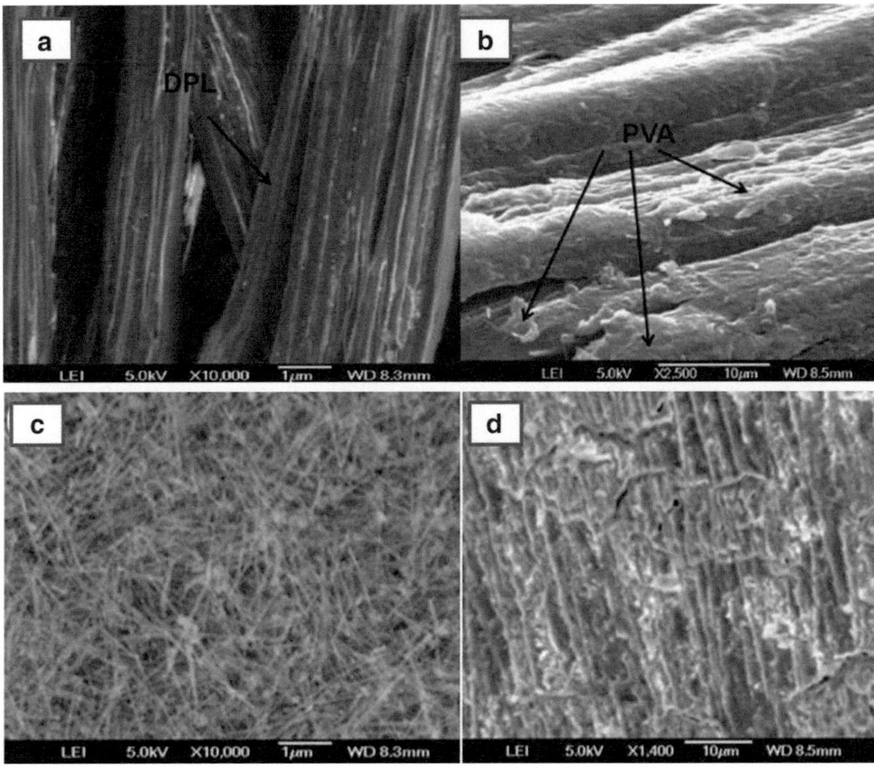

Fig. 14.7 (**a**, **b**) FESEM of DPL fibre, (**c**) PVA/DPL (28 wt%), and (**d**) PVP/DPL biocomposites [*Polym. Compos.* 15:1062–1070 (2013), *Fibres Polym.* 15, 1062–1070 (2014)]

14.7 Properties of Polymer Composites

In Table 14.3, the mechanical strength of DPL/PVA biocomposite with different wt% of DPL fibre loading is illustrated. As the DPL fibre content exceeds 28 wt%, the matrix coverage is appearing insufficient with the appearance of many voids leading to poor interfacial adhesion. According to common law of short fibre composites, it is noticed from Table 14.3 that the mechanical properties are observed more at 28 % of fibre content as discussed in the previous section. From Table 14.3, it is noticed that the Young's modulus was increased with an increase in fibre loading. A drastic rise in tensile properties is observed at 28 wt% of fibre loading of DPL fibre-reinforced PVA biocomposites. After crossing 28 wt% fibre loading, the modulus decreases with an increase in fibre loading as it is observed in the case of tensile strength. The reason behind the decrease in modulus after 28 wt% of DPL fibre content can be explained from the fact that at higher fibre content, the fibres act as flaws and crazing occurs, whereas the sharp increase of modulus at 28 wt% fibre content indicates that the matrix with reinforcement has become more stiff and can

Table 14.3 Mechanical properties of PVA/DPL biocomposites at different fibre loading

Fibre in PVA/DPL (wt%)	Yield stress (MPa)	Yield strain (%)	Young's modulus (MPa)	Elongation at break (%)	Flexural strength (MPa)	Impact strength (kJ m^{-2})
0	12.261 ± 0.5	3.1 ± 0.2	362.5 ± 26.8	1.734 ± 0.09	19.21 ± 0.86	3.79 ± 0.21
10	17.80 ± 0.65	0.48 ± 0.17	1,009.0 ± 33.0	1.620 ± 0.12	25.55 ± 1.70	3.97 ± 0.31
20	19.34 ± 1.09	0.376 ± 0.02	896.0 ± 46.0	1.786 ± 0.14	34.39 ± 2.02	4.22 ± 0.28
28	20.19 ± 1.06	0.484 ± 0.02	1,183.0 ± 50.0	1.910 ± 0.06	35.92 ± 1.28	4.31 ± 0.17
30	19.045 ± 1.42	0.471 ± 0.03	1,151.0 ± 52.0	1.854 ± 0.09	36.29 ± 1.92	4.30 ± 0.20
40	18.956 ± 1.48	1.102 ± 0.08	1,078.0 ± 56.0	1.580 ± 0.07	28.10 ± 1.20	3.89 ± 0.29

withstand higher stress at the same strain level. After 28 wt%, the further fibre loading causes a generation of stress concentration areas which lower the stiffness of biocomposite. Similar type of trend also occurs in case of flexural properties of the prepared biocomposites, and it is found that the flexural strength of the biocomposite increases initially compared to virgin PVA matrix. A competent stress transfer between PVA and natural fibre occurs due to the addition of date palm leaf fibres. The rise in peak value (36.290 MPa) happens at 28 wt% DPL fibre loading. After that, it decreases subsequently which may be realized from the fact that higher fibre loading encourages fibre-to-fibre interaction and leads to some sort of fibre agglomeration within PVA polymeric matrix.

The same trend like tensile and flexural strengths is also observed for Izod impact strength, with the maximum impact strength (i.e., 4.307 kJ m^{-2}) at 28 wt% of DPL fibre loading. After that, it decreases as fibre content increases. It is clear that dynamic mechanical properties of a composite material depend on the fibre content, compatibilizer, fibre orientation, and board of testing. An investigation is carried out to find out the effect of frequency on the dynamic mechanical analysis of DPL fibre-reinforced PVA biocomposite with 28 wt% of fibre; DMTA test was performed over a temperature range of 20–180 °C at a heating rate of 10 °C min^{-1}. From the dynamic mechanical thermal properties of the biocomposite, the damping peak (tan δ) was found to be 0.1090 delta at temperature of 73.87 °C (Fig. 14.8b). The thermal gravimetric analysis results of PVA and PVA/DPL biocomposites were compared in Fig. 14.8c. Storage modulus and loss modulus of the prepared biocomposite were measured, and it was found that the storage modulus of the biocomposite was 2,550.68 MPa at 41.41 °C, whereas the loss modulus was 190.6 MPa at a temperature of 64.99 °C (Fig. 14.8a).

For PVP/DPL biocomposite with dynamic mechanical and thermal properties, the damping peak (tan δ) of the biocomposite is found to be 0.3415 delta at a temperature of 124.72 °C, whereas for PVP matrix, it is 0.3915 delta at 107.09 °C (Fig. 14.8e). From the storage moduli and loss moduli results of both matrix and biocomposite, it is found that the storage modulus of the biocomposite is 2,893 MPa at 30.48 °C, whereas that of the PVP is 2,389 MPa at 30.35 °C. Further, the loss

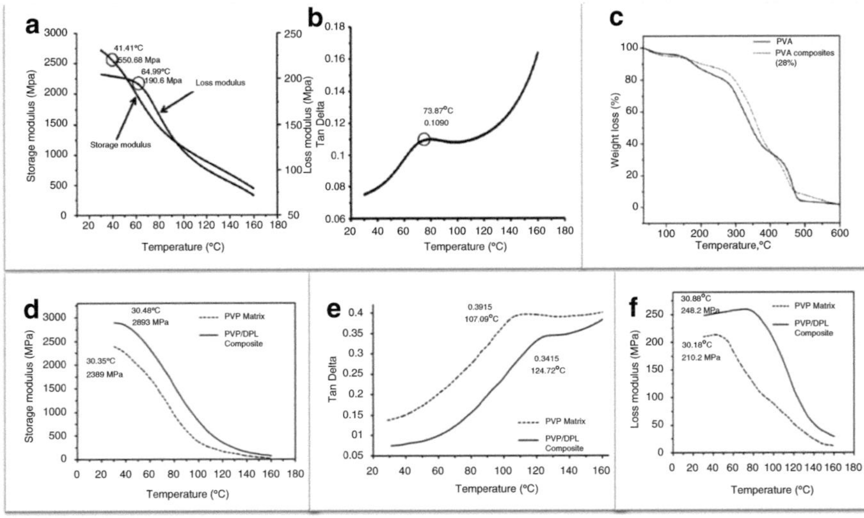

Fig. 14.8 (**a**, **b**, **c**) Temperature dependence of storage, loss modulus, and TGA of PVA/DPL biocomposites and (**d**, **e**, **f**) temperature dependence of storage and loss modulus of PVP/DPL biocomposites [*Polym. Compos.* 15:1062–1070 (2013), *Fibres Polym.* 15, 1062–1070 (2014)]

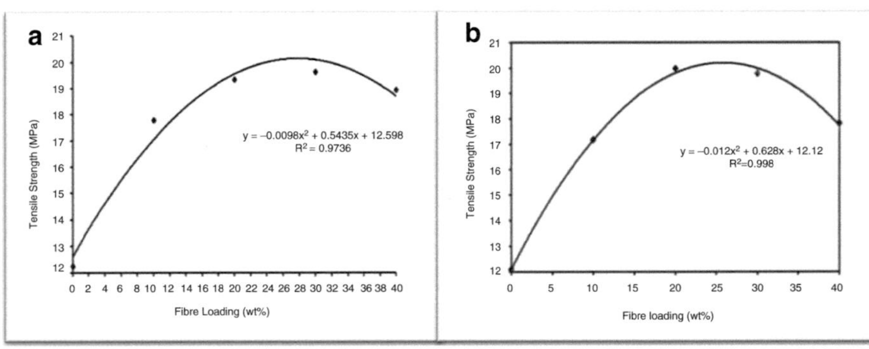

Fig. 14.9 (**a**) Second-degree polynomial curve in tensile test results of various wt% (**b**) and tensile strength of PVP/DPL biocomposites at different wt% of fibre loading [*Polym. Compos.* 15:1062–1070 (2013), *Fibres Polym.* 15, 1062–1070 (2014)]

modulus of biocomposite is 248.20 MPa at a temperature of 30.88 °C, whereas it is 210.2 MPa for PVP at a temperature of 30.18 °C (Fig. 14.8d, f).

Tensile strength data of prepared biocomposites were plotted against different weight percentage of acrylic acid-treated DPL fibres (shown in Fig. 14.5a). The actual wt% of DPL fibre at optimum strength has been obtained by regression analysis. The data fit very well by second-degree polynomial with regression coefficient (R^2-value) of 0.9736 (Fig. 14.9a), which gives the optimum strength (i.e., 20.19 ± 61.06 MPa) at 28 wt% of fibre loading, and this weight percentage of DPL fibre has been considered as the optimum fibre percentage for the development of PVA/DPL biocomposites for subsequent studies. From Fig. 14.5, it is noticed that the tensile strength increases first with DPL fibre loading and reaches a maximum value and then further decreases. Similarly, in Fig. 14.9b, the measured tensile strength data of prepared biocomposites have been plotted against the above weight percentage of acrylic acid-treated DPL fibres, and it is observed that the tensile strength first increases with fibre loading and further decreases after reaching a maximum value.

14.8 Conclusion

The DPL/PVP and DPL/PVA biocomposites have been fabricated by reinforcing chemically modified DPL fibres through melt mixing technique. The chemical interaction of polymers with fibres has been evidenced by FTIR which indicates the uniform distribution of fibres in PVP and PVA matrix. The optimum mechanical properties of PVP/DPL and PVA/DPL biocomposites have been observed at 26 and 28 wt% of fibre loading, respectively. The storage modulus, loss modulus, and tan δ values have been compared with neat polymers. From the study of the rheological properties, it can be stated that the improvement in the above properties may lead to easy processability of biocomposites. The conducting property of biocomposites has been compromised with substantial improvement in thermal and mechanical properties. Due to the hydrophobic and easily processable nature, prepared biocomposites can be utilized as water- and chemical-resistant packaging materials. Prepared DPL/PVA materials may be used in automotive, housing, and other industrial applications. These biocomposites can also act as a competent reinforcement for concrete composites to avoid accidental facture of concrete surface.

14.9 Future Prospective

Due to the lightweight nature and easy cease of chemical modification of date palm leaf fibres and its efficiency to introduce an improvement in tensile and flexture strength to the fibre-reinforced biocomposite, it can be effectively used as reinforcing component in different polymer-based biocomposites which can be used as an alternative to civil construction materials. In future perspective, other strong biofibres (elephant grass, sisal grass, sabai grass, coconut fibres) on the same type of chemical modification can be used as competent reinforcing component in the formation of lightweight and strong biocomposite which can be a suitable alternative for heavyweight civil engineering materials.

References

Agrawal R, Saxena NS, Sharma KB, Thomas S, Sreekala MS (2000) Activation energy and crystallization kinetics of untreated and treated oil palm fiber reinforced phenol formaldehyde composites. Mater Sci Eng A 277:77–82

Alawar A, Hamid MA, Kaabi A (2009) Characterization of treated date palm tree fiber as composite reinforcement. Compos Part B Eng 40:601–609

Al-Sulaiman FA (2003) Date palm fiber reinforced composite as a new insulating material. J Energy Res 27:1293–1297

Bledzki K, Reihmane S, Gassan J (1996) Properties and modification methods for vegetable fibers for natural fiber composites. J Appl Polym Sci 59:1329–1336

Joseph K, Thomas S (1996) Effect of chemical treatment on the tensile properties of short sisal fiber reinforced polyethylene composites. Polymer 37:5139–5149

Joseph K, Mattoso LHC, Toledo RD, Thomas S, de Carvalho LH, Pothen L, Kala S, James B, Carlos S (eds) (2000) Natural fiber reinforced thermoplastic composite. Embrapa, USP-IQSC, UNESP, Brazil, pp 159–201

Kaddmi H, Dufresne A, Khelifi B, Bendhaou A, Taourirte M, Raihane M, Issartel N, Sautereau H, Gerard J, Sami N (2006) Short palm tree fibers–thermoset matrices composites. Compos Part A Appl Sci Manuf 37:1413–1417

Kang JT, Kim SH (2011) Improvement in the mechanical properties of polyactide and bamboo fiber biocomposites by fiber surface modification. Macromol Res 19:789–796

Lei Y, Wu Q, Yao F, Xu Y (2007) Preparation and properties of recycled HDPE/natural fiber composites. Compos Part A Appl Sci Manuf 38:1664–1674

Mahdavi S, Kermanian H, Varshoei A (2010) Comparison of mechanical properties of date palm fiber polyethylene composites. BioResources 5:2391–2403

Mohanty JR, Das SN, Das HC, Swain SK (2013) Effect of chemically modified date palm leaf fiber on mechanical, thermal and rheological properties of polyvinylpyrrolidone. Polym Compos 15:1062–1070

Mohanty JR, Das SN, Das HC, Swain SK (2014) Effective mechanical properties of polyvinylalcohol biocomposites with reinforcement of date palm leaf fiber. Fibers Polym 15:1062–1070

Paul A, Joseph K, Thomas S (1997) Effect of surface treatments on the electrical properties of low density polyethylene composites reinforced with short sisal fiber. Compos Sci Technol 57:67–79

Pothan LA, George CN, Jacob M, Thomas S (2010) Mechanical performance of short banana/sisal hybrid fiber reinforced polyester composites. J Compos Mater 41:2371–2386

Rao KMM, Prasad AVR, Babu MNVR, Rao KM, Gupta AVSSKS (2007) Tensile properties of elephant grass fiber reinforced polymer composites. J Mater Sci 42:3266–3272

Chapter 15
Manufacturing of Coir Fibre-Reinforced Polymer Composites by Hot Compression Technique

Mahbub Hasan, M. Enamul Hoque, Samia Sultana Mir, N. Saba, and S.M. Sapuan

Abstract This present chapter describes the manufacturing technique and properties of coir fibre-reinforced polypropylene composites manufactured using a hot press machine. The effects of basic chromium sulphate and sodium bicarbonate treatment on the physical and mechanical properties were also evaluated. Chemical treatment and fibre loading generally improved the mechanical properties. Five-hour basic chromium sulphate and sodium bicarbonate-treated coir-polypropylene had the best set of properties among all manufactured composites. Chemical treatment also improved water absorption characteristics. This proves that chemical treatment reduced the hydrophilicity of the coir fibre. Overall the hot compression technique was proved to be successful in manufacturing good quality coir reinforced polypropylene composites.

Keywords Coir • Polymer composite • Hot compression technique • Chemical treatment • Scanning electron microscopy • Mechanical properties

M. Hasan • S.S. Mir
Department of Materials and Metallurgical Engineering, Bangladesh University of Engineering and Technology, Dhaka 1000, Bangladesh

M. Enamul Hoque (✉)
Department of Biomedical Engineering, University of King Faisal, Al-Hofuf, Al-Hassa, Saudi Arabia
e-mail: enamul1973@gmail.com

N. Saba
Laboratory of Biocomposite Technology, Institute of Tropical Forestry and Forest Products (INTROP), Universiti Putra Malaysia, 43400 Serdang, Selangor, Malaysia

S.M. Sapuan
Department of Mechanical and Manufacturing Engineering, Universiti Putra Malaysia, 43400 Serdang, Selangor, Malaysia

M.S. Salit et al. (eds.), *Manufacturing of Natural Fibre Reinforced Polymer Composites*, DOI 10.1007/978-3-319-07944-8_15

15.1 Introduction

Composites are made up of two or more physical or mechanically distinguishable materials. Polymer matrix composites are plastics within which fibres are embedded. The plastic is known as the matrix, and the fibres orientated within it are known as the reinforcement. Natural fibre-reinforced polymer composites are gaining paramount importance day by day in the composite research field and industry due to their versatile diversified nature and renewability. They possess a range of potential advantages, especially with regard to their environmental performance. When natural fibre composite waste is incinerated, they do not cause net emission of carbon dioxide to the environment. The matrix type, amount, and properties of the filler and fibre–matrix interaction mainly affect the properties of lignocellulosic natural fibre-reinforced polymer composites. The properties of fibre–matrix interface are different from the properties of the bulk matrix. The interface also transfers stress between fibre and matrix and controls the properties and durability of the composites. Natural fibres, obtained from lignocellulose, contain strongly polarized hydroxyl groups. As a result, they are generally hydrophilic. These hydroxyl groups are responsible for water absorption and deformation of the resultant products. As a result, the hydrophilic natural fibres are inherently incompatible with hydrophobic polymer. One way of increasing the compatibility is chemical treatment of natural fibre (Westerlind 1988; Haque et al. 2009).

Hot compression is a well-known technique of producing polymer matrix composites. During hot compression, both heat and pressure are applied; thus, it produces god quality composites. This present chapter describes the manufacturing process of coir fibre-reinforced polymer composites using hot compression technique. The characterization techniques and properties of the composites are also included in the chapter.

15.2 Experimental Procedure

15.2.1 Materials

Coir or coconut fibre (Fig. 15.1) is a hard structural commercial product that is obtained from the coconut husk. Coir is an outstanding substitute for cypress mulch or peat moss as it is a renewal resource, unlike peat and cypress. Its harvest does not cause environmental damage, as does peat mining, and it does not contain disease organisms transmittable to plants. Individual coir fibre has a length of 0.3–1.0 mm, diameter of 0.01–0.0.2 mm, and aspect ratio of 35. It has medium- to large-sized and polygonal-rounded or elliptic-shaped lumen. The vascular bundle is collateral, which is surrounded by thick sclerenchymatous sheath. Lignin and hemicelluloses that form the cementing materials of fibre cells increase, while the pectin decreases with an increase in the age of the fibre. The fibre becomes stiffer and tougher with

Fig. 15.1 Coir or coconut fibre

the lignin content. The fibre length determines its spinnability and commercial utility. Spinnability is the ease with which textile fibres may be twisted into continuous, uniform yarns, having commercially acceptable properties (Mir 2012). In the present chapter, basic chromium sulphate ($Cr_2(SO4)_3 \cdot 12(H_2O)$) and sodium bicarbonate ($NaHCO_3$) were used to modify coir fibre for increasing coir fibre and polypropylene compatibility in the resultant composites.

Polypropylene (PP) is a thermoplastic polymer extensively used throughout the world. PP is widely available with various reinforcing agents, for example, talc, mica, or calcium carbonate and chopped or continuous strand fibre. Additives are developed to enhance its thermal stability that minimizes degradation during processing. One of the main necessities during composite manufacturing is that PP should be relatively pure with no residual catalyst.

15.2.2 Coir Fibre Extraction

Initially, coir was collected from the local market. Then fibre was extracted from husk. At this stage, the fibre length was random. The fibre contained moisture after extraction. Thus, it was sun-dried before chemical treatment.

15.2.3 *Chemical Treatment of Coir Fibre*

Basic chromium sulphate and sodium bicarbonate treatments were conducted on coir fibre for increasing its compatibility with polypropylene. The chromium sulphate treatment was carried out for 3 h and called single-stage treatment (SST).On the other hand, the sodium bicarbonate was continued after the SST for two more hours. It was called double-stage treatment (DST). A pH of 2.5–3 was ensured during SST by preparing a 0.5 % chromium sulphate solution with 2/3 drops of HCl. The pH became 8–9 during the reaction. During SST, the coir was shaken for 3 h inside the prepared solution. The coir was then washed properly using distilled water. The same procedure, with an addition of 0.02 % $NaHCO_3$, was conducted for DST. The coir fibre was shaken for two more hours inside the new solution. Finally, the treated fibre was taken out from the solution and washed with distilled water properly.

15.2.4 *Composite Manufacturing*

15.2.4.1 Hot Compression Moulding Technique

The term compression moulding belongs to the type of methods in which desiccated fortification is permeated inside the liquefied resin through sealed cavity or in closed mould called liquid composite moulding (LCM). Replacement of synthetic fibres and reinforcements with natural plant fibres in the LCM processes significantly reduce the environmental influence; deliver supplementary economical, technological, ecological remunerations; and enhance the quality of composite parts prepared via customary open-mould processes. This technique offers a cost-effective substitute to prepreg techniques besides conserving quality of the major composite part (Walbran et al. 2013). A growing list of composite manufacturing processes through closed mould is listed below in Table 15.1.

In the compression moulding process, moulding of material in a confined dimension and shape is done by applying pressure and heat. When the mould gets occupied, curing of the thermoset polymer commences, and when the part gets hard (or stiff)

Table 15.1 The growing list of composite manufacturing processes through closed mould

Liquid-composite moulding (LCM)	Injection moulding (IM)
	Compression moulding (CM)
	Compression RTM (CRTM)
	Resin transfer moulding (RTM)
	RTM light moulding
	Resin infusion moulding (Vacuum Assisted RTM)
	Seemann composite resin infusion moulding process (SCRIMP)

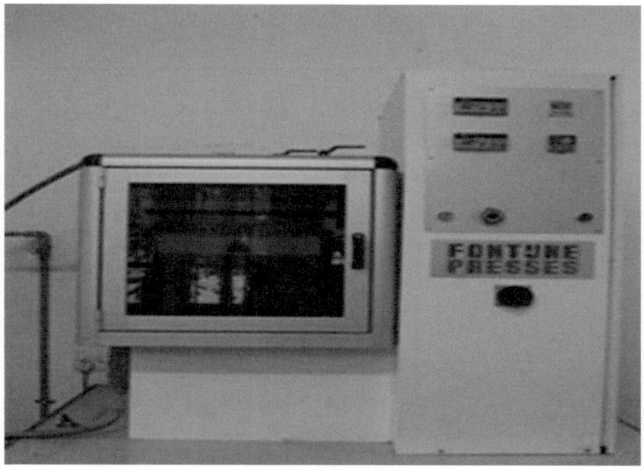

Fig. 15.2 Hot compression moulding machine

enough and is impassive from the mould, the curing ends. This process usually followed by the two step, involving preheating and pressurizing under hydraulic system. The hot compression moulding machine is shown in Fig. 15.2. Compression and flow compression moulding are the two commonly used methods which vary with respect to the types of semi-finished products used and its cutting. Hybrid fleeces and plane semi-finished products are frequently used in compression moulding process, whichever longer than the form or are expurgated precisely to desired size (Faruk et al. 2012).Usually resin is impregnated in the moulds by an alignment of vacuum-driven infusion and positive pressure, additionally conceivably assisted by compression-driven flow, which comprises thin, flexible, either semirigid or rigid membranes generally used in compression moulding (Bickerton and Abdullah 2003).

Currently the CM fabricating technique is extensively being used for fibre-reinforced polymer composite. In this technique, no damage to the fibre is observed as it is gently positioned within the mould without applying energetic motion and shear stress (Ho et al. 2012).Various researches have been focused on the probability of using cellulosic fibres as reinforcement mixed with renewable or biodegradable polymers to form biocomposites as a novel class of materials concluded by compression moulding process (Mir et al. 2013; Zhu et al. 2013).

15.2.4.2 Advantages

This closed mould pressure process is the alliance involving hot press and autoclave processes. The substantial advantages proposed by this process are comparable to open-mould manufacturing processes and hence are currently prevalent for the production of natural fibre composites and fibre-reinforced plastics (Walbran et al. 2013). Fibre mats or short fibres are premixed with the compounds for the fabrication of natural fibre composites in compression moulding; it would act as reinforcement to lower the contraction or shrinkages in composite materials (Ho et al. 2012). Some of the advantages are mentioned below:

- Reduced exposure to hazardous solvent emissions
- Produced products with better quality parts and properties
- Possess high reproducibility and low cycle time
- Shows economic and ecological benefits
- Damage of the fibre is minimal
- Composite having higher volume fracture is obtained from long fibres
- Suitable for thermoplastic, thermosetting and bio-polymeric materials

15.2.4.3 Applications

The compression moulding process is one of the utmost important techniques used for moulding the rubber into a variety of useful plastic products (door stopper, plastic rubber bands, popular cap, and jar cover designs of different sizes), as shown in Fig. 15.3. Extensive implementation of compression moulding process in appliance and automotive applications is governed by the production of high-strength sheet moulding

Fig. 15.3 Diverse plastic products made through hot compression moulding technique

compounds. Moreover, the turf of application broadens when the favourable understanding and control required for these processes expands (Walbran et al. 2013).

15.2.4.4 Improvement or Advancement

The compression moulding process can be improved to widen its applications through compression resin transfer moulding (CRTM) process, which pools the characteristics of CM with conventional RTM. In the production of high volume parts, CRTM manufacturing is regarded as advanced composite net shape process performing in a mould cavity. Figure 15.4 shows the scheme of various steps involved in CRTM process. The CRTM process is defined in three steps: (1) injection of resin into the gap, (2) gap closing, and (3) preform actual compression and resin distribution (Simacek et al. 2008). Some of the important research carried out on coir fibre reinforcement in polymeric resin by CM technique is summarized in Table 15.2.

15.2.4.5 Fabrication

Polypropylene composites reinforced with coir fibre were prepared using hot compression technique and a hot press machine, as shown in Fig. 15.2. The machine had two plates that could be operated separately. The pressure and temperature of both plates were set at required temperatures during composite manufacturing. The pressure application system was hydraulic. The total procedure of composite manufacturing is described below.

SST and DST coir fibre-reinforced composites were manufactured at 10, 15, and 20 wt.% fibre loading. Polypropylene was used as matrix material. The required coir

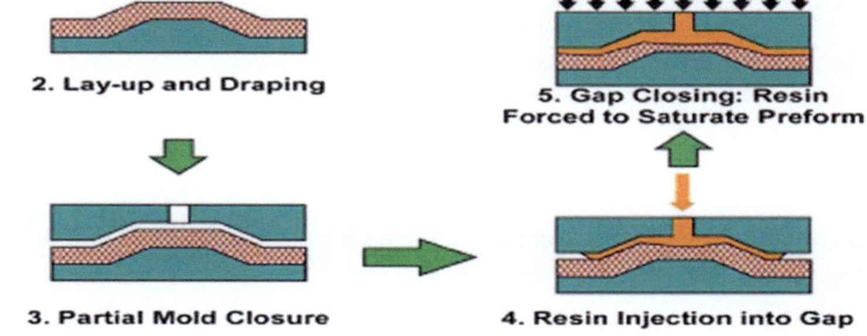

Fig. 15.4 Scheme of various steps involved in CRTM process

Table 15.2 Reported research work on the coir fibres reinforcement in polymeric resin by compression moulding process

Reinforcements/resin	Fabrication technique	Reference
Coir pith, nylon fabric/epoxy	Hand lay-up followed by compression moulding	Narendar et al. (2014)
Coir fiber/epoxy	Compression moulding	Romli et al. (2012)
Coconut (coir) fiber/wheat gluten	Compression moulding	Muensri et al. (2011)
Coir fiber/polyester	Compression moulding	Monteiro et al. (2008)
Untreated, treated coir fiber/ (PP)	Hot press/compression moulding	Mir et al. (2013)
Alkali treated coir fiber/poly (lactic acid)	Laminate fabrication	Dong et al. (2014)
Untreated, treated coconut sheath fiber/epoxy	Hand layup followed by compression moulding	Kumar et al. (2014)
Untreated, treated coconut fiber/wheat gluten	Compression moulding	Hemsri et al. (2012)
Modified coconut fiber/ polyester	Compression moulding	Mulinari et al. (2011)
Coir fiber/poly (lactic acid)	Commingled yarn method followed by compression moulding	Jang et al. (2012)
Chopped coconut fiber/low density polyethylene	Extrusion cum compression moulding process	Brahmakumar et al. (2005)
Coir fiber/poly (butylene succinate)	Hot press-compression moulding	Nam et al. (2011)
Coir fiber/(PP), (PVDF) (MAPP)	Compression moulding	Tran et al. (2013)
Coir fiber/(PP)	Compression moulding	Lai et al. (2005)
Coir fiber/(PP)	Single extruder and injection moulding machine.	Haque et al. (2010)
Alkali treated jute/coir fiber/ (PP)	Stacking followed by hot press	Hai et al. (2009)
Coir fiber/epoxy	Hand lay-up followed by hot compression moulding	Harish et al. (2009)
Coir fiber/maleated- and silanized-PP	Hot compression moulding	Santos et al. (2009)
Coir fiber/orthophthalic unsaturated polyester	Hot compression moulding	Yousif (2009)
Coir fiber/MMT nanoclay/ epoxy/(ENR)	Hand layup technique followed by compression moulding	Saw (2015)
Coir/alkali treated jute fiber/ (ENR)	Hand layup technique followed by compression moulding	Saw et al. (2012)
Coir/short bagasse fiber/ (ENR)	Hand lay-up followed by compression molding	Saw et al. (2011)
Coir fiber/shell particle/epoxy	Compression moulding	Kumar and Kumar (2012)
Coir yarn-jute fiber/(PP)	Compression moulding	Haydaruzzaman et al. (2010)
Coir/sisal fiber/(PP)	Fibers wounding onto metal plate followed by compression moulded (commingling technique)	Arya et al. (2015)

(continued)

Table 15.2 (continued)

Reinforcements/resin	Fabrication technique	Reference
Coir fiber/(PP)	Hot press	Ayrilmis et al. (2011)
Coir fiber/polyester	Compression moulding	Kuriakose et al. (2012)
Short coir/glass fiber/polyester	Mold with hydraulic press	Sindhu et al. (2007)

PP polypropylene, *PVDF* polyvinylidene fluoride, *MAPP* maleic anhydride-grafted polypropylene, *ENR* epoxy novolac, *MMT* montmorillonite

fibre was weighted. It was then cut into approximately 3 mm in size and dried in the oven at 100 °C for about 1 h for moisture evaporation. The required amount of commercial PP was weighted and dried in an oven at about 100 °C for 50 min for preventing voids, water bubbles, and poor fibre–matrix adhesion. The dried fibre and PP were uniformly mixed inside a beaker. Mould releasing agent was sprayed over carefully cleaned mould surfaces properly in order to removal of the composite easily. Picture of the mould is shown in Fig. 15.5. The uniform mixture of fibre and polypropylene was poured onto the female mould, whose depth was 3 mm. The female mould with randomly oriented fibre with polypropylene was covered by a male mould. Moulds were then placed in hot pressing machine in between the two hot press plates (Fig. 15.6) at 160 °C temperature, and 30 KN pressure was applied simultaneously for about 10 min. There were two conduction plates in between the mould and hot press plate. Then the temperature was increased to 180 °C and held for 10 min. After that, the whole system was slowly cooled using a water cooling system. Finally, the composite (Fig. 15.7) was discharged carefully from the mould.

15.2.5 Microstructural Analysis of Composites

FTIR spectra of raw, SST, and DST coir were taken under a Nicolet 380 spectrophotometer in the transmittance range of 370–4,000 cm^{-1}. A Philips XL 30 scanning electron microscope was employed for examining the interaction between coir and PP in manufactured composites. The coir composites were made conductive by applying gold coating with the help of sputtering technique. The gold coating propelled the electrons for interaction with the inner atomic shells of the composites.

15.2.6 Mechanical Testing of Composites

15.2.6.1 Tensile Test

Tensile properties of the composites were evaluated using a universal testing machine (Model: MSC-5/500, Agawn Seiki Company Limited, Japan) following ASTM D 638-01 (ADTM D 638-01).The crosshead speed used during the test was

Fig. 15.5 Pictures of mould used for composite manufacturing

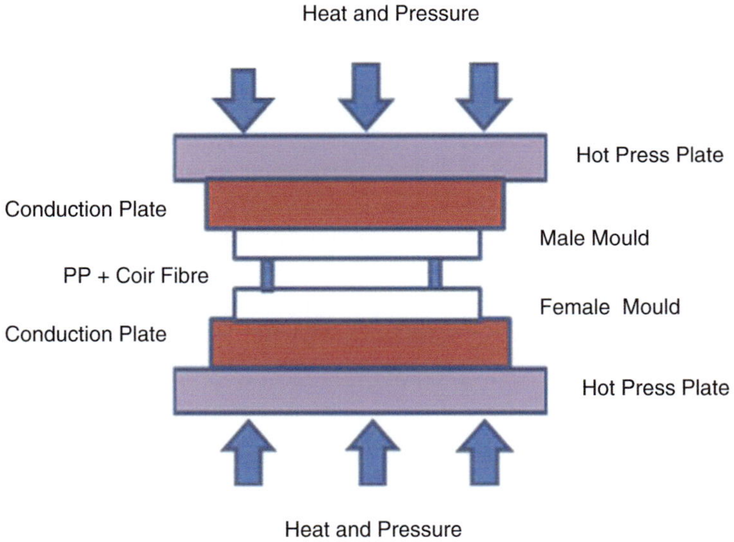

Fig. 15.6 Schematic representation of composite consolidation

Fig. 15.7 Picture of manufactured composites

10 mm/min. The composite specimen had the length, width, and thickness of 148 mm, 10 mm, and 4.1 mm, respectively.

15.2.6.2 Flexural Test

Static flexural test of the composites was carried out under the same testing machine using the same crosshead speed following ASTM D 790-00 (ADTM D 790-00). The specimen had a dimension of $79 \times 10 \times 4.1\ \text{mm}^3$. The flexural strength and flexural modulus were calculated using the following equations:

$$\text{Flexural strength}, \sigma_f = \frac{3PL}{2bd^2} \tag{15.1}$$

$$\text{Flexural modulus}, E = \frac{L^3 m}{4bd^3} \tag{15.2}$$

where P = maximum applied load, L = length of support span, m = slope of the tangent, b = width, and d = thickness of the specimen.

15.2.6.3 Impact Test

Dynamic Charpy impact test of the coir composites was conducted using a universal impact testing machine according to ASTM D 6110-97 (ASTM D 6110-97). Notched composite specimens of dimension of $79 \times 10 \times 4.1\ \text{mm}^3$ were used during the experiment.

15.2.6.4 Water Absorption Test

Rectangular composites with a dimension of $39 \times 10 \times 4.1$ mm^3 were prepared for measuring the water absorption characteristics. The composites were dried in an oven at 105 °C, cooled in a silica gel desiccator and immediately weighted. The dried and weighted composites were immersed in hot distilled water for 2 h according to ASTM D 570-99 (ASTM D 570-99). Excess water on the specimen surface was removed with soft cloth, and the specimens were weighted after immersion. The increase in the weight of the specimens was calculated using Eq. (15.3):

$$\text{Water absorption } (\%) = \frac{\text{Final Weight} - \text{Original Weight}}{\text{Original Weight}} \times 100 \qquad (15.3)$$

15.2.6.5 Hardness Test

A Shore hardness testing machine was utilized for measuring the hardness of the coir composites. The obtained hardness of the composites is described later.

15.3 Properties of Manufactured Composites

15.3.1 Tensile Properties

Variation of the tensile strength of raw, SST, and DST coir-polypropylene composites against fibre loading is shown in Fig. 15.8. The tensile strength decreased with fibre loading (Haque et al. 2009, 2010; Yang et al. 2004, 2006, 2007; Thwe and Liao 2002; Islam et al. 2010). Coir fibre and PP are, respectively, hydrophilic and hydrophobic in nature. Due to this reason, coir did not interact well with PP. The PP-coir interfacial area increased with fibre loading that resulted in decreased tensile strength. The tensile strength of all composites increased at 10 % fibre loading as compared to PP. After that, it decreased with further fibre loading. The chemical treatment involving coupling with basic chromium sulphate salt changed the structure of the cellulose unit of coir by reducing its hydroxyl groups. This in turn changed the overall tensile strength.

The above phenomenon is confirmed through FTIR analysis. The FTIR spectrum of treated coir indicates the absorption bands of –OH and C–O stretching near 3,479 cm^{-1} and cm^{-1}, respectively (Fig. 15.9). On the other hand, the FTIR spectrum of the raw coir indicates that there was an absorption band at the region of 1,732.7 cm^{-1}. This was due to carboxyl group of acetyl ester in cellulose and carboxyl aldehyde in lignin. The coupling reaction between hydroxyl groups of coir fibre and $CrSO_4$ salt (Fig. 15.10) increased the interaction and adhesion between the coir and PP. The cellulose anhydro-glucose unit of coir fibre generally contains

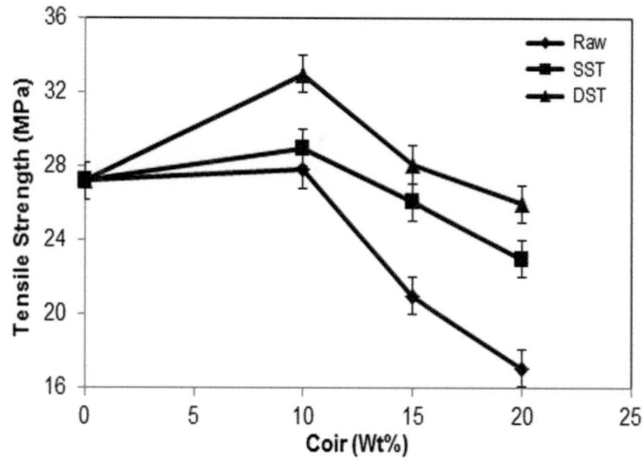

Fig. 15.8 Variation of tensile strength of PP composites reinforced with raw, SST, and DST coir

three hydroxyl groups. One is the primary hydroxyl group at C6, and the other two are secondary hydroxyl groups present at C2 and C3. Although C6 is more reactive than C2 and C3, the $CrSO_4$ salt broke the OH group of C6 and C2 during the reaction. The chrome complexes reacted with the coir cellulose carboxyl groups and shared the hydroxyl groups in SST with $CrSO_4$. The hydroxyl groups were fully shared by chromium atoms during DST with basic $CrSO_4$ and $NaHCO_3$. This resulted in a chrome product with cellulose by converting the hydroxyl groups into a chrome group (Figs. 15.9 and 15.10). Different degrees of coupling reaction occurred during SST and DST. Both of those treatments had common 3 h shaking in $CrSO_4$ reagent. Coir fibre mainly consists of cellulose and lignin. Cellulose is a hydrophilic glucan polymer containing hydroxyl groups, which reacted with chromium salt and changed the nature of coir from hydrophilic to hydrophobic. This in turn improved the interfacial adhesion between the coir and PP. Unlike cellulose, lignin does not possess any free hydroxyl groups for reaction with chromium salt. Two hours shaking in $NaHCO_3$ during DST dissolved the lignin component and increased the percentage of cellulose of coir. Sharing was incomplete during the initial stage in basic $CrSO_4$ media as the pH of the solution was 4. During DST, the reaction was completed as the addition of $NaHCO_3$ increased the pH to 10. As a result, treatment of 5 h during DST improved the interfacial bonding to the highest extent as compared to the basic $CrSO_4$ media treatment during SST. Thus, DST coir composites had the highest tensile strength, followed by SST and raw coir composites, respectively.

Figure 15.11 shows the values of the Young's modulus of various manufactured composites against fibre loading. Increase in fibre loading increased the Young's modulus of composites (Thwe and Liao 2002; Yang et al. 2004; Rana et al. 2003). Partially separated micro-spaces were created during tensile loading that obstructed stress propagation between PP and coir. The degree of obstruction increased with fibre loading that also increased the stiffness. The DST coir composites had the Young's modulus, followed by SST and raw coir fibre-reinforced PP composites, respectively.

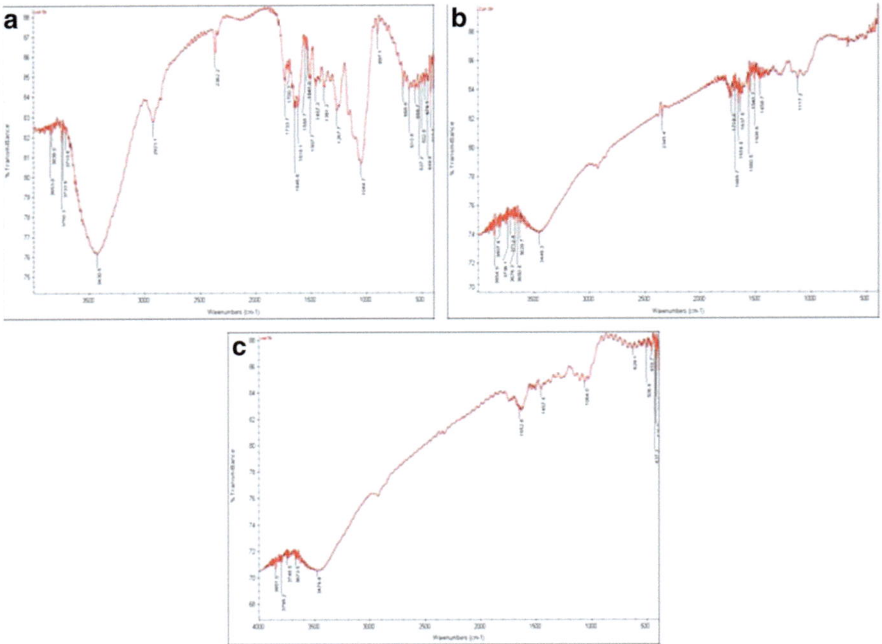

Fig. 15.9 FTIR spectra of (a) raw, (b) SST, and (c) DST coir fibre

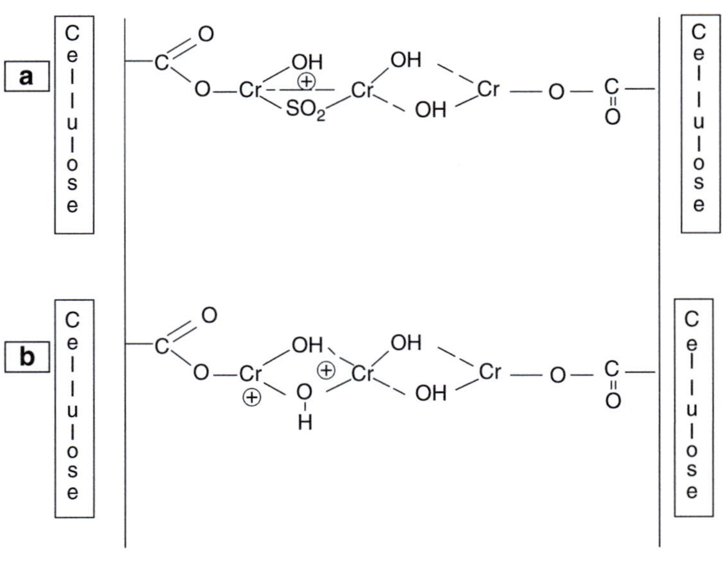

Fig. 15.10 Chemical reaction of cellulose of coir fibre with (a) basic $CrSO_4$ and (b) $NaHCO_3$

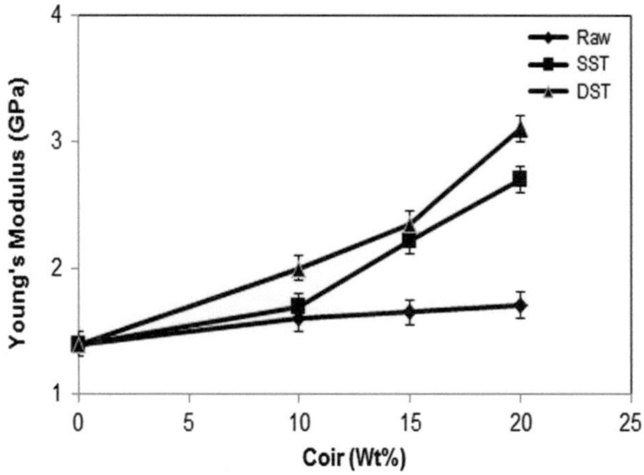

Fig. 15.11 Comparative Young's modulus of raw, SST, and DST coir fibre-reinforced PP composites

15.3.2 Flexural Properties

Figures 15.12 and 15.13 show the flexural properties of various manufactured composites against fibre loading. Higher fibre loading resulted in higher flexural strength. DST coir-reinforced composites had the highest flexural strength, followed by SST and raw coir composites, respectively. Polypropylene alone had the lowest flexural strength value. Like the flexural strength, higher fibre loading also increased the flexural modulus (Yang et al. 2004; Haque et al. 2010; Islam et al. 2010; Lin et al. 2006; Arrakhiz et al. 2013). Higher fibre content demanded higher stress for the same deformation due to the fact that coir is a higher modulus material. Thus, the addition of hard coir into the soft PP increased the flexural modulus. The DST coir composites had the highest flexural modulus, followed by SST and raw coir composites, respectively. SST and DST coir composites thus provide evidence of even coir distribution in PP and better coir–PP interaction.

15.3.3 Impact Strength

Figure 15.14 shows the variation of Charpy impact strength of various manufactured composites against fibre loading. Higher fibre loading increased the impact strength (Haque et al. 2009, 2010; Islam et al. 2010; Joseph et al. 2002; Romanzini et al. 2013). The impact strength of natural fibre-filled polymer composites mainly depends on the type of the fibre, polymer, and their interfacial interaction (Joseph et al. 2003). High fibre loading increases the probability of fibre agglomeration and formation of stress-concentrated regions, thus requiring less energy for crack propagation (Karmakar et al. 2007). Higher impact strength at higher fibre loading indicates that coir was capable of absorbing energy due to strong coir–PP interfacial bonding. Another reason of impact failure of a composite is fibre pull out. The impact strength of the DST coir composites was the highest, followed by SST and raw coir-reinforced composites, respectively.

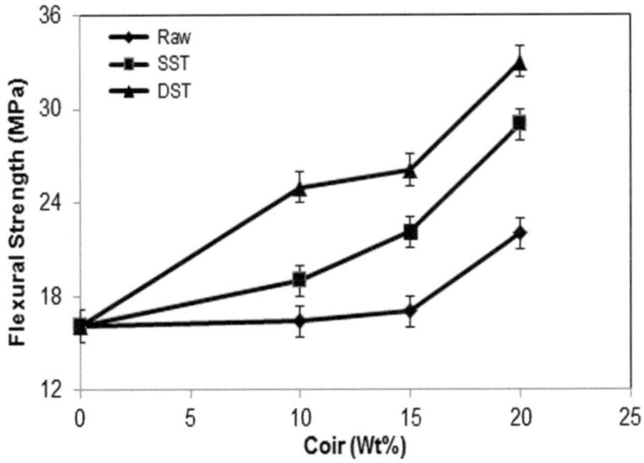

Fig. 15.12 Variation of flexural strength of PP composites reinforced with raw, SST, and DST coir

Fig. 15.13 Comparative flexural modulus of raw, SST, and DST coir fibre-reinforced PP composites

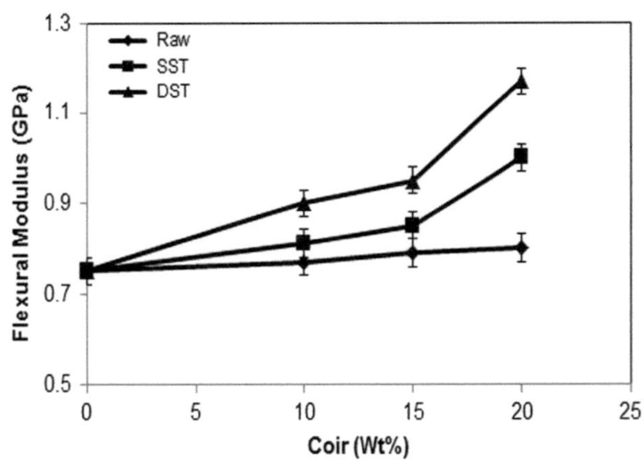

15.3.4 Hardness

Hardness is the resistance to deformation after application of force on it. Fibre distribution inside the matrix is the prime factor determining the hardness of a composite. Generally, the more flexible the matrix, the lower the hardness of composites (Premlal et al. 2002). Variation of hardness of various manufactured composites is shown in Fig. 15.15. The incorporation of raw, SST, and DST fibre into matrix reduced its flexibility that resulted in more rigid composites. The addition of coir into the PP also reduced the mobility of PP chain in the composites. As a result, SST and DST coir composites had higher hardness than the raw ones. This was due to better distribution and bonding of coir inside PP.

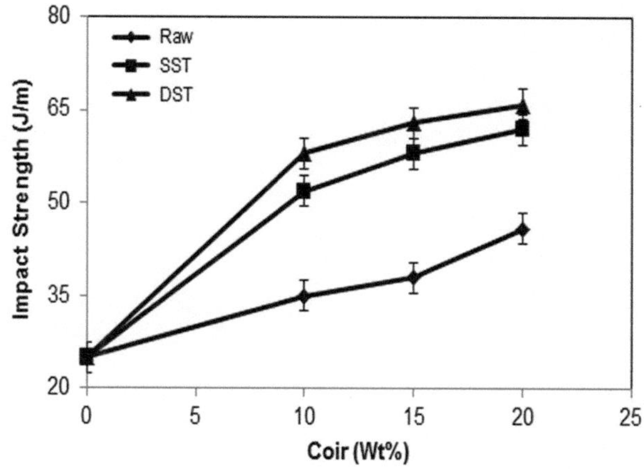

Fig. 15.14 Variation of impact strength of PP composites reinforced with raw, SST, and DST coir

15.3.5 Water Absorption Characteristics

Figure 15.16 presents the water absorption characteristics of various manufactured composites against fibre loading. Generally, raw natural fibre-reinforced polymer composites possess high water absorption as they contain a number of voids (Yang et al. 2007; Sanadi et al. 1995; Pasquini et al. 2008). Higher fibre loading increased the water absorption (%), as micro-voids and hydroxyl group (–OH) responsible for the water absorption increase with increase in fibre loading. Raw coir composites had the highest water absorption, followed by SST coir and DST coir–PP composites, respectively. Almost all hydrophilic –OH groups in the raw coir was fully converted to hydrophobic –O–Cr groups during DST. The DST coir-reinforced composites had the lowest water absorption, as hydrophobic –O–Cr groups have less affinity to water compared to –OH groups. However, SST only converted a few hydrophilic –OH groups into hydrophobic –O–Cr groups. As a result, water absorption of SST coir composites was higher than the DST composites. Micro-voids in the composites were substantially minimized in treated coir composites due to favourable interaction between coir and PP giving lower water uptake capacity.

15.3.6 SEM Morphology

Figure 15.17 shows the SEM morphology of tensile-fractured surfaces of 20 % coir-reinforced PP composites. A number of fibre agglomerations and fibre pullout marks are visible in the SEM image of the raw coir composite. This indicates poor interfacial bonding between PP and raw coir (Fig. 15.17a). Thus, raw coir fibres are clearly seen in the composite micrograph resulting in the worst set of mechanical properties for raw coir fibre-reinforced PP composite. Hydrophilic –OH groups in the raw coir were converted to hydrophobic groups during chemical treatment. As PP is hydrophobic, the interaction and adhesion between coir and PP increased

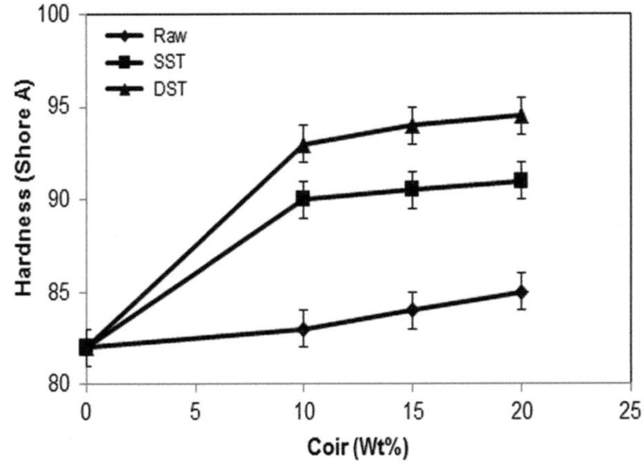

Fig. 15.15 Comparative hardness of raw, SST, and DST coir fibre-reinforced PP composites

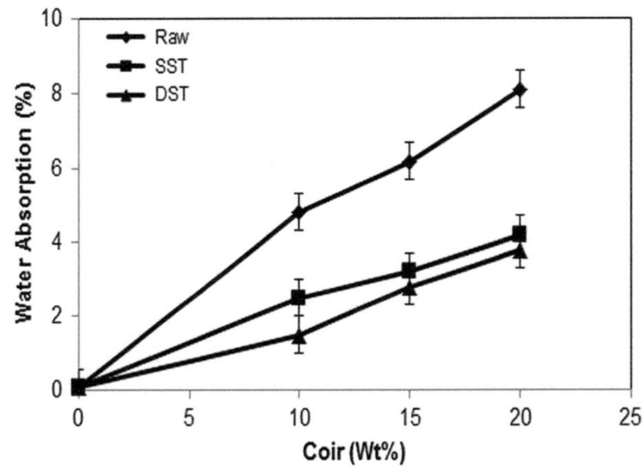

Fig. 15.16 Variation of water absorption of PP composites reinforced with raw, SST, and DST coir

after chemical treatment. Thus, the fibre and matrix are not clearly differentiable in SEM images of treated composites (Fig. 15.17b, c). Chemically treated composites also showed almost no signs of fibre agglomeration and micro-void formation. Furthermore, fibre pullout was substantially decreased in treated coir composites. Thus, the treated coir composites had better mechanical properties as compared to the raw coir composites.

15.4 Conclusions and Future Perspective

The present chapter describes hot press technique of coir fibre-reinforced polypropylene composite manufacturing. It also elaborates the effects of fibre treatment on the physico-mechanical properties of coir fibre–PP composites. The mechanical properties generally improved with fibre loading in coir–PP composites. The DST coir–PP composites had the best of mechanical properties, followed by SST and raw coir–PP

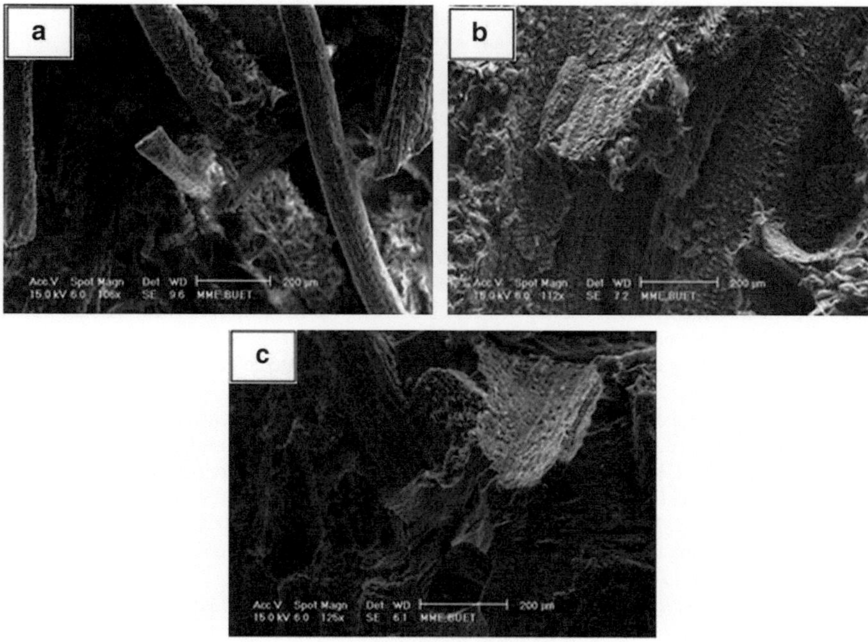

Fig. 15.17 SEM micrographs of 20 % (a) raw, (b) SST, and (c) DST coir fibre composites

composites, respectively. Water absorption behaviour also improved chemical treatment. This indicates basic $CrSO_4$ and $NaHCO_3$ treatment significantly reduced hydrophilic nature of coir. Besides, better interfacial interaction and minimization of micro-void at the coir–PP interface occurred upon chemical treatment. Overall the hot compression technique was successful in manufacturing good quality coir fibre-reinforced polymer composites.

References

Arrakhiz FZ, Achab ME, Malha M, Bensale MO, Fehri OF, Blouhfid R, Benmoussa K, Qaiss A (2013) Mechanical and thermal properties of natural fibers reinforced polymer composites: doum/low density polyethylene. Mater Des 43:200–205

Arya A, Tomlal JE, Gejo G, Kuruvilla J (2015) Commingled composites of polypropylene/coir-sisal yarn: effect of chemical treatments on thermal and tensile properties. e-Polym 15(3): 169–177. http://www.degruyter.com/

ASTM D 570-99 (2002) Standard test methods for water absorption of plastics. Annual Book of ASTM Standard, vol 08.01

ASTM D 6110-97 (2002) Standard test methods for determining the Charpy impact resistance of notched specimens of plastics. Annual Book of ASTM Standard, vol 08.03

ASTM D 638-01 (2002) Standard test methods for tensile properties of plastics. Annual Book of ASTM Standard, vol 08.03

ASTM D 790-00 (2002) Standard test methods for flexural properties of unreinforced and reinforced plastics and electrical insulating materials. Annual Book of ASTM Standard, vol 08.01

Ayrilmis N, Jarusombuti S, Fueangvivat V, Bauchongkol P, White RH (2011) Coir fiber reinforced polypropylene composite panel for automotive interior applications. Fiber Polym 12(7): 919–926

Bickerton S, Abdullah MZ (2003) Modelling and evaluation of the filling stage injection/compression molding. Compos Sci Technol 63:1359–1375

Brahmakumar M, Pavithran RM, Pillai RM (2005) Coconut fibre reinforced polyethylene composites: effect of natural waxy surface layer of the fibre on fibre/matrix interfacial bonding and strength of composites. Compos Sci Technol 65:563–569

Dong Y, Ghataura A, Takagi H, Haroosh HJ, Nakagaito AN, Lau KT (2014) Polylactic acid (PLA) biocomposites reinforced with coir fibres: evaluation of mechanical performance and multifunctional properties. Compos Part A 63:76–84

Faruk O, Bledzki AK, Fink HP, Sain M (2012) Biocomposites reinforced with natural fibers: 2000–2010. Prog Polym Sci 37:1552–1596

Hai NM, Kim BS, Lee S (2009) Effect of NaOH treatments on jute and coir fiber PP composites. Adv Compos Mater 18:197–208

Haque MM, Hasan M, Islam MS, Ali M (2009) Physico-mechanical properties of chemically treated palm and coir fiber reinforced polypropylene composites. Biores Technol 100:4903–4906

Haque MM, Islam MN, Hasan M, Islam MS, Islam MS (2010) Coir fiber reinforced polypropylene composites: physical and mechanical properties. Adv Compos Mater 19:91–106

Haque MM, Islam MS, Islam MS, Islam MN, Huque MM, Hasan M (2010) Physico-mechanical properties of chemically treated palm fiber reinforced polypropylene composites. J Reinf Plast Compos 29:1734–1742

Harish S, Michael DP, Bensely A, Lalb DM, Rajadurai A (2009) Mechanical property evaluation of natural fiber coir composite. Mater Charact 60:44–49

Haydaruzzaman, Khan AH, Hossain MA, Mubarak A, Khan MA, Ruhul A, Khan RA (2010) Mechanical properties of the coir fiber-reinforced polypropylene composites: effect of the incorporation of jute fiber. J Compos Mater 44(4):401–416

Hemsri S, Grieco K, Asandei AD, Parnas RS (2012) Wheat gluten composites reinforced with coconut fiber. Compos Part A 43:1160–1168

Ho MP, Wang H, Lee JH, Ho CK, Lau KT, Leng J, Hu D (2012) Critical factors on manufacturing processes of natural fibre composites. Compos B 43:3549–3562

Islam MN, Rahman MR, Haque MM, Huque MM (2010) Physico-mechanical properties of chemically treated coir reinforced polypropylene composites. Compos A 41:192–198

Jang JY, Jeong TK, Oh HJ, Youn JR, Song YS (2012) Thermal stability and flammability of coconut fiber reinforced poly (lactic acid) composites. Compos Part B 43:2434–2438

Joseph S, Sreekala MS, Oommen Z, Koshy P, Thomas S (2002) A comparison of mechanical properties of phenol formaldehyde composites reinforced with banana fibers and glass fibers. Compos Sci Technol 62:1857–1868

Joseph PV, Mathew G, Joseph K, Groeninckx G, Thomas S (2003) Dynamic mechanical properties of short sisal fiber reinforced polypropylene composite. Compos A 34:275–290

Karmakar A, Chauhan SS, Modak JM, Chanda M (2007) Mechanical properties of wood-fiber reinforced polypropylene composites. Compos A 38:227–233

Kumar S, Duraibabu SM, Subramanian DK (2014) Studies on mechanical, thermal and dynamic mechanical properties of untreated (raw) and treated coconut sheath fiber reinforced epoxy composites. Mater Design 59:63–69

Kumar S, Kumar B (2012) Study of mechanical properties of coconut shell particle and coir fibre reinforced epoxy composite. Int J Adv Eng Res 4(2):39–62

Kuriakose S, Varma D, Vaisakh VG (2012) Int J Emerging Technol Adv Eng 2(12):751–757

Lai CY, Sapuan SM, Ahmad M, Yahya N, Dahlan KZHM (2005) Mechanical and electrical properties of coconut coir fiber-reinforced polypropylene composites. Polym Plast Technol 44(4): 619–632

Lin JC, Chan LC, Nie MN, Ho HL (2006) Mechanical behavior of various nanoparticle filled composites at low-velocity impact. Compos Struct 27:30–36

Mir SS (2012) Manufacturing and characterization of chemically treated coir fiber reinforced polypropylene composites. Bachelor thesis, Bangladesh University of Engineering and Technology, Bangladesh

Mir S, Nafsin N, Hasan M, Hasan N, Hassan A (2013) Improvement of physico-mechanical properties of coir-polypropylene biocomposites by fiber chemical treatment. Mater Des 52:251–257

Monteiro SN, Terrones LAH, D'Almeida JRM (2008) Mechanical performance of coir fiber/polyester composites. Polym Test 27:591–595

Muensri P, Kunanopparat T, Menut P, Siriwattanayotin S (2011) Effect of lignin removal on the properties of coconut coir fiber/wheat gluten biocomposite. Compos Part A 42:173–179

Mulinari DR, Baptista CARP, Souza JVC, Voorwald HJC (2011) Mechanical properties of coconut fibers reinforced polyester composites. Procedia Eng 10:2074–2079

Nam TH, Ogihara S, Tung NH, Kobayashi S (2011) Effect of alkali treatment on interfacial and mechanical properties of coir fiber reinforced poly (butylene succinate) biodegradable composites. Compos Part B 42(6):1648–1656

Narendar R, Dasan KP, Nair M (2014) Development of coir pith/nylon fabric/epoxy hybrid composites: mechanical and ageing studies. Mater Design 54:644–651

Pasquini D, Teixeria EDM, Curvelo AADS, Belgacem MN, Dufresne A (2008) Surface esterification of cellulose fibres: processing and characterization of low-density polyethylene/cellulose fibres composite. Compos Sci Technol 68:193–201

Premlal HGB, Ismail H, Baharin A (2002) A comparison of the mechanical properties of rice husk powder filled polypropylene composites with talc filled polypropylene composites. Polym Test 21:833–839

Rana AK, Mandal A, Bandyopadhyay S (2003) Short jute fiber reinforced polypropylene composites: effect of compatibiliser, impact modifier and fiber loading. Compos Sci Technol 63:801–806

Romanzini D, Lavoratt A, Ornaghi HL, Amico SC, Zatte AJ (2013) Influence of fiber content on mechanical and dynamic mechanical properties of glass/ramie polymer composites. Mater Design 47:9–15

Romli FI, Alias AN, Rafie ASM, Abdul Majid DLA (2012) Factorial study on the tensile strength of a coir fiber reinforced epoxy composite. AASRI Procedia 3:242–247

Sanadi AR, Caulfield DF, Jacobson RE, Rowell RM (1995) Renewable agricultural fibers as reinforcing fillers in plastics: mechanical properties of kenaf fiber-polypropylene composites. Ind Eng Chem Res 34:1889–1896

Santos EF, Mauler RS, Nachtigall SMB (2009) Effectiveness of maleated- and silanized-PP for coir fiber-filled composites. J Reinf Plast Compos 28(17):2119–2129

Saw SK (2015) Static and dynamic mechanical analysis of coir fiber/montmorillonite nanoclay-filled novolac/epoxy hybrid nanocomposites. Adv Structural Mater 75:137–154

Saw SK, Sarkhel G, Choudhury A (2011) Dynamic mechanical analysis of randomly oriented short bagasse/coir hybrid fibre reinforced epoxy novolac composites. Fibers Polym 12:506–513

Saw SK, Sarkhel G, Choudhury A (2012) Preparation and characterization of chemically modified jute-coir hybrid fiber reinforced epoxy novolac composites. J Appl Polym Sci 125:3038–3049

Simacek P, Advani SG, Lobst LA (2008) Modeling flow in compression resin transfer molding for manufacturing of complex lightweight high-performance automotive parts. J Compos Mater 42:2523–2545

Sindhu K, Joseph K, Joseph JM, Mathew TV (2007) Degradation studies of coir fiber/polyester and glass fiber/polyester composites under different conditions. J Reinf Plast Compos 26(15): 1571–1585

Thwe MM, Liao K (2002) Effects of environmental aging on the mechanical properties of bamboo-glass fiber reinforced polymer matrix hybrid composites. Compos A 33:43–52

Tran LQN, Fuentes CA, Dupont-Gillain C, Vuure AWV, Verpoest I (2013) Understanding the interfacial compatibility and adhesion of natural coir fibre thermoplastic composites. Compos Sci Technol 80:23–30

Walbran W, Bickerton S, Kelly P (2013) Evaluating the shear component of reinforcement compaction stress during liquid composite moulding processes. J Compos Mater 47:513–528

Westerlind BS (1988) Surface energy of untreated and surface-modified cellulose fibre. J Appl Polym Sci 36:523–534
Yang HS, Kim HJ, Park HJ, Lee BJ, Hwang TS (2004) Rick-husk flour filled polypropylene composites; mechanical and morphological study. Compos Struct 63:305–312
Yang HS, Kim HJ, Park HJ, Lee BJ, Hwang TS (2006) Water absorption behavior and mechanical properties of lignocellulosic filler-polyolefin bio-composites. Compos Struct 72:429–437
Yang HS, Kim HJ, Park HJ, Lee BJ, Hwang TS (2007) Effect of compatibilizing agents on rice husk flour reinforced polypropylene composites. Compos Struct 77:45–55
Yousif BF (2009) Frictional and wear performance of polyester composites based on coir fibres. Proc Inst Mech Eng Part J: J Eng Tribol 223(1):51–59
Zhu J, Njuguna J, Abhyankar H, Zhu H, Perreux D, Thiebaud F, Chapelle D, Pizzi A, Sauget A, Larmina A, Nicollin A (2013) Effects of fibre configurations on mechanical properties of flaz/tannin composites. Indus Crops Prod 50:68–76

Chapter 16
Manufacturing of Natural Fibre-Reinforced Polymer Composites by Solvent Casting Method

I. Kong, K.Y. Tshai, and M. Enamul Hoque

Abstract Globally increasing environmental concern of petroleum-based material leads to finding the alternative renewable natural sources. Natural fibre-based composite is gaining immense interest not only because of its positive environmental impact but also its economic advantages. One of the very first and simplest processing techniques that have been used for preparing natural fibre-reinforced polymer composites is solvent casting method. In practice, the major advantage of solvent casting is its ease of fabrication without the need of specialized equipment. There are several factors that may influence solvent casting method and hence, the performance of the overall polymer composites. The present chapter provides a comprehensive overview on the manufacturing of natural fibre-reinforced polymer composites by solvent casting method. It comprises information on the factors that influence the method and the properties of the natural fibre-reinforced polymer composites prepared by this method as well as the possible applications.

Keywords Solvent casting • Natural fibre • Mechanical properties • Thermal properties

Abbreviations

DMA	Dynamic mechanical analysis
DMF	*N*, *N*-dimethylformamide
DSC	Differential scanning calorimetry

I. Kong (✉) • K.Y. Tshai
Department of Mechanical, Materials and Manufacturing Engineering, University of Nottingham Malaysia Campus, Jalan Broga, Semenyih 43500, Selangor, Malaysia
e-mail: ing.kong@nottingham.edu.my

M. Enamul Hoque
Department of Biomedical Engineering, University of King Faisal, Al-Hofuf, Al-Hassa, Saudi Arabia

M.S. Salit et al. (eds.), *Manufacturing of Natural Fibre Reinforced Polymer Composites*, DOI 10.1007/978-3-319-07944-8_16

HDS	Hexadecyltrimethoxy-silanes
MEK	Methyl ethyl ketone
MIBK	Methyl isobutyl ketone
MPS	γ-Methacryloxypropyltrimethoxy
MRPS	γ-Mercaptoproyltrimethoxy
PCL	Poly(ε-caprolactone)
PEG	Polyethylene glycol
PHBV	Polyhydroxybutyrate-co-valerate
PLA	Poly(lactic acid)
PMMA	Poly(methyl methacrylate)
Poly(S-co-BuA)	Poly(styrene-co-butyl acrylate)
PVA	Poly(vinyl alcohol)
PVAc	Polyvinyl acetate
SEM	Scanning electron microscope
T_g	Glass-transition temperature
THF	Tetrahydrofuran
TMA	Thermomechanical analysis
WPU	Waterborne polyurethane

16.1 Introduction

In the past decades, the use of natural fibre as reinforcing materials in both thermoplastic and thermoset matrix composites has been intensively studied due to their specific properties and clearly positive environmental impact, like partial/total independence on nonrenewable energy/material sources, lower greenhouse gas emissions, increased energy recovery, and biodegradability. Besides, the natural fibres provide additional benefits such as relatively less costly, lightweight, recyclable, and so on (Saheb and Jog 1999; Wambua et al. 2003). Until recently, there has also been an increasing interest in the research and development of natural fibre-filled biocomposites due to the growing awareness in sustainability and environmental credentials among scientists, industrialists, and consumers (Habib et al. 2011; Le Duigou et al. 2014; Gallo et al. 2013). Biopolymers are naturally occurring polymeric biomolecules that are found in all living organisms. They can be derived from plants and animals such as polysaccharides, polypeptides, lipids, and their composites (Phisalaphong et al. 2008). With the increasing concerns on global warming, price of crude oil, and depleting natural resources, natural fibre-based composites appear as an attractive alternative to synthetic fibre-reinforced polymer composites for use in various fields (Khwaldia et al. 2004).

The properties of natural fibre-reinforced composites can vary depending on the type and the treatment or modification to the fibre, resin matrix, material compositions, environmental parameters, and the processing methods for the composites (Faruk et al. 2012). A large proportion of research efforts have been done to investigate the effect of the abovementioned factors on the properties of natural fibre-reinforced composites (Messiry 2013; Shibata et al. 2005; Venkateshwaran et al. 2012; Kabir et al. 2012; Summerscales and Grove 2014; Rippon and Evans 2012). Among the various factors, processing methods have significant influences on the final properties of the composites. The affinity to moisture and the tendency for agglomeration during processing are the major concerns about the use of natural fibres as reinforcements in polymers (Saheb and Jog 1999; Abdelmouleh et al. 2007).

As a contribution to updating the state of knowledge, this chapter aims to provide the readers an overview on one of the very first and simplest processing techniques that has been used for preparing natural fibre-reinforced polymer composites, referred to as the solvent casting method. The chapter is organized as follows:

In the first section, a brief overview on the standard solvent casting method to fabricate polymer composites along with its advantages and drawbacks is discussed. The second section details the factors that influence the solvent casting process, which include but not limited to the choices of solvents, polymer molecular weight, etc. This is followed by the properties of polymer composites prepared by solvent casting method. The discussion will somewhat focus on mechanical, thermal, and morphology properties. Last but not least, some applications of natural fibre-reinforced polymer composites with emphasis on their properties are briefly outlined.

16.2 Solvent Casting

Solvent casting, sometimes called solution casting or wet processing method, has been popular over the years due to its simplicity and involvement of nonspecialized equipment. Generally, solvent casting is a manufacturing process that involves mixing of solubilized polymer matrix and filler under continuous agitation through mechanical stirring, followed by casting and solvent evaporation or drying steps. It has been extensively used to prepare coatings in a wide range of industrial packaging (Salehifar et al. 2013) and biological applications (Lieder et al. 2013; Yamauchi et al. 1996; Katoh et al. 2004). In the standard process of solvent casting as shown in Fig. 16.1, the first step involves preparation of polymer solution by dissolving polymer (in any form) in an appropriate solvent. Alcohol, water, or any organic solvent is commonly used in this processing method. To enhance the formation and to improve the properties of film, often the dissolved polymer solution is heated or the pH is adjusted. The resulting polymer solution is then cast into a mold or on a heated drum or onto a flat surface thus producing a film matrix, while the solvent is removed. The produced film may be subjected to thermal treatment for stress removal.

There are several prerequisites for the raw materials of the polymer solution casting process. One of the most important requirements is that the polymer must be soluble in a volatile solvent or water. To achieve better results, a stable solution with reasonable minimum solid content and viscosity should be formed. Another consideration is the possibility to form a homogeneous film and to release the film from the casting support.

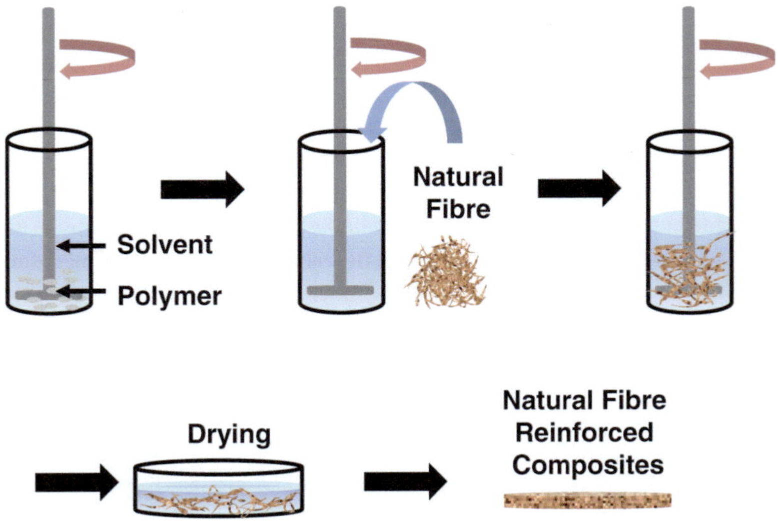

Fig. 16.1 Process of solution casting to fabricate polymer composites

16.2.1 Behavior and Mechanisms of Dissolution of Polymer

In general, there are two transport processes involved in the dissolution of a polymer into a solvent, namely, the disentanglement of polymer chains and the solvent diffusion through the outer layer at the polymer–solvent interface. The solvent will diffuse into the polymer when an uncross-linked (which is amorphous or glassy) polymer is in contact with a thermodynamically compatible solvent (Fig. 16.2). A gel-like swollen layer is produced in addition to two separate interfaces because of the plasticization effect of the polymer by the solvent. One interface is formed between the gel layer and glassy polymer, and the other interface is in between the solvent and the gel layer. As the process further progresses, diffusion of polymer chains into solvent reservoir takes place and the polymer eventually dissolves.

The first surface layer formation process was outlined in 1968 (Ueberreiter 1968). According to Ueberreiter, the solvent starts its aggression through pushing the swollen polymer into the solvent. With the progress of time, a more dilute upper layer is produced and moves toward the solvent stream direction. Continuous insertion of the solvent into the solid polymer further extends the swollen surface layer until the swelling time ends by reaching a quasi stationary state. Thus, further increase of the layer is restricted as the transportation of the macromolecules from the surface into the solution is prevented. During the dissolution process, the surface layers of glassy polymers are structured in between the pure polymer and pure solvent following a sequence such as infiltration layer, solid swollen layer, gel layer, and liquid layer (as shown in Fig. 16.3). A polymer in its glassy phase contains numerous channels and holes of molecular dimensions, thus having free volume. The diffusion process starts with the penetration of solvent molecules into these empty spaces forming the

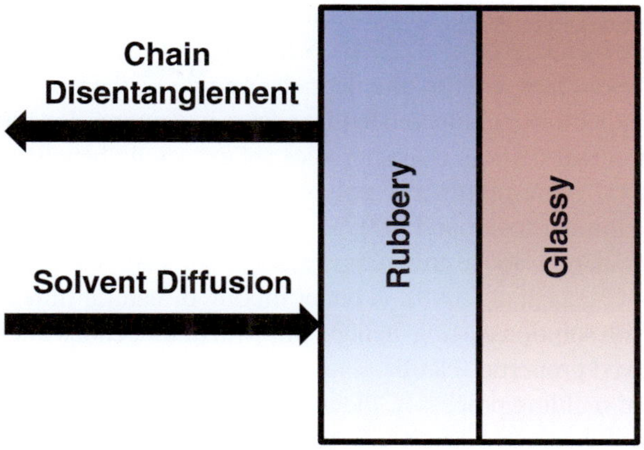

Fig. 16.2 A schematic of one-dimensional solvent diffusion and polymer dissolution (adapted from Miller-Chou and Koenig 2003)

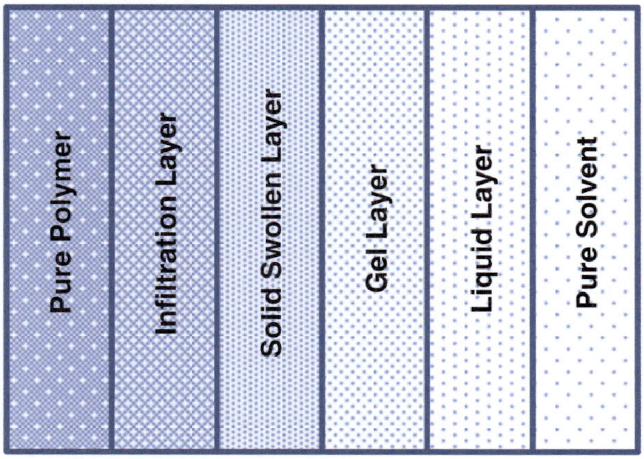

Fig. 16.3 A schematic of the composition of the surface layer (adapted from Miller-Chou and Koenig 2003)

first layer (i.e., infiltration layer). The next solid swollen layer is still in the glassy state followed by the gel layer, which is in a rubbery state, and the last layer is in the liquid state that encompasses every solid part of the polymer.

16.2.2 Advantages and Shortcomings

The prime advantage of solution casting is its ease of fabrication without the need of very specialized equipment (Gregory et al. 2001) that are essentially required for the extrusion or injection molding processes. The solution casting process also avoids fibre damage that usually happens in the melt-mixing process, while the fibre and polymer are blended (Li et al. 2000). Since the solvent casting fabrication process occurs at room temperature, the degradation rate of the solvent-cast polymer composites is rather lower compared to that occurs in compression molding or melt-compression processes. These characteristics are especially beneficial for the fabrication of natural fibre-reinforced polymer composites where the processing temperature is limited to around 200 °C as the lignocellulosic materials exhibit significant degradation above this temperature (e.g., 230 °C) (Azizi Samir et al. 2005; Sanchez-Garcia et al. 2008). A better mixing of natural fibre and polymer is also possible with solution casting, hence, film with homogenous thickness distribution and improved properties can be obtained (Dufresne et al. 1997). Compared to melt mixing and molding processes, the mixing of natural fibre in polymer solution greatly improved the degree of fibre dispersement. In order to further improve the fibre/polymer compatibility and their interfacial adhesion, the natural fibre can be pre-wetted in the same solvent used to dissolve the polymer prior to the incorporation into polymer solution.

Difficulties with solution casting usually are associated with the large amounts of organic solvent required to dissolve polymer, which is one of the reasons that this method is not ideal for industrial scale-up. In addition, the possibility of toxic solvent retention within the polymer is also a critical issue in solution casting technique. This can be overcome by total evaporation of residual solvent by allowing the polymer to completely dry under a vacuum environment. However, this process is very time-consuming. Another concern is that the use of solvents in such system often poses the risk of denaturation of drugs or proteins incorporated into the polymer, though this can also occur in film fabricated by other techniques such as compression molding. The denatured species are therapeutically inactive and might induce harmful side effects including immunogenicity etc. (Makadia and Siegel 2011). Since solution casting is not a continuous process, potential increases of batch-to-batch variation is expected. In cost-performance comparison, this method requires extra energy costs of solvent recovery and investments in facilities for handling solvent solutions. Despite the disadvantages, solution casting method with the highlighted benefits is still preferred among researchers mainly due to the performance and the high quality of solvent-cast films which cannot be achieved using other processes.

16.3 Factors That Influence Solvent Casting Method

The interactions between polymer and natural fibre are critically important in forming a continuous three-dimensional network for effective fabricating of a cohesive film. The interaction depends upon the polymers involved and the film-forming conditions. There are several aspects that play significant roles in polymer dissolution, which include the polymer molecular weight, stereochemistry, tacticity, polymer structure, composition and conformation, types of solvents and additives, environmental parameter, and processing conditions such as drying temperature, drying rate, agitation, and stirring frequency of the solvent. Among them, the effect of different solvents as well as the polymer molecular weight will be further discussed. The solvent nature influences the polymer structure in solution that could affect the final physical and mechanical properties and the aging characteristics of the produced polymer film.

16.3.1 Effect of Choices of Solvent

The polymer chain configuration in solution is governed by the solvent. Generally, solvents can be classified by their intermolecular force. If in a solution, the polymer–polymer attraction forces are lower than the polymer–solvent interactions, the chain segment will be extended by diffusion of solvent molecules, and thus inducing swelling by increasing in the free volume of the polymer matrix and loosening out from the coiled shape. On the contrary, contraction will occur (Fig. 16.4). Solvent with the capability to promote greater chain extension is considered a more effective solvent.

From the thermodynamics viewpoint, the Gibbs free energy of mixing controls the solvent dissolution of polymer; hence, such dissolution becomes a function dependent on the changes in enthalpy and entropy of the system (Grulke 1999). A spontaneous mixing process occurs when the overall Gibbs free energy is negative; otherwise, the mixing process may result in two or more phases. If the molecular weight is higher, the entropic driving force contribution for dissolution will be weaker

Fig. 16.4 The effects of solvent quality on the conformation of a polymer molecule in solution

(Holmberg et al. 2003). However, polymer dissolution is typically controlled by kinetics rather than thermodynamics, and an effective solvent must therefore be capable to overcome the low entropy gain by favorable solvent/polymer interactions.

Cooper et al. (1986) investigated the dissolution rate of poly(methyl methacrylate) (PMMA) film in the solvent that contained methyl ethyl ketone (MEK) as the major component. Besides, methanol, ethanol, 1-propanol, 2-propanol, ethylene glycol, and water were used as minor components. A significant increase in the dissolution rate of PMMA film was observed when small non-solvent molecules (water and alcohols are non-solvent for PMMA) were added to a good solvent (e.g., MEK). It was suggested that the improvement in dissolution rate was due to the fact that the small and rapidly diffusing non-solvent molecules induced plasticizing effect to the polymer films. It was also noticed that the film dissolution rate was reduced by the higher alcohols.

Ouano and Carothers (Ouano and Carothers 1980) used a critical angle microscopy technique to demonstrate the kinematics of polymer dissolution in real time. The dissolution of PMMA was studied using a number of solvents that include methyl acetate, methyl isobutyl ketone (MIBK), and tetrahydrofuran (THF). It was found that the cracks formed faster in the case of the smaller and stronger solvents (THF and methyl acetate) than in the case of a bulky and poorer solvent (MIBK) due to the higher diffusion rates and thus swelling power of these smaller solvent molecules.

Tang and coworkers (Tang et al. 2004) investigated the biocompatibility and surface properties of solvent-cast poly(ε-caprolactone) (PCL) films. Four solvent systems were used, namely, THF, acetone, chloroform, and ethyl acetate. The advancing contact angle for the surface exposed to atmosphere of the hydrophobic solvent systems (ethyl acetate and chloroform) during casting was observed to be higher than that for the surface in contact with the glass Petri dish, while with hydrophilic solvent systems (THF and acetone), the opposite phenomenon happened.

The dissolution properties of one polymer might vary in different solvents, and also the dissolution properties of different polymers might vary in the same solvent. Therefore, the selection of solvent deserves a careful consideration.

16.3.2 Effect of the Polymer Molecular Weight

Besides the solvent type, the polymer molecular weight can also influence the dissolution process. It would be worthy to comprehend how molecular weight affects penetration rate. It was observed that the dissolution rate of PMMA films in MIBK decreased with the increase of polymer molecular weight (Manjkow et al. 1987). Cooper et al. (1986) reported that the molecular weight affected the dissolution rates of PMMA films. For example, polymers with higher molecular weights dissolved more slowly compared to the polymers with lower molecular weights. It was considered that the longer chain in higher molecular weight polymer experiences a much slower relaxation rate due to greater chain entanglement, which retards flow and formation of the advancing gel boundary.

Papanu et al. (1990) studied how the swelling of PMMA thin film in low molecular weight alcohols is influenced by the polymer molecular weight. It was observed that the lower molecular weight polymer had a lower free volume and thus a lower dissolution rate. This indicated that the penetration rate significantly depended on molecular weight. The increase in penetration rate at higher molecular weight was considered to be because of the increased free volume that occurred from the inability of the higher molecular weight chains to contract rapidly during the ambient quench.

Hansen et al. (1991) studied the effects of solution on physical and mechanical properties of thermoplastic amorphous polymers used in conservation. They concluded that lower molecular weight polyvinyl acetate (PVAc) was more often used in conservation with the understanding that the lower molecular weight polymers caused less coiling and uncoiling of polymer chains.

16.4 Natural Fibre-Reinforced Polymer Composites

A natural fibre-reinforced polymer composite with outstanding combination of properties can be prepared by using appropriate processing method(s), compatibilizers/coupling agents, and fibre treatments. Owing to the abovementioned advantages, numerous studies on the fabrication of natural fibre-reinforced polymer composites through solution casting were reported. Table 16.1 summarizes the reported work on natural fibre composites.

16.5 Performance of Natural Fibre-Reinforced Polymer Composites

16.5.1 Mechanical Properties

The investigation of mechanical properties of natural fibre-reinforced polymer composites are mainly performed in the nonlinear (e.g., tensile tests) and linear range (e.g., dynamic mechanical analysis, DMA). DMA allows the determination of materials' mechanical characteristics in broad temperature/frequency range, which are closely related to the composites' morphological properties.

Dufresne et al. (1997) investigated the effect of processing technique on the mechanical properties of poly(S-co-BuA)/straw cellulose whiskers. It was found that the composite fabricated by solution casting method had superior mechanical characteristics compared to that produced by hot-pressing and freeze-drying with the same mixture. The solution cast samples showed a more pronounced reinforcing effect which could be explained by two origins: (a) the creation of a whisker network within the polymer matrix through the formation of strong interaction between the whiskers and (b) the deposition of the cellulose whiskers during evaporation, as shown by wide-angle X-ray scattering, scanning electron microscopy, and DMA.

Table 16.1 Reported work on natural-fibre reinforced polymer composites by solution casting method

Fibre	Matrix	Solvent	References
Wheat straw	Poly(styrene-co-butylacrylate) (poly(S-co-BuA))	Water	Dufresne et al. (1997)
Silk	Microcrystalline cellulose	Water	Noishiki et al. (2002)
Alfa (esparto grass)	Polyisoprene	Water	Abdelmouleh et al. (2007)
Pine			
Technocel-50 (commercial microcrystalline fibre)			
Technocel-2500			
Avicel (purified plant fibre)			
Avicel (purified plant fibre)	Natural rubber	Toluene	Ly et al. (2008)
Cotton linter	Polyvinyl acetate (PVAc)	Water	Roohani et al. (2008)
	PVA		
Purified cellulose	Polyhydroxybutyrate-co-valerate (PHBV)	Chloroform	Sanchez-Garcia et al. (2008)
	Polylactic acid (PLA)		
	PCL		
Microcrystalline cellulose	Polyvinyl alcohol (PVA)	Water	Lee et al. (2009)
Flax yarn	PLA	Chloroform	Liu et al. (2010)
Avicel (purified plant fibre)	PHBV	*N*, *N*-dimethylformamide (DMF)	Ten et al. (2010), Ten et al. (2012), and Jiang et al. (2008)
Hemp	PLA	Chloroform	Mustapa et al. (2013a, b)
Kenaf	Cellulose acetate	Acetone, dichloromethane	Pang et al. (2013)

Noishiki et al. (2002) prepared silk fibroin-microcystalline cellulose composite films by solution casting technique. They observed that the mixing ratio had direct influence on the composite's Young's modulus, while showing a maximum tensile strength of approximately 160 MPa at 70–80 % cellulose content. This could be due to the formation of β-structure of fibroin caused by contact with the highly ordered surface of cellulose whiskers.

Abdelmouleh et al. (2007) examined the effect of silane coupling agents and fibre loading on the natural rubber composites reinforced with short natural fibre and prepared by solution casting method. Four different cellulose fibres (e.g., Avicel, Technocel-50, Technocel-2500, alfa pulps and pine fibres) with different average lengths were utilized. The delignified cellulose fibres were treated with three different silane coupling agents such as γ-mercaptoproyltrimethoxy (MRPS), γ-methacryloxypropyltrimethoxy (MPS), and hexadecyltrimethoxy-silanes (HDS) that differ by the functionality of the radical moiety appended to silicon atoms, in order to improve their adhesion to polymer matrix. It was observed that the stiffness was higher for the composites reinforced with longer fibre (2.5 and 3.5 mm), and the mechanical properties prepared with MPS and MRPS displayed good mechanical performance. A year later, Thielemans et al. (Ly et al. 2008) prepared the same Avicel-reinforced natural rubber composites through film casting. The analysis of mechanical properties of the composites demonstrated that the storage modulus of the initial matrix was increased significantly from 0.5 to 3 MPa due to the incorporation of 30 % microcrystalline cellulose particles. The treatment of Avicel with MPS further improved the storage modulus (3.8 MPa at 30 %). The quality of the fibres/matrix interface was also improved by the chemical modification of Avicel that eventually enhanced the composite's mechanical properties.

Roohani et al. (2008) studied the mechanical and thermal properties of nanocomposite film produced from copolymer of PVA and PVAc reinforced with cotton cellulose whiskers. The length and average diameter of the cotton whiskers were approximately 171.6±48.2 and 14.6±3.9 nm, respectively. Whisker content and relative humidity of Young's modulus for nanocomposites were illustrated in Fig. 16.5. For each relative humidity condition, the storage modulus of the PVA matrix was enhanced by the incorporation of cotton whiskers. At a specific filler content, increase in the humidity caused a significant decrease in tensile modulus due to the lower glass-transition temperature, which is below room temperature.

In recent studies, Ten et al. (2010, 2012) observed that the composites fabricated by solution casting method showed good cellulose nanowhisker (CNW) dispersion and good mechanical properties compared to melt-processed composites which demonstrated reduced mechanical properties due to severe CNW agglomeration (Jiang et al. 2008). Both bulge test and tensile test were performed to characterize the mechanical properties of CNW-reinforced polymer composites. Bulge testing is a common method to investigate the mechanical performance of thin films such as

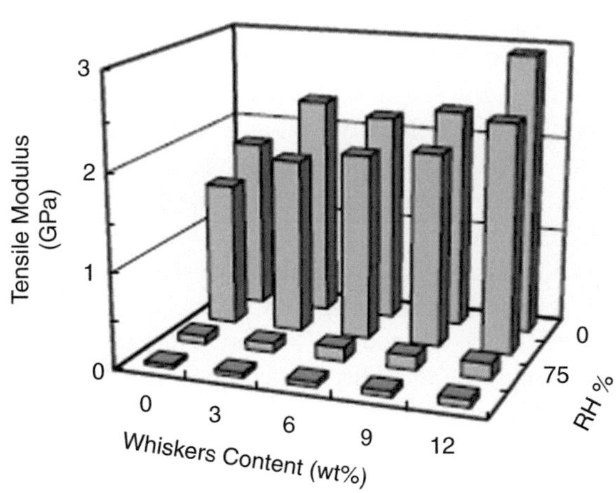

Fig. 16.5 Evolution of the Young's modulus as a function of the whisker content and relative humidity for cellulose whisker reinforced PVA copolymer nanocomposites (adapted from Roohani et al. 2008)

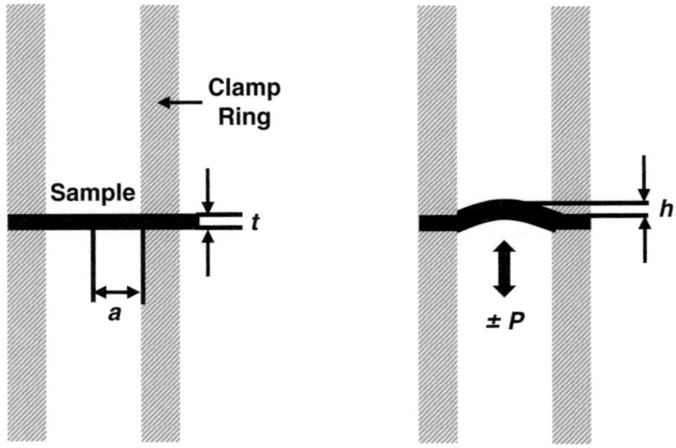

Fig. 16.6 Schematic representation of bulge test setup

elastic modulus and residual stresses as shown in Fig. 16.6 where, *P* represents gas pressure, *a* is bulge radius, *h* is the deflection, and *t* is the sample thickness. The results obtained from these two tests reasonably agreed each other. It has been previously proven that both tensile and bulge test results complement each other. The tensile test characterizes the linear and nonlinear behaviors, while the bulge test measures residual stresses (Edwards et al. 2004).

16.5.2 Thermal Properties

The glass-transition temperature, T_g, being an important parameter controls numerous properties of the composite, namely, its swelling behavior, matrix chains dynamics, and mechanical characteristics (Azizi Samir et al. 2005). Three different techniques can be used to measure T_g, e.g., differential scanning calorimetry (DSC), thermomechanical analysis (TMA), and DMA. DSC is considered to be the most commonly used technique for measuring T_g for most of the polymer composites. In DSC experiments, T_g is taken as the inflection point of the specific heat increment without latent heat of fusion at the transition of glass to rubber phase. DMA is probably the most sensitive technique for T_g analysis. The measured T_g by DMA is reported though three typical approaches, e.g., (a) peak of the tan delta curve, (b) onset of the storage modulus curve, and (c) peak of the loss modulus curve. All these approaches are practically viable though the different techniques may produce different results.

Thermal analysis of cellulose whisker-reinforced PVA copolymer nanocomposites was performed using DSC (Sanchez-Garcia et al. 2008). At different atmosphere (0, 35, 75, and 98 % relative humidity (RH)), the T_g values were found to decrease as the humidity content increases regardless of the composition of cellulose. This can be attributed to the well-known plasticizing effect of water molecules of PVA chains. Another interesting finding from this study was that, for moist samples (98 % RH), T_g increases significantly upon whisker addition regardless of the degree of hydrolysis of the matrix. It is probably due to the competitive interactions between the PVA, water, and cellulose whisker surface interfaces. The formation of a water layer at the interface resulted in a less plasticized PVA matrix by water.

Ten et al. (2010, 2012) investigated the effect of CNW content on the thermal properties of CNW composite film using both DSC and DMA techniques. It has been observed that the addition of ≤2.3 wt% CNW increased T_g by 4 °C. At higher CNW content (≥2.9 wt%), T_g decreased and was only 1.5 °C higher than that of the neat PHBV. The T_g of the composites indicated by the peak of tan δ curve was shifted to higher temperatures because of the restrained mobility of PHBV chains. Similar results were observed in their previous study (Ten et al. 2010). This could be explained that at temperatures above T_g, PHBV chains in the amorphous regions of the polymer were allowed for long-range cooperative motion. The higher transition temperatures indicate that higher energy is essential for such motion of the chains. This energy requirement change is probably because of the restraint of the PHBV and CNW spherulites, whose size decreased significantly upon incorporation of CNW.

Mustapa et al. (2013a, b) prepared hybrid composites of PLA with natural fibre (hemp) and inorganic filler (silica) and investigated the thermal properties of the composites. The T_g was determined using the loss modulus curves that demonstrated how the hemp fibre-PLA matrix affected on mobility of PLA chains as shown in Fig. 16.7. It was observed that with addition of silica the T_g of PLA composites was changed to a higher temperature. With the increase of silica content, the T_g increased and the highest T_g for PLA-hemp-nanosilica composite were found with 7.5 wt% of nanosilica at 59.0 °C compared to that of 57.4 °C for composite with nanosilica content of 2.5 wt% and 57.0 °C for PLA-hemp composite. Matthew et al. (2006) mentioned that the shifting of T_g to a higher temperature generally suggested limited molecular movement due to stronger fibre-matrix interaction. Besides, the α-relaxation involved polymer chain movements. The reinforcement behaves as physical cross-links leading to the formation of rigid amorphous phase that decreases the mobility and thus increases the stiffness of the composite.

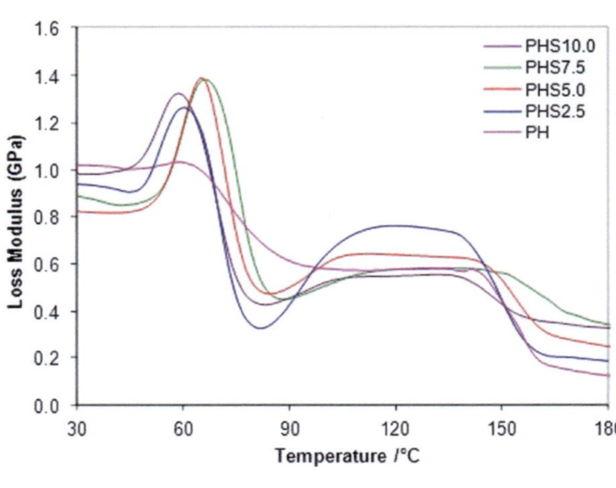

Fig. 16.7 Loss modulus of PLA-hemp and PLA-hemp-nanosilica composites (adapted from Mustapa et al. 2013a)

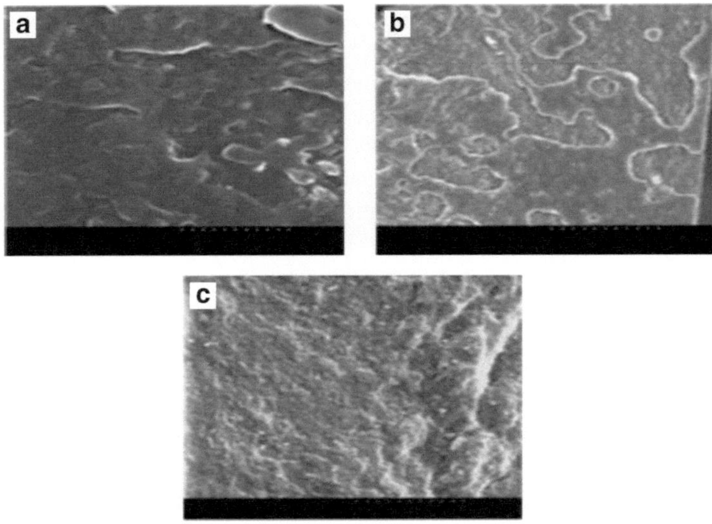

Fig. 16.8 SEM micrographs of the fracture surfaces of PHBV/CNW composites with (**a**) 0, (**b**) 2, and (**c**) 5 wt% of nanowhiskers prepared by solution casting (adapted from Jiang et al. 2008)

16.5.3 Morphology

Morphological investigations are most commonly performed by the scanning electron microscope (SEM) that often involves observation of fractured surfaces under liquid nitrogen environment. This investigation mainly observes the homogeneity of the composite, fibre configuration, aggregate dispersions, presence of voids, and so on. The natural fibre can easily be identified through comparison of the micrographs of fractured surfaces of neat polymer and composite. The SEM images of solution-casted (Fig. 16.8) and injection-molded (Fig. 16.9) composites based on PHBV and CNW demonstrated that more homogeneous whisker dispersion was obtained by the former process (Jiang et al. 2008).

16.6 Potential Applications

Natural fibre-reinforced polymer (thermoplastic or thermoset) composites have been used in many engineering applications that require high strength-to-weight ratio especially in the automotive and aerospace industries (Riedel and Nickel 2005). Many of the industrial composites have been prepared by the techniques that include extrusion, compounding, kneading, compression molding, injection molding, etc., whereas the use of solution casting method in preparing the industrial composites remains rather less explored. However, some applications of the solution-casted composites could be suggested as the homogeneous fibre dispersion can be achieved through this technique, which is essentially important for the enhancement of composite properties.

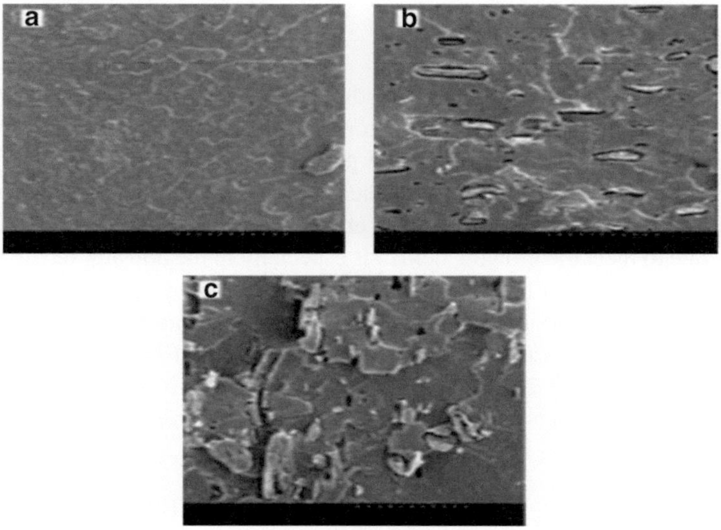

Fig. 16.9 SEM micrographs of the fracture surfaces of PHBV/CNW composites with (**a**) 0, (**b**) 2, and (**c**) 5 wt% nanowhiskers prepared by extrusion/injection molding (adapted from Jiang et al. 2008)

Table 16.2 Examples of interior and exterior automotive parts produced from natural materials

Vehicle part	Materials used
Interior	
Glove box	Wood/cotton fibres molded, flax/sisal
Door panels	Flax/sisal with thermoset resin
Seat coverings	Leather/wool backing
Seat surface/backrest	Coconut fibre/natural rubber
Trunk panel	Cotton fibre
Trunk floor	Cotton with PP/PET fibres
Insulation	Cotton fibre
Exterior	
Floor panels	Flax mat with PP

Source: Adapted from Holbery and Houston (2006)

16.6.1 Automotive

Natural fibre (kenaf, jute, hemp, flax, sisal, etc.)-reinforced polymer composites have been used as replacement of greater proportion of the mineral and glass filler-reinforced composites in numerous exterior and interior automotive parts. Table 16.2 summarizes various examples of exterior and interior automotive parts that are produced from numerous natural materials.

16.6.2 Construction

The application of natural fibre-reinforced composites is not limited to the automotive industry. With the progressive environmental awareness, the tendency toward utilization of eco-friendly materials and products is ever increasing. The government along with individuals is putting much emphasis on the quality of the products by reducing its economic, social, and environmental impacts, while increasing delivery rate to meet the growing demands especially in the housing sector. Roofing panels consist of natural fibre and polymer resin that can be used to substitute corrugated iron roofing or traditional plantation roofing. The natural fibre sandwich has improved thermal and acoustic insulation properties and it is cheaper to manufacture (Riedel and Nickel 2005). Other typical applications include building panels, roofing sheets, partition boards, doors, windows, tiles, and so on (Singh et al. 2011).

16.6.3 Biomedical

Nanocellulose and CNW are fillers that could be easily dispersed in polymer matrix by solution casting technique. Matthew and Oksman (2010) studied the potential of cellulose nanofibre-based material as tendon and ligament substitute. This study indicated that the nanofibres derived from softwood pulp by mechanical fibrillation make cellulose-based nanocomposites an excellent candidate for load-bearing components in biomedical applications. The cellulose-based nanocomposites inherit numerous favorable biomedical properties that include low or no toxicity, biocompatibility, and biodegradability with superior biomechanical properties.

16.7 Future Perspective

As mentioned earlier in Sect. 16.2.2, one of the main drawbacks of using solution casting method to produce natural fibre-reinforced composites is that this method involves a large amount of organic solvent to dissolve polymer. This makes it to be a less preferred method compared to extrusion, compression molding, and injection molding among the composite manufacturers. However, the high-quality plastic films can easily be manufactured by solution casting method rather than any other technique. The use of solution casting method for large-scale industrial production is still under progression. In the near future, the production of composites using solution casting method could be directed toward the development of high-volume continuous solution cast technology for the production of films with extremely high quality. The idea of replacing toxic solvents with greener alternatives (nontoxic, benign to environment and human health) might help to promote solution casting method as eco-friendly and competitive.

16.8 Conclusions

Solution casting being one of the oldest polymer-based composite processing techniques offers the unique process of drying the fabricated composite's surface without applying further mechanical or thermal stress. With careful selection of solvent and controllable fabrication technique, a natural fibre-reinforced composite with excellent properties can be obtained. The real challenge for the scientist is to improve the solution casting technology resolving the constraints associated with large-scale manufacturing as well as the environmental issues that arise from the use of toxic solvents.

References

Abdelmouleh M, Boufi S, Belgacem MN, Dufresne A (2007) Short natural-fibre reinforced polyethylene and natural rubber composites: effect of silane coupling agents and fibres loading. Compos Sci Technol 67(7–8):1627–1639. doi:10.1016/j.compscitech.2006.07.003

Azizi Samir MAS, Alloin F, Dufresne A (2005) Review of recent research into cellulosic whiskers, their properties and their application in nanocomposite field. Biomacromolecules 6(2): 612–626. doi:10.1021/bm0493685

Cooper WJ, Krasicky PD, Rodriguez F (1986) Dissolution rates of poly(methyl methacrylate) films in mixed solvents. J Appl Polym Sci 31(1):65–73. doi:10.1002/app.1986.070310107

Dufresne A, Cavaillé J-Y, Helbert W (1997) Thermoplastic nanocomposites filled with wheat straw cellulose whiskers. Part II: effect of processing and modeling. Polym Compos 18(2): 198–210. doi:10.1002/pc.10274

Edwards RL, Coles G, Sharpe WN Jr (2004) Comparison of tensile and bulge tests for thin-film silicon nitride. Exp Mech 44(1):49–54. doi:10.1007/BF02427976

Faruk O, Bledzki AK, Fink H-P, Sain M (2012) Biocomposites reinforced with natural fibers: 2000–2010. Prog Polym Sci 37(11):1552–1596. doi:10.1016/j.progpolymsci.2012.04.003

Gallo E, Schartel B, Acierno D, Cimino F, Russo P (2013) Tailoring the flame retardant and mechanical performances of natural fiber-reinforced biopolymer by multi-component laminate. Compos Part B Eng 44(1):112–119. doi:10.1016/j.compositesb.2012.07.005

Gregory ER, Cheryl AM, Surya KM (2001) Processing of polymer scaffold: solvent casting. In: Anthony A, Robert PL (eds) Methods of tissue engineering. Academic Press, California, pp 681–686

Grulke EA (1999) Solubility parameter values. In: Brandrup J, Immergut EH, Grulke EA (eds) Polymer handbook. Wiley-Interscience, New York

Habib M, Guessasma S, Bassir D, Benseddiq N (2011) Interfacial damage in biopolymer composites reinforced using hemp fibres: finite element simulation and experimental investigation. Compos Sci Technol 71(11):1419–1426. doi:10.1016/j.compscitech.2011.05.015

Hansen EF, Derrick MR, Schilling MR, Garcia R (1991) The effects of solution application on some mechanical and physical properties of thermoplastic amorphous polymers used in conservation: poly(vinyl acetate)s. J Am Inst Conserv 30(2):203–213. doi:10.1179/019713691806066764

Holbery J, Houston D (2006) Natural-fiber-reinforced polymer composites in automotive applications. JOM 58(11):80–86. doi:10.1007/s11837-006-0234-2

Holmberg K, Jönsson B, Kronberg B, Lindman B (2003) Introduction to surfactants. In: Holmberg K, Jönsson B, Kronberg B, Lindman B (eds) Surfactants and polymers in aqueous solution, 2nd edn. Wiley, West Sussex

Jiang L, Morelius E, Zhang J, Wolcott M, Holbery J (2008) Study of the poly(3-hydroxybutyrate-co-3-hydroxyvalerate)/cellulose nanowhisker composites prepared by solution casting and melt processing. J Compos Mater 42(24):2629–2645. doi:10.1177/0021998308096327

Kabir MM, Wang H, Lau KT, Cardona F (2012) Chemical treatments on plant-based natural fibre reinforced polymer composites: an overview. Compos Part B Eng 43(7):2883–2892. doi:10.1016/j.compositesb.2012.04.053

Katoh K, Shibayama M, Tanabe T, Yamauchi K (2004) Preparation and physicochemical properties of compression-molded keratin films. Biomaterials 25(12):2265–2272. doi:10.1016/j.biomaterials.2003.09.021

Khwaldia K, Perez C, Banon S, Desobry S, Hardy J (2004) Milk proteins for edible films and coatings. Crit Rev Food Sci Nutr 44(4):239–251. doi:10.1080/10408690490464906

Le Duigou A, Baley C, Grohens Y, Davies P, Cognard J-Y, Créach'cadec R, Sohier L (2014) A multi-scale study of the interface between natural fibres and a biopolymer. Compos Part A Appl Sci Manuf 65(0):161–168. doi:10.1016/j.compositesa.2014.06.010

Lee S-Y, Mohan DJ, Kang I-A, Doh G-H, Lee S, Han S (2009) Nanocellulose reinforced PVA composite films: effects of acid treatment and filler loading. Fibers Polym 10(1):77–82. doi:10.1007/s12221-009-0077-x

Li Y, Mai Y-W, Ye L (2000) Sisal fibre and its composites: a review of recent developments. Compos Sci Technol 60(11):2037–2055. doi:10.1016/S0266-3538(00)00101-9

Lieder R, Darai M, Orlygsson G, Sigurjonsson O (2013) Solution casting of chitosan membranes for in vitro evaluation of bioactivity. Biol Proc Online 15(1):11. doi:10.1186/1480-9222-15-11

Liu DY, Yuan XW, Bhattacharyya D, Easteal AJ (2010) Characterisation of solution cast cellulose nanofibre – reinforced poly(lactic acid). Express Polym Lett 4(1):26–31. doi:10.3144/expresspolymlett.2010.5

Ly B, Thielemans W, Dufresne A, Chaussy D, Belgacem MN (2008) Surface functionalization of cellulose fibres and their incorporation in renewable polymeric matrices. Compos Sci Technol 68(15–16):3193–3201. doi:10.1016/j.compscitech.2008.07.018

Makadia HK, Siegel SJ (2011) Poly lactic-co-glycolic acid (PLGA) as biodegradable controlled drug delivery carrier. Polymers 3(3):1377–1397. doi:10.3390/polym3031377

Manjkow J, Papanu JS, Hess DW, Soane DS, Bell AT (1987) Influence of processing and molecular parameters on the dissolution rate of poly-(methyl methacrylate) thin films. J Electrochem Soc 134(8):2003–2007. doi:10.1149/1.2100807

Mathew AP, Oksman K (2010) Cellulose nanofiber based composites for use as ligament or tendon substitute. In: International conference on nanotechnology for the forest products industry 2010, Otaniemi, Espoo, Finland, 2010. TAPPI Press, Norcross, pp 980–982

Mathew AP, Oksman K, Sain M (2006) The effect of morphology and chemical characteristics of cellulose reinforcements on the crystallinity of polylactic acid. J Appl Polym Sci 101(1):300–310. doi:10.1002/app.23346

Messiry ME (2013) Theoretical analysis of natural fiber volume fraction of reinforced composites. Alexandria Eng J 52(3):301–306. doi:10.1016/j.aej.2013.01.006

Miller-Chou BA, Koenig JL (2003) A review of polymer dissolution. Prog Polym Sci 28(8):1223–1270. doi:10.1016/S0079-6700(03)00045-5

Mustapa IR, Shanks RA, Kong I (2013a) Melting behaviour and dynamic mechanical properties of poly(lactic acid) hemp nanosilica composites. Asian Trans Basic Appl Sci 3(2):29–37. doi:ATBAS-30320028

Mustapa IR, Shanks RA, Kong I (2013b) Poly(lactic acid)-hemp-nanosilica hybrid composites: thermomechanical, thermal behavior and morphological properties. Int J Adv Sci Eng Technol 3(1):192–199

Noishiki Y, Nishiyama Y, Wada M, Kuga S, Magoshi J (2002) Mechanical properties of silk fibroin–microcrystalline cellulose composite films. J Appl Polym Sci 86(13):3425–3429. doi:10.1002/app.11370

Ouano AC, Carothers JA (1980) Dissolution dynamics of some polymers: solvent-polymer boundaries. Polym Eng Sci 20(2):160–166. doi:10.1002/pen.760200208

Pang C, Shanks RA, Ing K, Daver F (2013) Plasticised cellulose acetate-natural fibre composite. World J Eng 10(5):405–409. doi:10.1260/1708-5284.10.5.405

Papanu JS, Hess DW, Soane DS, Bell AT (1990) Swelling of poly(methyl methacrylate) thin films in low molecular weight alcohols. J Appl Polym Sci 39(4):803–823. doi:10.1002/app.1990.070390404

Phisalaphong M, Suwanmajo T, Sangtherapitikul P (2008) Novel nanoporous membranes from regenerated bacterial cellulose. J Appl Polym Sci 107(1):292–299. doi:10.1002/app.27118

Riedel U, Nickel J (2005) Applications of natural fiber composites for constructive parts in aerospace, automobiles, and other areas. In: Biopolymers online. Wiley-VCH Verlag GmbH & Co. KGaA. doi:10.1002/3527600035.bpola001

Rippon JA, Evans DJ (2012) Improving the properties of natural fibres by chemical treatments. In: Kozlowski R (ed) Handbook of natural fibres: processing and applications, vol 2. Woodhead Publishing, Cambridge, UK, pp 63–140

Roohani M, Habibi Y, Belgacem NM, Ebrahim G, Karimi AN, Dufresne A (2008) Cellulose whiskers reinforced polyvinyl alcohol copolymers nanocomposites. Eur Polym J 44(8):2489–2498. doi:10.1016/j.eurpolymj.2008.05.024

Saheb DN, Jog JP (1999) Natural fiber polymer composites: a review. Adv Polym Technol 18(4):351–363. doi:10.1002/(sici)1098-2329(199924)18:4<351::aid-adv6>3.0.co;2-x

Salehifar M, Beladi Nejad MH, Alizadeh R, Azizi MH (2013) Effect of LDPE/MWCNT films on the shelf life of Iranian Lavash bread. Pelagia Res Libr 3(6):183–18

Sanchez-Garcia MD, Gimenez E, Lagaron JM (2008) Morphology and barrier properties of solvent cast composites of thermoplastic biopolymers and purified cellulose fibers. Carbohydr Polym 71(2):235–244. doi:10.1016/j.carbpol.2007.05.041

Shibata S, Cao Y, Fukumoto I (2005) Press forming of short natural fiber-reinforced biodegradable resin: effects of fiber volume and length on flexural properties. Polym Test 24(8):1005–1011. doi:10.1016/j.polymertesting.2005.07.012

Singh B, Gupta M, Tarannum H, Randhawa A (2011) Natural fiber-based composite building materials. In: Kalia S, Kaith BS, Kaur I (eds) Cellulose fibers: bio- and nano-polymer composites. Springer, Berlin, Heidelberg, pp 701–720. doi:10.1007/978-3-642-17370-7_24

Summerscales J, Grove S (2014) Manufacturing methods for natural fibre composites. In: Hodzic A, Shanks R (eds) Natural fibre composites: materials, processes and applications. Woodhead Publishing, Cambridge, UK, pp 176–215

Tang ZG, Black RA, Curran JM, Hunt JA, Rhodes NP, Williams DF (2004) Surface properties and biocompatibility of solvent-cast poly[ε-caprolactone] films. Biomaterials 25(19):4741–4748. doi:10.1016/j.biomaterials.2003.12.003

Ten E, Turtle J, Bahr D, Jiang L, Wolcott M (2010) Thermal and mechanical properties of poly(3-hydroxybutyrate-co-3-hydroxyvalerate)/cellulose nanowhiskers composites. Polymer 51(12):2652–2660. doi:10.1016/j.polymer.2010.04.007

Ten E, Bahr DF, Li B, Jiang L, Wolcott MP (2012) Effects of cellulose nanowhiskers on mechanical, dielectric, and rheological properties of poly(3-hydroxybutyrate-co-3-hydroxyvalerate)/cellulose nanowhisker composites. Ind Eng Chem Res 51(7):2941–2951. doi:10.1021/ie2023367

Ueberreiter K (1968) The solution process. In: Crank J, Park GS (eds) Diffusion in polymers. Academic Press, New York, pp 219–257

Venkateshwaran N, Elayaperumal A, Sathiya GK (2012) Prediction of tensile properties of hybrid-natural fiber composites. Compos Part B Eng 43(2):793–796. doi:10.1016/j.compositesb.2011.08.023

Wambua P, Ivens J, Verpoest I (2003) Natural fibres: can they replace glass in fibre reinforced plastics? Compos Sci Technol 63(9):1259–1264. doi:10.1016/S0266-3538(03)00096-4

Yamauchi K, Yamauchi A, Kusunoki T, Kohda A, Konishi Y (1996) Preparation of stable aqueous solution of keratins, and physiochemical and biodegradational properties of films. J Biomed Mater Res 31(4):439–444. doi:10.1002/(SICI)1097-4636(199608)31:4<439::AID-JBM1>3.0.CO;2-M

Chapter 17
Processability of Wood Fibre-Filled Thermoplastic Composite Thin-Walled Parts Using Injection Moulding

M.D. Azaman, S.M. Sapuan, S. Sulaiman, E.S. Zainudin, and A. Khalina

Abstract A review of research on injection-moulded wood fibre-filled thermoplastic composites is presented in this chapter. Brief description of injection-moulding compounding process (including drying and mixing and extrusion) as well as injection-moulding process itself is also presented. A review on wood fibre-reinforced thermoplastic composites is also reported. An in-depth discussion is presented on thin-part moulding and the formation of residual stresses, volumetric shrinkage and warpage using injection moulding for composite products. Simulation work and statistical analysis related to the topic are also reviewed. Finally, some ideas about further research on wood-filled thermoplastic composite are proposed.

Keywords Wood polymer composites • Thin-walled part • Injection moulding • Residual stresses • Shrinkages • Warpage

M.D. Azaman • E.S. Zainudin
School of Manufacturing Engineering, Universiti Malaysia Perlis, 02600 Arau, Perlis, Malaysia

Department of Mechanical and Manufacturing Engineering, Universiti Putra Malaysia (UPM), 43400 Serdang, Selangor, Malaysia

S.M. Sapuan (✉)
Department of Mechanical and Manufacturing Engineering, Universiti Putra Malaysia (UPM), 43400 Serdang, Selangor, Malaysia

Laboratory of Biocomposite Technology, Institute of Tropical Forestry and Forest Products (INTROP), Universiti Putra Malaysia (UPM), 43400 Serdang, Selangor, Malaysia
e-mail: sapuan@upm.edu.my

S. Sulaiman
Department of Mechanical and Manufacturing Engineering, Universiti Putra Malaysia (UPM), 43400 Serdang, Selangor, Malaysia

A. Khalina
Department of Biological and Agricultural Engineering, Universiti Putra Malaysia (UPM), 43400 Serdang, Selangor, Malaysia

Aerospace Malaysia Innovation Centre, Universiti Putra Malaysia (UPM), 43400 Serdang, Selangor, Malaysia

M.S. Salit et al. (eds.), *Manufacturing of Natural Fibre Reinforced Polymer Composites*, DOI 10.1007/978-3-319-07944-8_17

17.1 Introduction

Composites have become very important materials in the recent years due to advantages that they offered including light weight, part integration, comparable stiffness and strength properties to some metals and corrosion resistant (Sapuan and Abdalla 1998; Hambali et al. 2009, 2010). They have been widely used in many industries such as automotive, construction, marine, furniture, aerospace, telecommunications and sport industries. The development of green-composite engineering materials made from natural resources has rapidly increased in product development and manufacturing today due to people's awareness concerning environmental issues (Bachtiar et al. 2010; Leman et al. 2008; Rashdi et al. 2009; Ishak et al. 2010; Sapuan et al. 2005). Wood residue resources have a high potential for use as fillers in producing green composites, which have been extensively investigated over the past few years. However, limited research efforts were conducted on the processability and quality of parts after processing, especially in the injection-moulding process.

Processed natural fibres that are ready for compounding can be categorized into two categories: fibres and particulates. The difference between the two types can be noticed by the ratio of the length to diameter (normally known as aspect ratio) of the fibres. Particulates generally have fibre aspect ratio close to unity. For this type of fibre, generally strengthening mechanism is almost nonexistence; however, it has demonstrated good stiffness properties. Fibres can be short or long depending on whether or not the fibres are in the form of particulates or continuous fibres. This fibre exhibits more significant strengthening and has the potential to replace conventional fibres such as glass fibres as reinforcement in thermoplastic composites.

Thermoplastic materials are more promising than thermoset materials because of their lower melting point and they have greater potential to be recycled (Sanadi et al. 1998). The low melting points are necessary to minimize the fibre degradation when exposed to heat during the manufacturing processing (Sanadi et al. 1998). Therefore, the suitability of polymers as matrices for natural fibre depends on the thermal stability properties. Bledzki and Gassan (1999) showed that polyethylene (PE) and polypropylene (PP) are the most suitable compounds with wood flour/fibre, whose processing temperature does not exceed 230 °C.

17.2 Wood Fibre-Reinforced Thermoplastic Composites

Natural fibres are normally demonstrated comparable specific properties to conventional fibres like glass fibres, abundantly available and generally biodegradable that can be used to achieve durability without using toxic chemicals (El-Shekeil et al. 2012; Suriani et al. 2007; Sastra et al. 2006; Ishak et al. 2012; Yusoff et al. 2010). Furthermore, natural fibre thermoplastic composites can be used as alternatives to conventional fibre composites. They can be regarded as environmentally friendly materials. They are also known as 'eco-composites' or 'green composites', which is suitable for semi- or nonstructural applications.

From a technical viewpoint, natural fibre-reinforced thermoplastic composites exhibit higher specific strength and stiffness than glass, reduced material weight and specific weight, high dimensional stability and high possible filling levels and are nonabrasive in nature. From an ecological viewpoint, natural fibre thermoplastic composites offer low cost, safe fibre handling, unlimited sources, low energy consumption, a wide variety of fibre types (i.e. wood, kenaf, bamboo, etc.), biodegradability and annual renewability. The future prospects of wood-reinforced thermoplastic composites are generally determined by many parameters such as customer requirements, introduction of newer products, high-quality components and research and development findings that act as inputs to these factors. In addition, products being produced must be affordable, recyclable and environmentally friendly (Ishak et al. 2010).

According to La Mantia and Morreale (2011), wood flour and fibres have been widely used reinforcements in polymer composites. Wood flour can be made available at low cost and in abundance from sawmill wastes after undergoing sieving process. Wood fibres were normally obtained from thermo-mechanical processing of sawmill wastes. Other important features of wood flour and fibres include dimensional stability and good stiffness properties. However, they suffered from some drawbacks such as low-strength and poor impact properties, poor matrix and filler or fibre bonding and thermal decomposition at temperatures above 200 °C (Netravali and Chabba 2003; Carroll et al. 2001).

Li and Wolcott (2004) studied the behaviour of high-density polyethylene (HDPE) wood fibre composites with two different filler loadings (40 and 60 wt.%) and two different types of wood (maple and pine) using capillary rheometer. They found that wall slip phenomena existed and they were mainly related to filler loadings. The analysis of flow material flow showed that the viscosity depended mainly upon the filler loading.

Ashori (2008) revealed that the most important criterion affecting the strength of wood composites is fibre size. Short-length fibres with the average size of 0.24–0.35 mm were proposed. These fibres demonstrated larger specific surface area, and the fibres distribution seemed to be more homogeneous in comparison with long fibre composites, thus promoted better fibres and polymer interfacial bonding. In addition, fibre swelling and breakage were minimally found.

Wood residues are currently gaining importance worldwide. This processed wood is either flours which have normally length to diameter ratio of 1/1 to 5/1 or fibres which have the ratio of slightly above 10/1. Wood flour can be considered as filler rather than fibres in composites (Rangaprasad 2003).

La Mantia and Morreale (2011) briefly noted some recent trends in natural fibre composites:

1. The development of 'green composites' with focus on making fully 'green' materials by having biodegradable polymers as matrices in place of synthetic polymers. Their findings are supported by other researchers (Netravali and

Chabba 2003; Lodha and Netravali 2002; Marsh 2003; Gould 2002; Nishino et al. 2003; Lee and Wang 2006; Bax and Müssig 2008; Huda et al. 2005; Plackett et al. 2003).
2. The problem concerning dispersion of fibres in matrices and polymer-fibre interfacial bonding (Scaffaro et al. 2009; Morreale et al. 2008a, b; Willett 1994; Coutinho et al. 1997; Nickel and Riedel 2003; Li et al. 2001).
3. The development of nanofillers as replacements for traditional micro fillers, thus to create 'green nanocomposites' field (Samir et al. 2005; Hubbe et al. 2008; Eichhorn et al. 2009).

Concerning the second issue, the problem in the manufacturing process should be considered as one of the worst problems, especially for producing products using natural fibre-reinforced thermoplastic composites. Researchers must have in-depth knowledge and the ability to predict some problems concerning the properties occurring during the manufacturing process (i.e. residual stresses, volumetric shrinkage and warpage distribution), especially when moulding thin-walled parts using the plastic injection-moulding process.

17.3 Challenges in Injection-Moulding Process for Wood Composites

In this section, issues related to challenges in injection-moulding process are discussed. They include thin-walled moulding process, minimizing the wall thickness and research on thin-walled parts.

17.3.1 Injection Moulding for Wood Composites

Injection moulding is a polymer and polymer composite manufacturing process in which injection-moulding compounds (in the form of granule) are injected into the mould through a hopper. Injection-moulding compound granules are fed into the mould via a heated barrel. Melting of the granules is caused by the heated barrel as well as the shearing action of the reciprocating screw inside the machine (Zainudin et al. 2001; Shaharuddin et al. 2006). Figures 17.1, 17.2, 17.3, 17.4, 17.5, 17.6, 17.7 and 17.8 portray the compounding process of injection-moulding compounds for wood fibre-reinforced thermoplastic composites. Figures 17.9, 17.10, 17.11 and 17.12 show machines, specimens and modelling used in injection-moulding process.

Research areas that are normally being focused in moulded part by natural filler-reinforced thermoplastic composites include processability and properties (residual stress properties, material warpage and material shrinkages) of injection-moulded thin-walled part.

Fig. 17.1 Virgin polypropylene

Fig. 17.2 Wood powder

Fig. 17.3 Additives

Fig. 17.4 Drying and mixing processes

Fig. 17.5 Twin screw extruder

Fig. 17.6 Granulate carrier

Fig. 17.7 Front view of carrier

Fig. 17.8 Fine granulate crusher machine

Fig. 17.9 Typical injection-moulding machine

Fig. 17.10 Injection-moulding machine at INTROP, Universiti Putra Malaysia

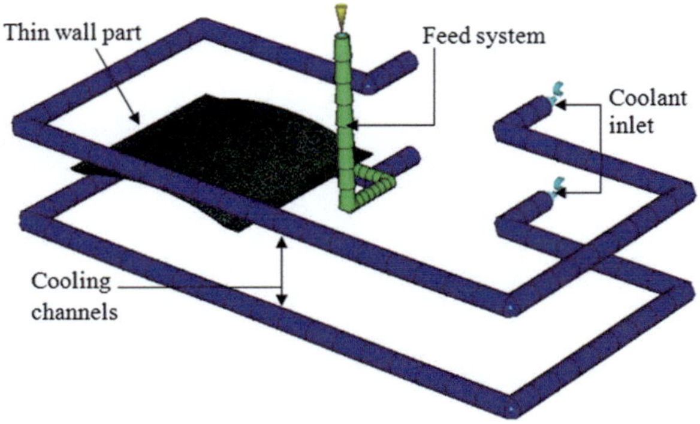

Fig. 17.11 Justification of the optimum mould design (Moldflow simulation on thin-walled parts)

Fig. 17.12 Specimens of wood polymer composites

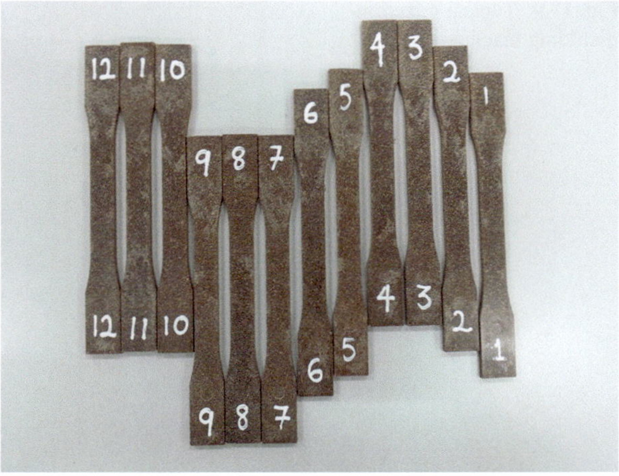

17.3.2 Thin-Walled Moulding Process

Nowadays, many thin applications use nominal wall thicknesses less than 1 mm compared to previous definition of thin-walled part (i.e. 1.5 mm) (Timothy 2001). However, Shoemaker (2006) stated that thin-walled materials are called 'advanced' for the thickness of 1.2–2 mm and 'leading edge' if the thickness is below 1.2 mm. He is of the opinion that the thin-walled definitions are considered on the basis of length to thickness ratio of the wall ranging from 100:1 to 150:1 or more. In addition, thin-walled moulding process also as known as when the part thickness smaller than 1.5 mm and (L:T) is greater than 100 (Chen et al. 2007). Sometimes, if the (L:T) is above 100 or 150 with thicknesses of 1 mm or thinner and a surface area of at least 50 cm^2, it is called ultra-thin-walled part and it was classified by Song et al. (2007).

17.3.3 Minimizing the Wall Thickness

Why do we want to reduce the wall thickness? The answer is 'cost'. Today, concern about economics and the environment causes the industries to focus on thin-moulding processes in their manufacturing. The specifications of a product can be made lighter, more compact and less expensive with sufficient mechanical properties and shorter moulding cycles (Anonymous 1995).

Therefore, manufacturers involving in research and development to produce 3C (computer, communication and consumer) products will be more concerned with product properties such as lightness, thinness, shortness and smallness (Shen et al. 2008).

According to Fischer (2003), decreasing the products' wall thickness means reducing plastic consumption and moulding with shorter moulding cycles. Hence, more than 70 % of the part's cost can be reduced. In addition, reducing the wall thickness also reduces the cooling time, which provides greater potential benefits, particularly when the cooling time is a significant portion of the cost in the moulding process.

Moving towards 'thin-walled technology', designers and manufacturers have increasingly explored the potential of thin-walled technology as a cost-reducing methodology that can reduce the moulding cycle time and material used (Slaviero et al. 2001). Preparing a smaller component enables the productivity rates to be increased, which reveals the manufacturer's attempt to rapidly compete in the world market (Rotheiser 2002).

17.3.4 Research on Thin-Walled Parts

Shen et al. (2008) stated that by looking at various injection-moulding conditions, the most appropriate number moulding gates during material filling can be determined. The geometry model analysed was a battery cover with a wall thickness on four sides of 1.0 mm and a base central region of 0.2 mm. Ozcelik and Sonat (2009) carried out the way to optimize the strength and warpage of a consumer product (handphone cover) and the thickness being studied were 0.9 mm, 1 mm and 1.1 mm and the materials used were carbon fibre-reinforced polycarbonate ABS blend composites.

Tang et al. (2007) fabricated a mould for a thin-walled plate part with dimensions of 120 mm × 50 mm × 1 mm to study the optimization of the moulding parameters on warpage. Their research was performed to investigate the effect of injection-moulding parameters such as filling time, melt temperature and packing pressure and time on the warpage behaviour of thin-walled plate parts.

17.4 Formation of Residual Stresses in Injection-Moulding Process

Struik (1978) classified residual stresses into two categories: those induced by molecular orientation and those induced by rapid inhomogeneous cooling during solidification in the cooling stage. However, Williams (1981) divided residual stress in injection moulding into three major sources: two of these sources are similar to those presented by Struik (1978); in addition, Williams (1981) proposed residual stresses that were caused by the difference of crystallinity levels between skin and core layers.

Most researchers accepted that residual stress of injection moulding took place from the sources of residual flow stresses and residual thermal stresses (Baaijens 1991; Hastenberg et al. 1992; Kabanemi and Crochet 1992; Boitout et al. 1995; Liu 1996; Zoetelief et al. 1996; Chen et al. 2000; Gu et al. 2001; Lu and Khim 2001; Kamal et al. 2002; Lee et al. 2002a, b; Kim et al. 2005; Isayev et al. 2006; Park et al. 2006).

Residual stress is normally referred to as the stresses that remained in a moulded component and this took place when external loads were not present; only temperature or thermally induced loads and pressure or flow-induced loads existed. Thermally induced loads were caused by the moulding component that was not cooled homogenously (Wang and Young 2005) and the parameters associated with cooling like mould and matrix temperatures should be given proper attention. Flow-induced stress is normally lower than thermal stress (Zoetelief et al. 1996). In injection-moulding process, residual stress occurred either during moulding or demoulding of the component (Choi and Im 1999).

In the moulding process, residual stresses were distributed along the thickness of a component, and Kim et al. (2002) reported that on the surface and in the middle of the parts, the stress is in tension, while on other locations, the stress is in compression. Zhou et al. (2008) reported different findings, i.e. residual stresses were distributed equally across the part thickness. When the parts were being demoulded, warpage and part shrinkage were observed due to residual stress (Zhou and Li 2005a, b).

17.5 Formation of Shrinkage and Warpage in Injection-Moulding Process

Problems of accuracies in dimensions and related to mechanical properties and optical properties can be associated with the residual stress in a part (Michaeli and Potsch 1995; Hastenberg et al. 1992; Isayev and Crouthamel 1984; Zhang et al. 2002; Weng 2010). These common problems depend on the source of residual stresses, which are either thermally induced or flow induced. Normally, thermally induced residual stresses produced part warpage, environmentally related stress crack and part shrinkage (Isayev and Crouthamel 1984; Baaijens 1991; Powell and Housz 1998; Maxwell 2005).

Jacques (1982) reported that inhomogeneous cooling of a part may lead to stress uneven distributed in the part and in turn may lead to part warpage. This is particularly true when the wall is very thin. Cheng et al. (2009) suggested ways to minimize this problem by paying particular attention to shape of the part, innovation in cooling and moulding technologies. Similar solutions were proposed by Tang et al. (2007). Constant mould temperature during cooling is important (Chiang 2007). Cheng et al. (2009) suggested that in addition to temperature, packing pressure had the major concern to warpage. Studies are being conducted to determine which factors had the greatest effect in reducing warpage: packing time, mould temperature and filling pressure. Ozcelik and Sonat (2009) had carried out ways to reduce shrinkage and warpage in the plastic products that they developed. One way that they proposed is to use composite materials.

17.6 Optimization and Analysis of the Effect of Processing Parameters on Residual Stress, Shrinkage and Warpage

In research involving the injection-moulding process, most researchers employ certain statistical tools to optimize the variables for injection-moulding parameters such as temperature (resin, mould), time (packing, cooling, filling), and pressure (injection, packing) and to investigate the effect of the processing parameters on the quality of product. Giboz et al. (2007) reported that shrinkage and warpage normally depend upon injection-moulding processing parameters. They suggested ways to minimize warpage and shrinkage and they also suggested that attention should be paid to controlling injection-moulding parameters.

Researchers had used mould simulation package like (Moldflow and C-MOLD) as well as statistical tools to simulate the behaviour of composite materials during warpage and shrinkage. Examples of statistical tool used were Taguchi (including orthogonal array and S/N ratio), *F*-test and ANOVA methods. Among the findings includes the determination of the optimum process parameters affecting warpage of composite-walled composite products. Process parameters being considered include time, pressure and temperature.

Choi and Im (1999) studied the influence of injection-moulding conditions on residual stress formation inside moulded parts for use in accurate prediction against shrinkage and warpage. They observed that increases in the packing pressure and mould temperature resulted in more significant effects on the residual stresses. The increase was due to slow propagation of solid-liquid boundary when temperature of the injection mould was high. As the packing pressure increases, the difference of stresses (in tension and compression) becomes large, while the melt temperature and injection velocities have no effect on the residual stresses.

Sen and Bhattacharya (2000) showed that the distribution of residual stresses in starch/synthetic polymer blends consists of compressive stresses on the surface and tensile stresses in the middle of the part. The profile was affected by the injection and packing pressures, mould temperature, ageing and the amount of starch in the blend. However, packing time and melt temperature did not greatly affect the distribution of residual stresses. The magnitudes of both the compressive and tensile stresses across the thickness decrease with decreasing packing pressure. The rise in temperature of the mould and injection pressure had reduced tensile stresses. However, it did not significantly affect compressive stresses.

The formation of internal stresses depends on the results of cooling time diversification in moulded parts. Rapidly cooled and solidified surface layers create a stiff layer, which limits the shrinkage of a still-liquid core. The result of such a limitation is the existence of tensile stresses in the outside layer and compressive stresses in the inner areas. Rapidly increasing the pressure during the packing stages and an excessive clamping pressure cause packing of the particles; this results in high tensile stresses in the solidifying skin layers of the mould piece surface. As an adverse effect, the moulded part may be cracked when ejected from mould (Kansal et al. 2001; Arif and Fazal 2003).

Wang and Young (2005) studied the effect of the moulding condition on residual stresses of thin-walled parts. The mould temperature was observed to yield a significant distribution of residual stresses. There were different types of residual stresses at different melt temperatures. However, they observed that high-packing pressure range did not significantly affect residual stresses compared to low-packing pressure counterpart.

A review of past research showed that the study of moulded thin-walled parts made from natural fibre-reinforced thermoplastic composites in the injection-moulding process was not widely carried out. Research works have only concentrated on unreinforced thermoplastic and glass fibre-reinforced thermoplastic composites in making thin-walled parts during the investigation on the relations between the moulding parameters and part quality.

17.7 Potential Industrial Applications

Concerning industrial applications, several paths have been undertaken by worldwide researchers to expand and continuously improve the production of high-quality natural fibre composites (Davoodi et al. 2011). A mobile phone casing made of polylactide (PLA) reinforced with kenaf fibres was launched by the NEC Corporation using a modified PLA developed by UNITIKA Ltd (Zini and Scandola 2011). The most suitable industrial applications of thin-walled parts focus on the electronics industry. The French moulding technology company RocTool has recently claimed that the thin-walled technology is going to expand and predicts increases in electronic applications. RocTool is the only company capable of producing electronic housing using engineering plastics with fibre contents that do not require painting; the thickness is 0.8 mm or less than 1 mm (Chris 2014a). The Taiwanese electronics industry moulder Ju Teng cooperated with RocTool to produce the latest smartphone and tablet casing component of less than 0.5 mm thick using glass- and carbon-reinforced thermoplastic composites. The Chinese mobile internet device and smartphone maker called Xiaomi is also exploring electronic packing applications with thin-walled casings produced for manufacturing its ultra-slim Mi2A device (Chris 2014b). Therefore, the best opportunities for lignocellulosic thermoplastic composites appear to be focused on electronic packaging. For how long will the idea of lignocellulosic thermoplastic composites be suitable for the thin-walled moulding process? In the initial stage, approaches using numerical simulations helped to identify the lignocellulosic thermoplastic composites that are suitable for thin-walled injection-moulding processes specifically for applications of similar casing parts. Based on these findings, this research became a platform for future exploration to expand the application of lignocellulosic thermoplastic composites in the development of thin-walled smartphone and tablet casings.

17.8 Conclusions

The development of wood-reinforced thermoplastic composites has been a hot topic recently due to the plan to reduce material costs and increasing environmental awareness. Specifically, wood flour-reinforced thermoplastic composites have a bright future for the aggressive exploration of new research scopes and applications. Wood fillers are also being used based on the reasons that were highlighted in the beginning of this review chapter. Therefore, the designer or engineer proposes the thin-walled part in their design, which is injection moulded using wood fibre-reinforced thermoplastic composites. However, in quality control of the part, the formation of residual stresses, shrinkage and warpage of the moulded part should be considered. This chapter provides a comprehensive review on previous research in material compounding, the properties of wood thermoplastic composites and the trend towards thin-moulding processing and the processability and properties (residual stresses, shrinkage and warpage) of injection-moulded thin-walled parts.

Acknowledgements This project is funded by Universiti Putra Malaysia through Research University Grant Scheme with project number of UPM/700-1/2/RUGS/05-02-12-1917RU. The authors also acknowledge financial support from Universiti Malaysia Perlis for the principal author to carry out his PhD work at Universiti Putra Malaysia. The contribution of SIRIM Berhad and INTROP, UPM, is highly appreciated.

References

Anonymous (1995) Thinwall®. In technical guidebook for electronics applications. GE Plastics.

Arif A, Fazal M (2003) Numerical prediction of plastic deformation and residual stresses induced by laser shock processing. J Mater Process Technol 136:120–138

Ashori A (2008) Wood-plastic composites as promising green-composites for automotive industries! Bioresour Technol 99:4661–4667

Baaijens FPT (1991) Calculation of residual stresses in injection molded products. Rheol Acta 30:284–299

Bachtiar D, Sapuan SM, Hamdan MM (2010) Flexural properties of alkaline treated sugar palm fibre reinforced epoxy composites. Int J Automot Mech Eng (IJAME) 1:79–90

Bax B, Müssig J (2008) Impact and tensile properties of PLA/Cordenka and PLA/flax composites. Compos Sci Technol 68:1601–1607

Bledzki AK, Gassan J (1999) Composites reinforced with cellulose based fibres. Prog Polym Sci 24:221–274

Boitout F, Agassant JF, Vincent M (1995) Elastic calculation of residual stresses in injection molding – influence of mold deformation and pressure in the liquid. Int Polym Process 10:237–242

Carroll DR, Stone RB, Sirignano AM, Saindon RM, Gose SC, Friedman MA (2001) Structural properties of recycled plastic/sawdust lumber decking planks. Resour Conserv Recycl 31:241–251

Chen X, Lam YC, Li DQ (2000) Analysis of thermal residual stress in plastic injection molding. J Mater Process Technol 101:275–280

Chen CS, Chen TJ, Chien RD, Chen SC (2007) Investigation on the weldline strength of thin-wall injection molded ABS parts. Int Commun Heat Mass Tran 43(4):448–455

Cheng XM, Zhou L, Sheng NY, Wu YD (2009) Injection molding and warpage of thin-walled parts based on simulated deformation. IEEE 2006:3–6

Chiang KT (2007) The optimal process conditions of an injection-molded thermoplastic part with a thin shell feature using grey-fuzzy logic: a case study on machining the PC/ABS cell phone shell. Mater Des 28:1851–1860

Choi D, Im Y (1999) Prediction of shrinkage and warpage in consideration of residual stress in integrated simulation of injection molding. Compos Struct 47:655–665

Chris S (2014a) Roctool claims thin wall gains in electronics. Injection World by Applied Market Information Ltd, p 47

Chris S (2014b) Xiaomi adopts LNP compounds for ultra-slim smartphone. Injection World by Applied Market Information Ltd, p 46

Coutinho FMB, Costa THS, Carvalho DL (1997) Polypropylene–wood fiber composites: effect of treatment. J Appl Polym Sci 65:1227–1235

Davoodi MM, Sapuan SM, Ahmad D, Aidy A, Khalina A, Jonoobi M (2011) Concept selection of car bumper beam with developed hybrid bio-composite materials. Mater Des 32:4857–4865

Eichhorn SJ, Dufresne A, Aranguren M, Marcovich NE, Capadona JR, Rowan SJ, Peijs T (2009) Review: current international research into cellulose nanofibres and nanocomposites. J Mater Sci 45:1–33

El-Shekeil YA, Sapuan SM, Abdan K, Zainudin ES (2012) Influence of fiber content on the mechanical and thermal properties of kenaf fiber reinforced thermoplastic polyurethane composites. Mater Des 40:299–303

Fischer JM (2003) Handbook of molded part shrinkage and warpage. Plastics Design Library William Andrew, Inc., USA

Giboz J, Copponnex T, Mélé P (2007) Microinjection molding of thermoplastic polymers: a review. J Micromech Microeng 17:R96–R109

Gould P (2002) Exploiting spider's silk. Materialstoday, (December), 42–47

Gu Y, Li H, Shen C (2001) Numerical simulation of thermally induced stress and warpage in injection-molded thermoplastics. Adv Polym Technol 20:14–21

Hambali A, Sapuan SM, Ismail N, Nukman Y (2009) Application of analytical hierarchy process in the design concept selection of automotive composite bumper beam during the conceptual design stage. Sci Res Essay 4:198–211

Hambali A, Sapuan SM, Ismail N, Nukman Y (2010) Material selection of polymeric composite automotive bumper beam using analytical hierarchy process. J Cent South Univ Technol 17:244–256

Hastenberg CHV, Wildervanck PC, Leenen AJH (1992) The measurement of thermal stress distribution along the flow path in injection-molded flat plates. Polym Eng Sci 32:506–515

Hubbe MA, Rojas OJ, Lucia LA, Sain M (2008) Cellulosic nanocomposites: a review. BioResources 3:929–980

Huda MS, Mohanty AK, Drzal LT, Schut E, Misra M (2005) "Green" composites from recycled cellulose and poly(lactic acid): physico-mechanical. J Mater Sci 40:4221–4229

Isayev AI, Crouthamel DL (1984) Residual stress development in the injection moldings of polymers. Polym Plast Technol Eng 22:177–232

Isayev AI, Shyu GD, Li CT (2006) Residual stresses and birefringence in injection molding of amorphous polymers: simulation and comparison with experiment. J Polym Sci B Polym Phys 44:622–639

Ishak MR, Leman Z, Sapuan SM, Edeerozey AMM, Othman IS (2010) Mechanical properties of kenaf bast and core fibre reinforced unsaturated polyester composites. IOP Confer Series: Mater Sci Eng 11:012006

Ishak MR, Sapuan SM, Leman Z, Rahman MZA, Anwar UMK (2012) Characterization of sugar palm (*Arenga pinnata*) fibres. J Therm Anal Calorim 109:981–989

Jacques MS (1982) An analysis of thermal warpage in injection molded at parts due to unbalanced cooling. Polym Eng Sci 22:241–245

Kabanemi KK, Crochet MJ (1992) Thermoviscoelastic calculation of residual stresses and residual shapes of injection molded parts. Int Polym Process 7:60–70

Kamal MR, Lai-Fook RA, Hernandez-Aguilar JR (2002) Residual thermal stresses in injection moldings of thermoplastics: a theoretical and experimental study. Polym Eng Sci 42:1098–1114

Kansal G, Rao PN, Atreya SK (2001) Study: temperature and residual stress in an injection moulded gear. J Mater Process Technol 108:328–337

Kim SK, Lee SW, Youn JR (2002) Measurement of residual stresses in injection molded short fiber composites considering anisotropy and modulus variation. Korea Aust Rheol J14:107–114

Kim KH, Isayev AI, Kwon K, van Sweden C (2005) Modeling and experimental study of birefringence in injection molding of semicrystalline polymers. Polymer 46:4183–4203

La Mantia FP, Morreale M (2011) Green composites: a brief review. Compos A Appl Sci Manuf 42:579–588

Lee SH, Wang S (2006) Biodegradable polymers/bamboo fiber biocomposite with bio-based coupling agent. Compos A Appl Sci Manuf 37:80–91

Lee YB, Kwon TH, Yoon K (2002a) Numerical prediction of residual stresses and birefringence in injection/compression molded center-gated disk. Part I: Basic modeling and results for injection molding. Polym Eng Sci 42:2246–2272

Lee YB, Kwon TH, Yoon K (2002b) Numerical prediction of residual stresses and birefringence in injection/compression molded center-gated disk. Part II: Effects of processing conditions. Polym Eng Sci 42:2273–2292

Leman Z, Sapuan SM, Azwan M, Ahmad MMHM, Maleque MA (2008) The effect of environmental treatments on fiber surface properties and tensile strength of sugar palm fiber-reinforced epoxy composites. Polym Plast Technol Eng 47:606–615

Li TQ, Wolcott MP (2004) Rheology of HDPE–wood composites. I. Steady state shear and extensional flow. Compos A Appl Sci Manuf 35:303–311

Li TQ, Ng CN, Li RKY (2001) Impact behavior of sawdust/recycled – PP composites. J Appl Polym Sci 81:1420–1428

Liu SJ (1996) Modeling and simulation of thermally induced stress and warpage in injection molded thermoplastics. Polym Eng Sci 36:807–818

Lodha P, Netravali AN (2002) Characterization of interfacial and mechanical properties of "green" composites with soy protein isolate and ramie fiber. J Mater Sci 37:3657–3665
Lu X, Khim LS (2001) A statistical experimental study of the injection molding of optical lenses. J Mater Process Technol 113:189–195
Marsh G (2003) Next step for automotive materials. Mater Today April:36–43
Maxwell AS (2005) Measurement of residual stress in plastics. DEPC (MN) 027
Michaeli W, Potsch G (1995) Injection molding: an introduction. Hanser Publishers, New York
Morreale M, Scaffaro R, Maio A, La Mantia FP (2008a) Effect of adding wood flour to the physical properties of a biodegradable polymer. Compos A Appl Sci Manuf 39:503–513
Morreale M, Scaffaro R, Maio A, La Mantia FP (2008b) Mechanical behaviour of Mater-Bi®/wood flour composites: a statistical approach. Compos A Appl Sci Manuf 39:1537–1546
Netravali AN, Chabba S (2003) Composites get greener. Mater Today April:22–29
Nickel J, Riedel U (2003) Activities in biocomposites. Mater Today April:44–48
Nishino T, Hirao K, Kotera M, Nakamae K, Inagaki H (2003) Kenaf reinforced biodegradable composite. Compos Sci Technol 63:1281–1286
Ozcelik B, Sonat I (2009) Warpage and structural analysis of thin shell plastic in the plastic injection molding. Mater Des 30:367–375
Park K, Kim B, Yao D (2006) Numerical simulation for injection molding with a rapidly heated mold, Part II: Birefringence prediction. Polym Plast Technol Eng 45:903–909
Plackett D, Andersen LT, Pedersen BW, Nielsen L (2003) Biodegradable composites based on L-polylactide and jute fibres. Compos Sci Technol 63:1287–1296
Powell CP, Housz IJA (1998) Engineering with polymers. Stanley Thornes Publishers, London
Rangaprasad R (2003) Wood plastic composites – an overview. In: IPI seminar on synthetic wood, pp. 8 http://www.ipiindia.org/index.php?option=com_zoo&task=callelement&format=raw&item_id=1010&element=cf0577f3-519f-4fe3-9d25-658179076aff&method=download&Itemid=328
Rashdi AAA, Sapuan SM, Zainudin ES, Khalina A (2009) Water absorption and tensile properties of soil buried kenaf fibre reinforced unsaturated polyester composites (KFUPC). J Food Agric Environ 7:908–911
Rotheiser JI (2002) Controlling warpage for the decorating and assembly of plastic parts. In: ANTEC, annual technical conference proceedings, conference 60, pp 3020–3023
Samir MASA, Alloin F, Dufresne A (2005) Review of recent research into cellulosic whiskers, their properties and their application in nanocomposite field. Biomacromolecules 6:612–626
Sanadi A, Caulfield DF, Rowell RM (1998) Lignocellulosic/plastic composites. Technology Summaries, Madison, pp 8–13
Sapuan SM, Abdalla HS (1998) A prototype knowledge-based system for the material selection of polymeric-based composites for automotive components. Compos A Appl Sci Manuf 29:731–742
Sapuan SM, Zan MNM, Zainudin ES, Arora PR (2005) Tensile and flexural strengths of coconut spathe-fibre reinforced epoxy composites. J Trop Agric 43:63–65
Sastra HY, Siregar JP, Sapuan SM, Hamdan MM (2006) Tensile properties of *Arenga pinnata* fiber-reinforced epoxy composites. Polym Plast Technol Eng 45:149–155
Scaffaro R, Morreale M, Re GL, Mantia FPL (2009) Effect of the processing techniques on the properties of ecocomposites based on vegetable oil-derived mater-bi and wood flour. J Appl Polym Sci 114:2855–2863
Sen A, Bhattacharya M (2000) Residual stresses and density gradient in injection molded starch/synthetic polymer blends. Polymer 41:9177–9190
Shaharuddin SIS, Salit MS, Zainudin ES (2006) Review of the effect of moulding parameters on the performance of polymeric composite injection moulding. Turk J Eng Environ Sci 30:23–34
Shen YK, Wu CW, Yu YF, Chung HW (2008) Analysis for optimal gate design of thin-walled injection molding. Int Commun Heat Mass Tran 35:728–734
Shoemaker J (2006) Moldflow design guide (a resource for plastic engineer). Hanser Gardner Publications Inc. 6915 Valley Avenue, Cincinnati, Ohio 45244–3029, USA, p 256

Slaviero C, Weiss K, Woodman D (2001) Thinwall injection molding for instrument panels. In: SAE Technical Paper Series

Song MC, Liu Z, Wang MJ, Yu TM, Zhao DY (2007) Research on effects of injection process parameters on the molding process for ultra-thin wall plastic parts. J Mater Process Technol 187–188:668–671

Struik LCE (1978) Orientation effects and cooling stresses in amorphous polymers. Polym Eng Sci 18:799–811

Suriani MJ, Hamdan MM, Sastra HY, Sapuan SM (2007) Study of interfacial adhesion of tensile specimens of *Arenga pinnata* fiber reinforced composites. Multidiscip Model Mater Struct 3:213–224

Tang SH, Tan YJ, Sapuan SM, Sulaiman S, Ismail N, Samin R (2007) The use of Taguchi method in the design of plastic injection mould for reducing warpage. J Mater Process Technol 182:418–426

Timothy AP (2001) Specializes molding techniques. In: Heim HP, Potente H (eds) 10 common pitfalls in thin wall plastic part design. Plastic Design Library, Norwich, p 107

Wang TH, Young WB (2005) Study on residual stresses of thin-walled injection molding. Eur Polym J 41:2511–2517

Weng C (2010) Modelling and simulation of residual stresses and birefringence in the precision injection moulding of microlens arrays. UMI Dissertation Publishing, The Hong Kong Polytechnic University

Willett JL (1994) Mechanical properties of LDPE/granular starch composites. J Appl Polym Sci 54:1685–1695

Williams JG (1981) On the prediction of residual stresses in polymers. Plast Rubber Process Appl 1:369–377

Yusoff MZM, Salit MS, Ismail N, Wirawan R (2010) Mechanical properties of short random oil palm fibre reinforced epoxy composites. Sains Malaysiana 39:87–92

Zainudin ES, Sapuan SM, Sulaiman S, Ahmad MMHM (2001) Fiber orientation of short fiber reinforced injection molded thermoplastic composites: a review. J Inject Mold Technol 6:1–10

Zhang X, Cheng X, Stelson KA, Bhattacharya M, Sen A, Voller VR (2002) Approximate model of thermal residual stress in an injection molded part. J Therm Stresses 25:523–538

Zhou H, Li D (2005a) Numerical simulation and experimental study of warpage of injection-molded parts. Polym Plast Technol Eng 44:603–617

Zhou H, Li D (2005b) Residual stress analysis of the post-filling stage in injection moulding. Int J Adv Manuf Technol 25:700–704

Zhou H, Xi G, Liu F (2008) Residual stress simulation of injection molding. J Mater Eng Perform 17:422–427

Zini E, Scandola M (2011) Green composites: an overview. Polym Compos 32:1905–1915

Zoetelief WF, Douven LFA, Housz AJI (1996) Residual thermal stresses in injection molded products. Polym Eng Sci 36:1886–1896

Chapter 18
Filament Winding Process for Kenaf Fibre Reinforced Polymer Composites

S. Misri, M.R. Ishak, S.M. Sapuan, and Z. Leman

Abstract In this chapter, a study on filament winding process for kenaf fibre reinforced polymer composite manufacture is presented. Filament winding process for conventional fibre composites is discussed. Advantages and disadvantages of filament winding process are also briefly covered. A review on filament winding from various research is also performed. Brief review of filament winding of natural fibre composites is provided. The chapter also describes work on improvement of existing filament winding process and fabrication of hollow shaft made from kenaf fibre reinforced composites. The main contribution was the use of drum-type resin bath and surfacing veil, and these had facilitated ease of fabrication of kenaf yarn fibre reinforced unsaturated polyester composite hollow shafts.

Keywords Filament winding • Natural fibre composites • Kenaf • Biocomposites • Hollow shafts

S. Misri • Z. Leman
Department of Mechanical and Manufacturing Engineering, Universiti Putra Malaysia (UPM), 43400 Serdang, Selangor, Malaysia

M.R. Ishak
Aerospace Malaysia Innovation Centre, Universiti Putra Malaysia (UPM), 43400 Serdang, Selangor, Malaysia

Laboratory of Biocomposite Technology, Institute of Tropical Forestry and Forest Products (INTROP), Universiti Putra Malaysia (UPM), 43400 Serdang, Selangor, Malaysia

S.M. Sapuan (✉)
Department of Mechanical and Manufacturing Engineering, Universiti Putra Malaysia (UPM), 43400 Serdang, Selangor, Malaysia

Laboratory of Biocomposite Technology, Institute of Tropical Forestry and Forest Products (INTROP), Universiti Putra Malaysia (UPM), 43400 Serdang, Selangor, Malaysia

Aerospace Manufacturing Research Centre (AMRC), Universiti Putra Malaysia (UPM), 43400 Serdang, Selangor, Malaysia

Materials Technology and Processing Laboratory, Institute of Advanced Technology (ITMA), Universiti Putra Malaysia (UPM), 43400 Serdang, Selangor, Malaysia
e-mail: sapuan@upm.edu.my

M.S. Salit et al. (eds.), *Manufacturing of Natural Fibre Reinforced Polymer Composites*, DOI 10.1007/978-3-319-07944-8_18

18.1 Introduction

Advances and innovation in composite material have made it a very important class of material, and it has been accepted as a major material in many industries such as in aerospace, automotive, oil and gas, marine, telecommunication, furniture, sport and leisure and construction industries. The advantages from using this material are enormous such as light weight, corrosion resistance, part integration, aesthetically pleasing and comparable strength and stiffness properties (Sapuan and Abdalla 1998; Davoodi et al. 2008; Hambali et al. 2009; Sapuan et al. 2002, 2005). Nowadays, composite technology has gone beyond conventional glass and fibre reinforced polymer composites. The development of ceramic matrix composites (CMC) and metal matrix composites (MMC) has now reached the commercialization stage. Nanocomposites (including carbon nanotubes, halloysite nanotubes and graphene-based composites), biomimetics, biocomposites, functionally graded materials and smart materials have become new important elements in composite technology, and they can be regarded as tomorrow's materials.

In the recent years, there is a growing interest in using natural fibres as reinforcements or filler materials for polymer-based composites due to many advantages that they offer such as abundance, environmentally friendly, low cost, low density, renewable, recyclable, low energy consumption, biodegradable, posing no harmful effect to the health and having comparable specific strength and stiffness properties (El-Shekeil et al. 2012; Anwar et al. 2009; Sastra et al. 2006; Yusoff et al. 2010; Ishak et al. 2010, 2012; Davoodi et al. 2011).

Nowadays, natural fibre reinforced polymer composites have been used to make numerous consumer and engineering products, and their applications are mainly restricted to semi- or nonstructural components. Examples of their use include furniture, food packaging, mulch film, drain cover, door or window panels for houses and offices, boat, automotive door panel, household appliances and food containers. Polymer system can be in the form of synthetic (thermoplastic and thermosetting polymers) or biopolymers (commercially available or in-house made). Among the natural fibres that are normally used as reinforcements for polymer composites include kenaf (Fig. 18.1), oil palm, sugar palm, cocoa pod, betel nut, pineapple leaf, sugarcane, abaca, banana stem, jute, sisal, hemp, Roselle and coconut. Manufacturing processes associated with natural fibre composites are normally the same processes used to produce components from traditional glass and carbon fibre composites such as injection moulding, pultrusion, vacuum bagging, hand lay-up, resin transfer moulding and filament winding. In this chapter, the manufacturing process of natural fibre composite products is presented and discussed.

Fig. 18.1 Kenaf plants

18.2 Filament Winding Process

Filament winding is a manufacturing process for fabrication of fibre reinforced composite components. Peters and Humphrey (1987) reported that the filament winding method was first presented in the form of patent since 1963. In 1964, a monograph describing the filament winding process was published. An automated filament placement technology was introduced in 1990 (Anon 2011). The filament winding process involves drawing continuous fibres through a container of resin mixture or bath and winding the continuous resin impregnated fibres around a rotating mandrel to achieve the desired shape (Fig. 18.2). The fibres are placed on the rotating mandrel by means of a horizontal carrier. The fibre orientation is controlled by adjusting the speed of the carrier. Mandrel can be removable, can be sacrificial or can form as part of the component. Thereafter, the component is cured in an oven for a certain period of time at a suitable temperature.

Defects are normally found in products made from filament winding process such as in the forms of voids, fibre wrinkles and delaminations (Mallick 2008). There are three common types of winding patterns of filament winding process, namely, hoop (circumferential), helical and polar. Typical products that are normally made from filament winding process include drive shafts, pipes, rods, hollow and solid tubes, space frames, tanks, gas cylinders, water filters, aerospace components, lamp posts, fishing rods and high-pressure vessels (Meijer and Ellyin 2008;

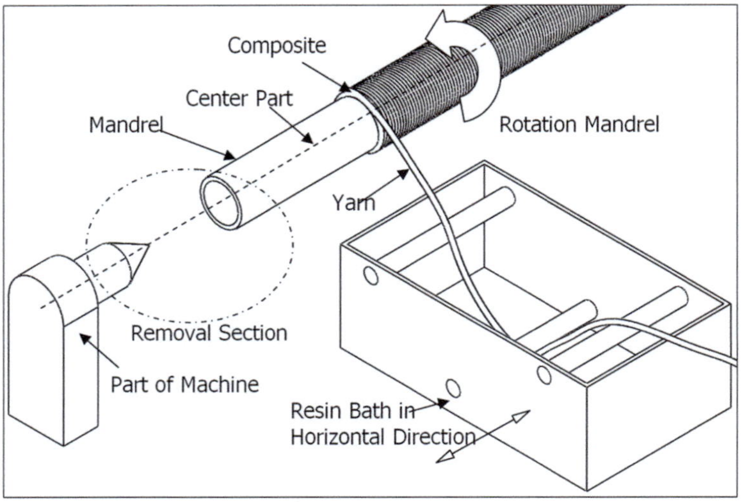

Fig. 18.2 3D schematic diagram of filament winding process

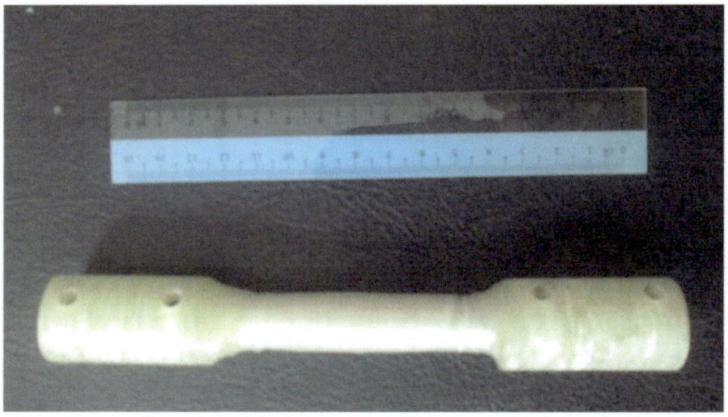

Fig. 18.3 Shaft made from filament winding process

Shaw-Stewart 1985). Figure 18.3 shows an example of a drive shaft made from filament winding method.

The polymer matrices used for filament winding process are mainly thermosetting polymers such as epoxy and unsaturated polyesters. The synthetic fibres such as glass, aramid and carbon fibres are the main reinforcing fibres used in the filament winding process, and so far, research and development work on filament winding process employing natural fibre composites is very limited. Advantages and disadvantages of filament winding process are presented in Table 18.1.

Table 18.1 Advantages and disadvantages of filament winding process (Zu 2012; Krishnamurthy and Idkan 2014; Buragohain and Velmurugan 2011)

Advantages	Disadvantages
1. Winding time is short because of simplified tooling concept 2. Mandrel preparation time is short 3. Availability of raw materials 4. Relatively low cost of raw materials (matrices and reinforcements) 5. Relatively low tooling costs 6. Polymers can easily be formulated, and the formulation can easily be changed according to individuals' needs 7. The process is reproducible or repetitive 8. Continuous fibres can be used for the entire components 9. High fibre volume is achievable 10. Fibres can be oriented in the loading direction 11. There is no need to use autoclave 12. Part size is not limited by the size of oven 13. Process can be automated and cost savings can be achieved 14. Limited use of capital equipment	1. Component must facilitate removal of mandrel 2. High cost of mandrel and may be complex 3. Winding reverse curvature is not possible 4. Problem in placing fibres parallel to the mandrel axis 5. The need for external mandrel surface treatment for surface evenness

18.3 Review of Filament Winding Process

A review of filament winding process based on published literature is presented in this section. Gemi et al. (2009) had carried out an experimental work on fatigue analysis of filament wound fibre reinforced polymer composites. The fibre used in the study was glass and the matrix used was epoxy. The hoop winding pattern was employed. In this study, composite pipe was subjected to internal pressures at different fibre loadings. Modes of failures at low and high fibre loadings were found to be different. Mutasher et al. (2012) had reported their work on the fabrication of a small filament winding machine used for fabrication of polymer composite parts for the purposes of teaching and research at Swinburne University, Malaysia branch. Mechanical, electrical and electronic control system was designed in an integrated manner to facilitate ease of use. Winding angles of 40–80° were achievable using this control system.

Wang et al. (2004) developed an improvement method during curing process in filament winding process. A model called 'helical tow model' was used to improve curing of the composites. Numerical problem was solved using finite volume method. The work was performed using computer code development and was used to investigate on-line thermal curing. Akkus et al. (2008) developed a control system that can optimize the pulling force of the resin impregnated fibre reinforcement. The mechatronics-based pretensioning system for filament winding process can be used for any types of fibres such as carbon, Kevlar and glass fibres, and it can ensure constant pulling force is achieved. This in turn can enhance the quality of finished filament wound composite parts.

Development in polymer technology had also improved the performance of filament winding process. New epoxy matrices were used with carbon fibres in filament winding process as reported by Chen et al. (2007). These new matrices had demonstrated good interfacial bonding with carbon fibres. Results of scanning electron microscopy (SEM), Fourier transform infrared spectroscopy (FTIR) and atomic force microscopy (AFM) revealed that these new epoxy matrices are superior to conventional epoxy matrices. Another development in composite materials for filament winding was the use of metal prepreg as reinforcement (Gordon 2006). It is a form of MMC. The performance of this MMC-based filament wound composites is comparable with polymer-based filament wound composites. In tensile properties (transverse direction), non-linear effects were detected, and these made the metal composites (MC) to be different from polymer-based composites.

Velosa et al. (2009) had developed new pressure vessels from filament wound glass fibre reinforced polymer composites. The liner was made from high-density polypropylene (HDPE), and the polymer matrix used in the composites was unsaturated polyester resin. The fibres used in this research were continuous E-glass rovings. Design analysis was conducted using a commercial finite element analysis software (Abaqus). The major parameter being studied was mainly internal burst pressure. The results of the study on internal burst pressures from finite element analysis and experimental work were compared and they have shown good agreement.

Hernández-Moreno et al. (2008) had carried out a study on the effect of winding pattern and architecture of impregnated glass fibres on mechanical properties of filament wound polymer composites. Winding pattern that was chosen for the research is helical winding. Detailed work on buckling characteristics at different winding patterns (e.g., one- and five-unit cell patterns) was carried out. Implosion pressure tests were carried out on the specimens. Results of implosion pressure tests and buckling analysis revealed that two different winding patterns did not significantly influence the implosion pressure and buckling behaviour of the polymer composites. Similar work on winding pattern issues in filament winding was presented by Park et al. (2002). The focus was on the study of through thickness direction. Helical winding was considered in this study. Winding angle and winding thickness were two major parameters being investigated in detail. The work on water pressure test was performed both numerically and experimentally for comparison purposes, and they have shown good agreement. A commercial finite element analysis software package (Abaqus) was used in the study. Ha and Jeong (2005) also perform similar work, but the component being studied was composite rings.

Morozov (2006) had carried out a research work on the influence of winding pattern on the mechanical properties of filament wound polymer composite products. The winding pattern that was considered in this study is called mosaic pattern. The products being investigated were thin-walled shells. The investigation was done with numerical analysis software (MSC Nastran). Different mosaic patterns had given different values of mechanical properties. Buragohain and Velmurugan (2011) had carried a study to determine the performance of filament wound polymer composite structures. The structures being investigated were in cylindrical form. The structures were stiffened by means of few ribs. The cylindrical structures were fabricated from filament winding process. Experimental study on compression and the finite element analysis were performed and their results were compared and discussed in details.

18.4 Filament Winding with Natural Fibre Composites: A Short Review

Very limited number of literature had reported on the use of filament winding of natural fibre composites. There is still a large scope of research in this topic.

Lehtiniemi et al. (2011) had studied the mechanical properties of flax fibre yarn-reinforced polymer composite tubes, and they had compared the results with the properties of glass fibre reinforced polymer composite tubes. Composites were produced using the filament winding process. The major problems associated with filament wound composites made from flax fibres and polymer matrix are poor interfacial bonding between fibres and matrix. Mahdi et al. (2002) had used filament winding process to fabricate polymer composite specimens from oil palm frond fibre reinforced epoxy composites. The composite products were circular, cylindrical and conical shells. Static crushing tests and crushing energy absorption tests were performed. The results were compared with polymer composites made from glass and carbon fibres.

Misri et al. (2014) had reported their work on fabrication of kenaf yarn fibre (Fig. 18.4) reinforced unsaturated polyester hollow tube composites using filament winding technique. Split-disk tests were performed for the specimens being wound with two different winding angles: 45° and 90°. The effect of filament wound processing parameter, namely, winding angle, was evaluated by determining the tensile properties of the composites using split-disk method (see Fig. 18.5). A recent work by the authors was concerned with the study of torsional behaviour of kenaf fibre reinforced polymer composite hollow shafts fabricated using filament winding process (Misri et al. 2015). The resin used was unsaturated polyester. Experimental work was conducted to study the torsional properties of the polymer composites. Figure 18.6 shows a kenaf fibre reinforced polymer composite specimen tested for torsion. The experimental work was compared with finite element analysis (Abaqus). Torsional properties of kenaf composites (with and without aluminium mandrels) at two different winding angles (45° and 90°) were determined. There was a good agreement between both sets of results. Nadia and Ishak (2014) had carried out a study on various aspects of kenaf fibre hollow composite shaft. They reported that changing the orientation angle would change the fatigue life of the filament wound composites. They concluded the kenaf composites can be used for light-weight and low-cost engineering application.

Fig. 18.4 Kenaf yarn fibre

Fig. 18.5 Split-disk test for a filament wound kenaf composite specimen

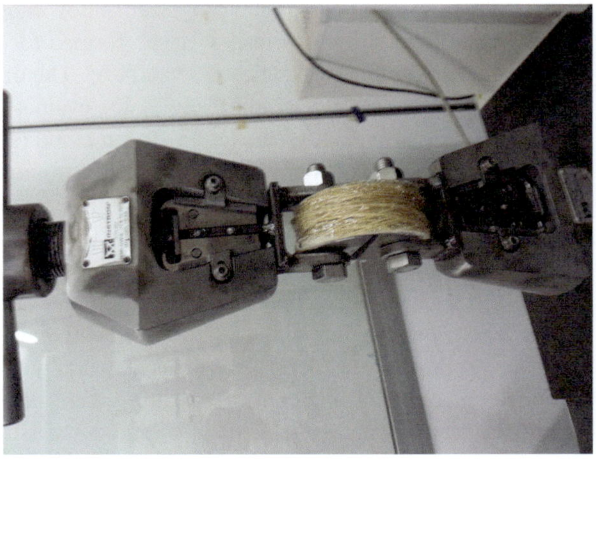

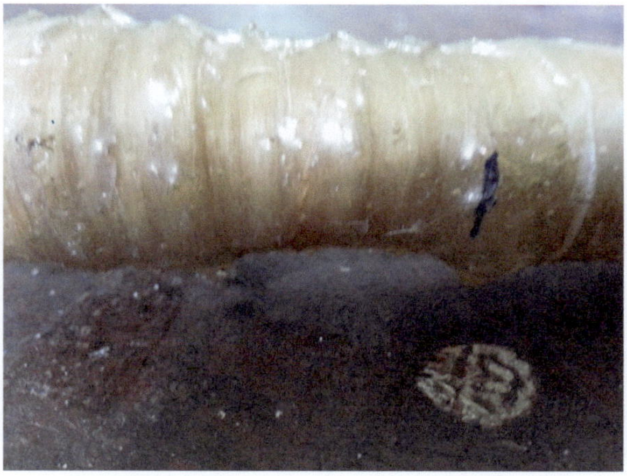

Fig. 18.6 Kenaf composite specimen in torsion test

18.5 Improvement of Filament Winding Machine

In this section, improvement of filament winding machine was made. The existing machine (Abdalla et al. 2007) was a lathe machine, and improvements were made to cater for the use in kenaf fibre reinforced polymer composite part development.

18.5.1 Innovation in Rotating Drum Design

Figure 18.7 shows the schematic diagram of the concept of fibre-dip type of resin bath, and the concept follows the work of Mutasher et al. (2007). This was the resin bath normally used for winding carbon and glass fibre rovings. In this process, excessive pulling force caused the fibres such as kenaf fibres to break as the fibres demonstrate lower tensile strength compared to glass and carbon fibres. To solve the problem of excessive pulling force, drum-type resin bath was proposed by Misri et al. (2014) as shown in Fig. 18.8. Three-dimensional representations of the resin bath are shown in Figs. 18.9 (schematic) and 18.10 (photo). A servomotor with the capacity of 12 V 20 A was attached at the drum. The drum rotating at the speed of 95 rpm was covered with surfacing veil made of polyester mat. The resin bath provided the impregnation mechanism for the fibre by wetting both the soaked drum surface and the static polyester mat (attached at the guide rollers) when the drum rotated. The impregnated yarn fibres were then passed through a guide roller to wipe off the excess resin before winding them over the rotating mandrel. The innovative rotating drum design helped to eliminate yarn fibre breakage by decreasing the excessive pulling load from the rotating mandrel during the manufacturing.

Hoop winding took place when the mandrel and fibre yarn were in perpendicular direction, while the helical winding took place when the mandrel and fibre yarn

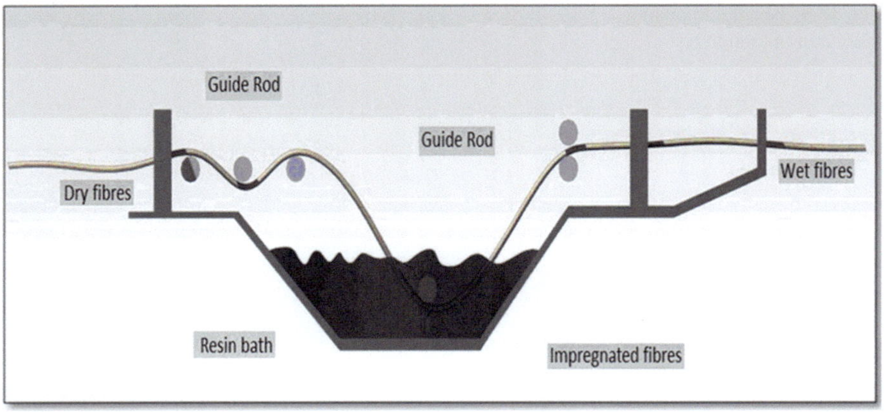

Fig. 18.7 The existing concept of resin bath (fibre-dip type)

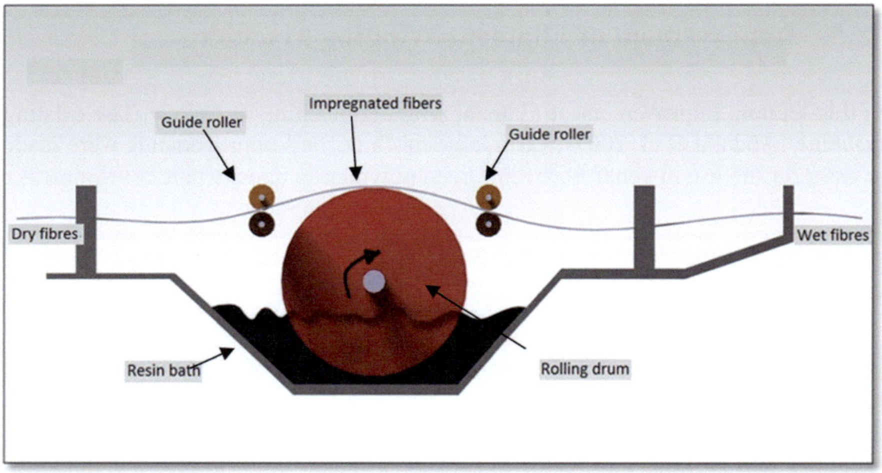

Fig. 18.8 The improved concept of resin bath (drum type)

were at certain angles lower than 90° (e.g., 45°) to each other. In this experiment, kenaf yarns were wound at 90° and 45° fibre angle orientations. It means that the kenaf yarn underwent the helical and hoop winding processes. The yarns were wound onto the rotation mandrel, which was made of hollow circular aluminium rod and PVC, and they covered all the mandrel surfaces. The filament winding process was performed by pulling the kenaf fibres through a resin bath.

Once the kenaf fibre reinforced unsaturated polyester composite hollow shafts were fabricated, it was cured in an oven at a temperature of 80 °C for 5 h. The surface of composite shaft was surface finished using a lathe machine as shown in Fig. 18.11. In this process, vernier caliper was used to control their thickness with tolerance of ±0.02 mm.

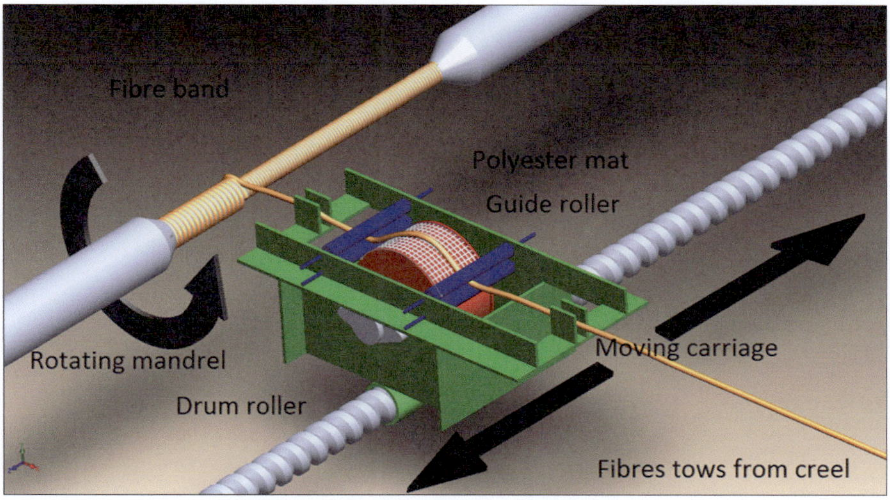

Fig. 18.9 A three-dimensional representation of the resin bath (drum type)

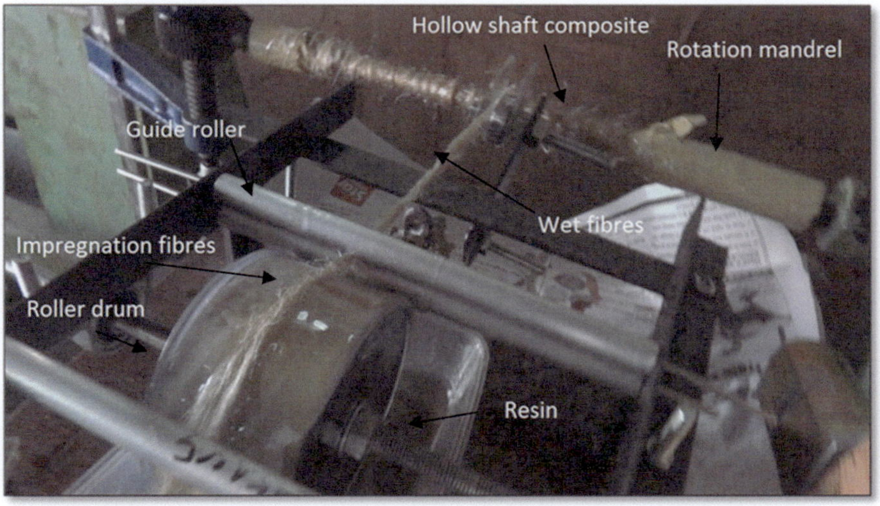

Fig. 18.10 Innovative rotating drum

18.5.2 Fabrication of Kenaf Composite Hollow Shaft Using Improved Method

Figure 18.12 shows the composite hollow shafts made by different types of raw materials (carbon, glass and kenaf fibres) that were fabricated by filament winding method. It shows the carbon and glass composite shafts were fabricated using the

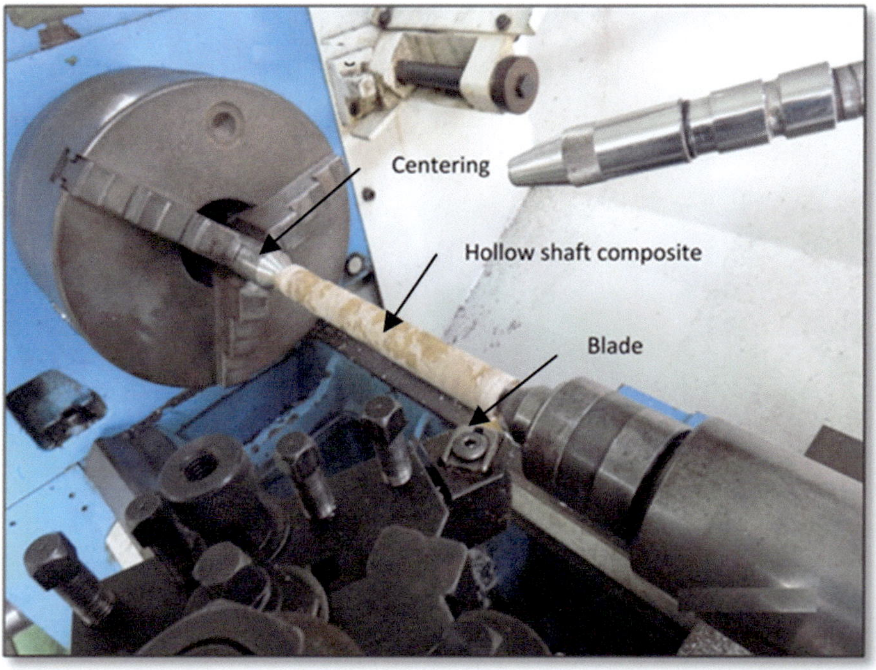

Fig. 18.11 Finishing process

Fig. 18.12 Carbon, glass and kenaf fibre hollow shaft composites

Fig. 18.13 Kenaf fibre composite hollow shafts fabricated using the improved method

regular method of resin bath fibre-dip type (as shown in Fig. 18.7), while the kenaf hollow shaft (Fig. 18.13) was fabricated using the improved method (resin bath drum type) (as shown in Fig. 18.8).

18.6 Conclusions

In this chapter, a review of filament winding process with conventional and natural fibre composites was presented. From the review, it is concluded that there is a huge scope of research and development in the area of natural fibre reinforced polymer composites using filament winding method. The emphasis of this study is on kenaf fibre composites. Various work on kenaf composites such as torsion and split-disk test are reported. Improvement in the existing fabrication process using filament winding had produced good kenaf composite specimens and parts. Drum type of resin bath and with the use of surfacing veil has helped in the fabrication and the study of kenaf fibre reinforced unsaturated polyester composite hollow shafts.

Acknowledgements The authors wish to thank the Universiti Putra Malaysia for the funding support through Research University Grant Scheme (RUGS 2-2012), project number: 05-02-12-1916RU. The support provided by staff of the Institute of Tropical Forestry and Forest Products (INTROP), Universiti Putra Malaysia, is highly appreciated.

References

Abdalla FH, Mutasher SA, Khalid YA, Sapuan SM, Hamouda AMS, Sahari BB, Hamdan MM (2007) Design and fabrication of low cost filament winding machine. Mater Des 28:234–239

Akkus N, Genc G, Girgin C (2008) Control of the pretension in filament winding process. Acta Mech Autom 2:5–10

Anon (2011) Overview of ATL/AFP, filament winding, Composite World. (Online), Gardner Publications, Inc, http://www.compositesworld.com/zones/atl-afp-filamentwinding

Anwar UMK, Paridah MT, Hamdan H, Sapuan SM, Bakar SE (2009) Effect of pressing time on physical and mechanical properties of bamboo strip fibre reinforced phenolic composites. Ind Crop Prod 29:214–219

Buragohain M, Velmurugan R (2011) Study of filament wound grid-stiffened composite cylindrical structures. Compos Struct 93:1031–1038

Chen W, Yu Y, Li P, Wang C (2007) Science and effect of new epoxy matrix for T800 carbon fiber/epoxy filament wound composites. Compos Sci Technol 67:2261–2270

Davoodi MM, Sapuan SM, Yunus R (2008) Conceptual design of a polymer composite automotive bumper energy absorber. Mater Des 29:1447–1452

Davoodi MM, Sapuan SM, Ahmad D, Aidy A, Khalina A, Jonoobi M (2011) Concept selection of car bumper beam with developed hybrid bio-composite material. Mater Des 32:4857–4865

El-Shekeil YA, Sapuan SM, Abdan K, Zainudin ES (2012) Influence of fiber content on the mechanical and thermal properties of kenaf fiber reinforced thermoplastic polyurethane composites. Mater Des 40:299–303

Gemi L, Tarakçioğlu N, Akdemir A, Şahin ÖS (2009) Progressive fatigue failure behavior of glass/epoxy $(\pm 75)_2$ filament-wound pipes under pure internal pressure. Mater Des 30:4293–4298

Gordon B (2006) Metal prepreg filament winding. SAMPE J 42:6–11

Ha SK, Jeong JY (2005) Science and effects of winding angles on through-thickness properties and residual strains of thick filament wound composite rings. Compos Sci Technol 65:27–35

Hambali A, Sapuan SM, Ismail N, Nukman Y (2009) Application of analytical hierarchy process in the design concept selection of automotive composite bumper beam during the conceptual design stage. Sci Res Essays 4:198–211

Hernández-Moreno H, Douchin B, Collombet F, Choqueuse D, Davies P (2008) Influence of winding pattern on the mechanical behavior of filament wound composite cylinders under external pressure. Compos Sci Technol 68:1015–1024

Ishak MR, Leman Z, Sapuan SM, Edeerozey AMM, Othman IS (2010) Mechanical properties of kenaf bast and core fibre reinforced unsaturated polyester composites. IOP Conf Series: Mater Sci Eng 11:012006

Ishak MR, Sapuan SM, Leman Z, Rahman MZA, Anwar UMK (2012) Characterization of sugar palm (*Arenga pinnata*) fibres. J Therm Anal Calorim 109:981–989

Krishnamurthy TN, Idkan M (2014) Fabrication of low cost filament winding machine. Int J Recent Trend Electr Electron Eng 4:30–39

Lehtiniemi P, Dufva K, Berg T, Skrifvars M, Järvelä P (2011) Natural fiber-based reinforcements in epoxy composites processed by filament winding. J Reinf Plast Compos 30:1947–1955

Mahdi E, Hamouda AMS, Sahari BB, Khalid YA (2002) Effect of material and geometry on crushing behaviour of laminated conical composite shells. Appl Compos Mater 9(5):265–290

Mallick PK (2008) Fiber-reinforced composites: materials, manufacturing, and design, 3rd edn. CRC Press, Boca Raton

Meijer G, Ellyin F (2008) A failure envelope for ±60° filament wound glass fibre reinforced epoxy tubulars. Compos A Appl Sci Manuf 39:555–564

Misri S, Ishak MR, Sapuan SM, Leman Z (2014) Split disk properties of kenaf-reinforced unsaturated polyester composites processed by filament winding. In: Colloquium on biocomposite technology 2014. Serdang, Malaysia. 24 November 2014, Book of Abstract, p 5

Misri S, Sapuan SM, Leman Z, Ishak MR (2015) Torsional behaviour of filament wound kenaf yarn fibre reinforced unsaturated polyester composite hollow shafts. Mater Des 65:953–960

Morozov EV (2006) The effect of filament-winding mosaic patterns on the strength of thin-walled composite shells. Compos Struct 76:123–129

Mutasher SA, Sahari BB, Hamouda AMS, Sapuan SM (2007) Experimental study of bending fatigue characteristics of a hybrid aluminum/composite drive shaft. J Compos Mater 41: 2267–2288

Mutasher S, Mir-nasiri N, Lee CL (2012) Small-scale filament winding machine for producing fiber composite products. J Eng Sci Technol 7:156–168

Nadia K, Ishak MR (2014) Bending fatigue – kenaf fibre hollow shaft composite: a review. In: The postgraduate symposium on composites science and technology 2014 & the 4th postgraduate seminar on natural fibre composites 2014, 28 January 2014, Putrajaya, pp 193–199

Park J, Hong C, Kim C, Kim C (2002) Analysis of filament wound composite structures considering the change of winding angles through the thickness direction. Compos Struct 55:63–71

Peters ST, Humphrey WD (1987) Filament winding. In: Dostal CA (ed) Engineered materials handbook, volume 1: composites. ASM International, Materials Park

Sapuan SM, Abdalla HS (1998) A prototype knowledge-based system for the material selection of polymeric-based composites for automotive components. Compos A Appl Sci Manuf 29: 731–742

Sapuan SM, Jacob MSD, Mustapha F, Ismail N (2002) A prototype knowledge-based system for material selection of ceramic matrix composites of automotive engine components. Mater Des 23:701–708

Sapuan SM, Maleque MA, Hameedullah M, Suddin MN, Ismail N (2005) A note on the conceptual design of polymeric composite automotive bumper system. J Mater Process Technol 159:145–151

Sastra HY, Siregar JP, Sapuan SM, Hamdan MM (2006) Tensile properties of *Arenga pinnata* fiber-reinforced epoxy composites. Polym Plast Technol Eng 45:149–155

Shaw-Stewart D (1985) Filament winding. Mater Eng 6:140–144

Velosa JC, Nunes JP, Antunes PJ, Silva JF, Marques AT (2009) Development of a new generation of filament wound composite pressure cylinders. Compos Sci Technol 69:1348–1353

Wang X, Lou DYS, Zhang N (2004) A helical tow model and numerical simulation of on-line thermal curing of thermoset composites in filament winding. Int J Heat Mass Transfer 47: 4807–4820

Yusoff MZM, Salit MS, Ismail N, Wirawan R (2010) Mechanical properties of short random oil palm fibre reinforced epoxy composites. Sains Malaysiana 39:87–92

Zu L (2012) Design and optimization of filament wound composite pressure vessels. Ph.D. Thesis, Delft University, Netherlands

About the Editors

Prof. Mohd Sapuan Salit, Ph.D. is a senior professor of composite materials at Department of Mechanical and Manufacturing Engineering, Universiti Putra Malaysia (UPM). He is currently head of Laboratory of Biocomposite Technology, INTROP, UPM, and also head of composites Technology Research Program at UPM. He is the vice president and honorary member of Asian Polymer Association; Fellow of Institute of Materials, Malaysia, Plastic and Rubber Institute, Malaysia, Malaysian Scientific Association and Society of Automotive Engineers International, USA. His research interests include natural fibre composites, composite materials selection and concurrent engineering for composites. Professor Mohd Sapuan Salit has published more than 450 journal papers and 400 conference papers. He has authored and edited 15 books in engineering. He has guided more than 40 Ph.D. students and 40 M.Sc. students. He is the recipient of various awards such as ISESCO Science Prize, Alumni Award, University of Newcastle, Australia, Rotary Research Award, Khwarizmi International Award, Plastic and Rubber Institute Malaysia Fellowship Award and Vice Chancellor Fellowship Prize, UPM.

Dr. Mohammad Jawaid, Ph.D. is currently working as Fellow Researcher (Associate Professor), at Biocomposite Technology Laboratory, Institute of Tropical Forestry and Forest Products (INTROP), Universiti Putra Malaysia, Serdang, Selangor, Malaysia and also Visiting Professor at Department of Chemical Engineering, College of Engineering, King Saud University, Riyadh, Saudi Arabia since June 2013. Previously he worked as Visiting Lecturer, Faculty of Chemical Engineering, Universiti Teknologi Malaysia (UTM) and also worked as Expatriate Lecturer under UNDP project with Ministry of Education of Ethiopia at Adama University, Ethiopia. He received his Ph.D. from Universiti Sains Malaysia, Malaysia. He has more than 10 years of experience in teaching, research and industries. His area of research interests include Hybrid Reinforced/Filled Polymer Composites, Advance Materials: Graphene/Nanoclay/Fire Retardant, Lignocellulosic Reinforced/Filled Polymer Composites, Modification and Treatment of Lignocellulosic Fibres and Solid Wood, Nano Composites and

M.S. Salit et al. (eds.), *Manufacturing of Natural Fibre Reinforced Polymer Composites*, DOI 10.1007/978-3-319-07944-8

Nanocellulose fibres, Polymer blends. So far he has published 4 books, 12 book chapters, and more than 100 international journal papers and 4 published review papers under Top 25 hot articles in science direct during 2013–2014. He is also the Deputy Editor-in-Chief of *Malaysian Polymer Journal* and reviewer of 30 high impact ISI journals.

Dr. Nukman Bin Yusoff, Ph.D. is an Associate Professor in the Department of Mechanical Engineering, University of Malaya (UM), Malaysia. Appointed as Head of Department of Engineering Design & Manufacture in 2007 till 2009 and actively doing research in the field of Laser Materials Processing, CAD/CAM, CNC and Manufacturing Process and Management. He acquired his Ph.D. from Loughborough University in 2009 and Master in Engineering Science, M.Sc. (Mechatronics) from DeMontfort University, Leicester United Kingdom in 1998. He acquired his first degree in CAD/CAM from University of Central England in Birmingham. Since then, he has been teaching undergraduate in the field of CAD/CAM and Manufacturing degrees and graduate level teaching since 2007. He published many papers in the field of manufacturing and articles in different international, regional and national journals, conference proceedings and bulletins and co-author of several technical books.

Dr. M. Enamul Hoque, Ph.D. is an Associate Professor in the Department of Mechanical, Materials & Manufacturing Engineering and served as Head for the Bioengineering Research Group at the University of Nottingham Malaysia Campus. He received his Ph.D. in Mechanical Engineering (major in Bioengineering) from NUS, Singapore in 2007. He has been intensively involved in numerous research projects that are collaborated locally and internationally. He has authored 3 books, edited 2 books and co-authored 7 book chapters, and published about 120 technical papers in high impact refereed journals and international conference proceedings. He serves as an editor for 5 journals, as member of editorial board for 5 journals, and technical reviewer for about 25 journals. He has established an international reputation in the areas of Materials/ Biomedical Engineering. He is an Associate Fellow to the Higher Education Academy, UK. Besides, he is a member of several professional scientific bodies including Tissue Engineering and Regenerative Medicine International Society (TERMIS), Institute of Materials, Minerals and Mining (IOM3), UK; Institute of Materials Malaysia (IMM) and Tissue Engineering Society Malaysia (TESMA). His research interests include the areas of Rapid Prototyping Technology, Nanotechnology, Biomaterials, Tissue Engineering, Stem Cells, Nanomaterials, Composite Materials, Bioenergy and Food Technology.

Index

M.S. Salit et al. (eds.), *Manufacturing of Natural Fibre Reinforced Polymer Composites*, DOI 10.1007/978-3-319-07944-8

F

G

H

MIX
Papier aus verantwortungsvollen Quellen
Paper from responsible sources
FSC® C105338
FSC
www.fsc.org

If you have any concerns about our products,
you can contact us on
ProductSafety@springernature.com

In case Publisher is established outside the EU,
the EU authorized representative is:
Springer Nature Customer Service Center GmbH
Europaplatz 3, 69115 Heidelberg, Germany

Printed by Libri Plureos GmbH
in Hamburg, Germany